Lecture Notes in Computer Science 5038

Commenced Publication in 1973
Founding and Former Series Editors:
Gerhard Goos, Juris Hartmanis, and Jan van Leeuwen

Catherine C. McGeoch (Ed.)

Experimental Algorithms

7th International Workshop, WEA 2008
Provincetown, MA, USA, May 30–June 1, 2008
Proceedings

 Springer

Volume Editor

Catherine C. McGeoch
Department of Mathematics
and Computer Science
Amherst College
Amherst, MA, USA
E-mail: ccm@cs.amherst.edu

Library of Congress Control Number: 2008927191

CR Subject Classification (1998): F.2.1-2, E.1, G.1-2, I.3.5, I.2.8

LNCS Sublibrary: SL 1 – Theoretical Computer Science and General Issues

ISSN   0302-9743
ISBN-10   3-540-68548-0 Springer Berlin Heidelberg New York
ISBN-13   978-3-540-68548-7 Springer Berlin Heidelberg New York

Springer is a part of Springer Science+Business Media

springer.com

© Springer-Verlag Berlin Heidelberg 2008
Printed in Germany

Typesetting: Camera-ready by author, data conversion by Scientific Publishing Services, Chennai, India
Printed on acid-free paper   SPIN: 12275052   06/3180   5 4 3 2 1 0

Preface

The Workshop on Experimental Algorithms, WEA, is intended to be an international forum for research on the experimental evaluation and engineering of algorithms, as well as in various aspects of computational optimization and its applications. The emphasis of the workshop is the use of experimental methods to guide the design, analysis, implementation, and evaluation of algorithms, heuristics, and optimization programs.

WEA 2008 was held at the Provincetown Inn, Provincetown, MA, USA, on May 30 – June 1, 2008. This was the seventh workshop of the series, after Rome (2007), Menorca (2006), Santorini (2005), Rio de Janiero (2004), Asconia (2003), and Riga (2001).

This volume contains all contributed papers accepted for presentation at the workshop. The 26 contributed papers were selected by the Program Committee on the basis of at least three referee reports, some contributed by trusted external referees.

In addition to the 26 contributed papers, the program contained two invited talks. Camil Demetrescu, of the University of Rome "La Sapienza," spoke on "Visualization in Algorithm Engineering." David S. Johnson of AT & T Labs – Research, gave a talk on "Bin Packing: From Theory to Experiment and Back Again."

We would like to thank the authors who responded to the call for papers, our invited speakers, the members of the Program Committee, the external referees, and the Organizing Committee members for making this workshop possible.

March 2008   Catherine C. McGeoch

Organization

Program Committee

Catherine C. McGeoch (Chair)	Amherst College (USA)
Lars Arge	University of Aarhus (Denmark)
Jon Bentley	Avaya Laboratories (USA)
Gerth Stolting Brodal	University of Aarhus, BRICS (Denmark)
Adam Buchsbaum	AT & T Labs (USA)
Camil Demetrescu	University of Rome "La Sapienza" (Italy)
Thomas Erlebach	University of Leicester (UK)
Irene Finocchi	University of Rome "La Sapienza" (Italy)
Andrew Goldberg	Microsoft (USA)
Mark Goldberg	Rensselaer Polytechnic Institute (USA)
Michael Goodrich	University of California, Irvine (USA)
Richard Ladner	University of Washington (USA)
Ian Munro	University of Waterloo (Canada)
Stefan Naeher	University of Trier (Germany)
Sotiris Nikoletseas	University of Patras and CTI (Greece)
Luis Paquete	University of Algarve (Portugal)
Mike Preuss	University of Dortmund (Germany)
Mauricio G.C. Resende	AT & T Labs (USA)
Celso Ribeiro	Universidade Federal Fluminense (Brazil)
Steve Skiena	Stony Brook University (USA)
Matt Stallmann	North Carolina State University (USA)
Cliff Stein	Columbia University (USA)
Thomas Steutzle	IRIDIA, Université Libre de Bruxelles (Belgium)
Roberto Tamassia	Brown University (USA)
Stefan Voss	University of Hamburg (Germany)
Dorothea Wagner	Universität Karlsruhe (Germany)

Referees

Mohammad Abam	Daniel Delling	Allan Jorgensen
Reinhard Bauer	Paola Festa	Alexis Kaporis
Vincenzo Bonifaci	Bob Fraser	Marcus Krug
Saverio Caminiti	Marco Gaertler	Veli Makinen
Ioannis Caragiannis	Mark Goldberg	Caserta Marco
Marco Chiarandini	Robert Görke	Sascha Meinert
Albert Choi	Michael Hirsch	Thomas Moelhave
Razaul Alam Chowdhury	David Johnson	Alberto Moraglio

Gabriel Moruz   Franz Rendl   Renato Werneck
Pat Nicholson   Peter Sanders   Ke Yi
Richard Peng   Srinivasa Rao Satti   Martin Zachariasen
Giuseppe Persiano   Frank Schwartz   Christos Zaroliagis
Rajeev Raman   Silvia Schwarze
Christoforos Raptopoulos Matthew Skala

WEA Steering Committee

Edoardo Amaldi   Politecnico di Milano (Italy)
David A. Bader   Georgia Institute of Technology (USA)
Josep Diaz   T.U. of Catalonia (Spain)
Guiseppe F. Italiano   University of Rome "Tor Vergata" (Italy)
David Johnson   AT & T Labs (USA)
Klaus Jansen   Universität Kiel (Germany)
Kurt Mehlhorn   Max-Plank-Institut für Informatik (Germany)
Ian Munro   University of Waterloo (Canada)
Sotiris Nikoletseas   University of Patras and CTI (Greece)
Jose Rolim (Chair)   University of Geneva (Switzerland)
Pablos Spirakis   University of Patras and CTI (Greece)

Table of Contents

Reducing Splaying by Taking Advantage of Working Sets

Timo Aho, Tapio Elomaa, and Jussi Kujala

Department of Software Systems, Tampere University of Technology
P.O. Box 553 (Korkeakoulunkatu 1), FI-33101 Tampere, Finland
{timo.aho,tapio.elomaa,jussi.kujala}@tut.fi

Abstract. Access requests to keys stored into a data structure often exhibit locality of reference in practice. Such a regularity can be modeled, e.g., by working sets. In this paper we study to what extent can the existence of working sets be taken advantage of in splay trees. In order to reduce the number of costly splay operations we monitor for information on the current working set and its change. We introduce a simple algorithm which attempts to splay only when necessary. Under worst-case analysis the algorithm guarantees an amortized logarithmic bound. In empirical experiments it is 5% more efficient than randomized splay trees and at most 10% more efficient than the original splay tree. We also briefly analyze the usefulness of the commonly-used Zipf's distribution as a general model of locality of reference.

1 Introduction

Many search trees facilitate efficient access to the stored items by keeping the tree in balance using rotations [1]. The balance invariant is maintained independent of the sequence of access requests observed. Splay trees [2], on the other hand, manage to do without any invariant, but need to splay in connection of each access and update. Splay trees lose the provably logarithmic worst-case bounds of individual operations, but still behave well under amortized analysis. The need for (expensive) splaying can be reduced by randomizing the decision of whether to splay or not in connection of an operation [3,4] as well as by heuristic limit-splaying algorithms [2,5,6].

Several theoretical results indicate that splay trees should work particularly well when there is locality of reference in the request sequence [2]. However, some empirical studies [6,7,8] have indicated that they could be actually at their best in highly dynamic environments, where the focus of locality drifts over time. Moreover, despite careful implementation basic splay tree variations have empirically been observed to be less efficient than red-black trees (RBTs), standard binary search trees (BSTs), and hashing at least in some situations [6,7]. Randomized adaptive data structures can do better [4,6], but only heuristic limit-splaying has been competitive in practice [6]. However, some recent studies [9,10] have demonstrated that in some settings splay trees may be more efficient

C.C. McGeoch (Ed.): WEA 2008, LNCS 5038, pp. 1–13, 2008.

than other BSTs. We hope that paying better attention to the properties of the input — splaying only when necessary and useful — would lead to more efficient performance.

Randomized splay trees achieve some practical savings without giving in on asymptotic efficiency. Nevertheless, they still do not pay any attention to the properties of the request sequence. The best advantage of randomization has been shown for fixed probability distributions, while request sequences with high dynamic locality of reference benefit from randomization only slightly [4]. Our aim in this paper is to study whether the amount of splay operations could be reduced more efficiently. In other words, we investigate how large savings (if any) can be achieved by monitoring the actual input that is seen.

In particular, we examine access request sequences that exhibit locality of reference in the form of *working sets* [11]. This means that, at any time interval, most accesses refer only to a small portion of all keys — the current working set. Real-world situations often conform with this assumption. Of course, we cannot afford to implement a too complicated request sequence monitoring method, because then we would be destined to lose in time consumption to the in-practice efficient (randomized) splay trees.

We will introduce and analyze a quite straightforward version of the splay tree which takes the existence of working sets in the request sequence into account. The algorithm maintains a (discounted) counter to monitor the (average) depths of recent searches in the tree. Low average search depth indicates that a working set exists near the root of the tree and is being actively used. Occasional deep searches do not change the situation. Only when the searches are constantly deep, is there need to update the splay tree.

The remainder of this paper is organized as follows. In Section 2 we briefly recapitulate splay trees and their relation to working sets. Section 3 presents the main idea of conditional adaptation of a binary search tree studied in this paper. The splay tree algorithm based on this idea is introduced in Section 4. An empirical evaluation of the algorithm is reported and analyzed in Sections 5 and 6. Finally, Section 7 gives the concluding remarks of the paper.

2   Adaptive Data Structures and Working Sets

A splay tree is a BST with the keys in symmetric order. In it the accessed item is elevated to the root of the tree using splay rotations [2]. These operations keep the tree pretty well in balance. Because no other balancing is enforced, a splay tree does not contain any additional information and, thus, does not require extra storage space. Because of its strategy, a splay tree also keeps recently accessed items very near the root. Thus, it automatically handles also working sets quite efficiently. On the downside, the accessed key is splayed to the root even if it is accessed only once during the whole access sequence. Also unnecessary splaying is executed even if the current working set is near the root and, thus, already efficiently accessible.

Sleator and Tarjan [2] proved many interesting bounds and properties for the time consumption of splay trees. In this paper we need only one of these results. The formulation of the following theorem comes from Albers and Karpinski [4]. Throughout this paper we denote by m the number of access requests in the operation sequence and by n the number of keys mentioned (all of them stored in the tree).

Theorem 1 (Balance Theorem). *The total access time incurred by an access sequence is at most* $3m \log_2 n + m + C$, *where* $C = n \log_2 n$.

There are two basic implementations for splay trees: top-down and bottom-up splaying. Asymptotically their access times are the same, but the practical efficiency of these implementations has been under some controversy [6]. In our experiments top-down splaying was always more efficient than the bottom-up version. Therefore, we report our results using top-down splaying with all the algorithms.

An interesting theoretical examination of splay trees has been presented by Subramanian [12]. He proposed a more general group of trees possessing similar properties as splay trees. The splay heuristic has additionally been applied in other tree structures [13]. Iacono [14] has also discussed the topic of splay trees and working sets. He presented new distribution sensitive data structures consisting of multiple trees and proved very interesting features for them.

After the submission of this paper we learned that independent of us Lee and Martel [10] have proposed an algorithm very similar to ours for cache efficient splaying. Their algorithm uses a sliding window of accesses. The algorithm executes splaying if a too large portion of the accesses is deeper than a predefined limit depth. They also present experiments somewhat similar to ours. However, their test setting is more static.

For self-organizing lists a heuristic with some reminiscence with the splaying operation is the move-to-front rule [15]. Amer and Oommen [16] have recently examined the effect of locality of reference in self-organizing lists.

3   Conditional Adaptation of a Binary Search Tree

Before introducing the BST algorithm intended to cope with working sets, we give the basic philosophy of the algorithm: how to identify a working set and its change.

We execute splay operations only when the active working set is not near the root or when the whole tree is unbalanced. Thus, unnecessary splaying is avoided in accessing a single item outside the working set as well as an item of the working set already near the root. The latter is useful especially when the access operations are approximately uniformly distributed among the items in a working set; in other words, there are not several layers of working sets. However, assuming this would be very restrictive and, thus, we do not use it in the paper.

In order to gain knowledge of the input we simply maintain a discounted depth counter, which gives us information about the relation between on-going access operations and the current working set. If the last few access operations are deep, we can conclude that the working set has changed or the tree is not in balance. Thus, there is need to splay to correct the situation.

For our counter we need an approximate value for the size of working sets w. See e.g. [11,17] for techniques on approximating w. With this value we can calculate the limit depth $limit_w$, which represents the acceptable average depth for access operations in working sets. We set $limit_w = a \log_2(w + 1)$, where the multiplier a is a constant chosen suitably for the current environment.

The value of the *condition counter* is updated in connection of an access operation to *depth* as follows:

$$counter \leftarrow d \cdot counter + depth - limit_w,$$

where the *discounting factor* d, $0 \leq d < 1$, is a constant regulating the impact of the history of access operations on the current value. The difference $depth - limit_w$ tells us how much (if at all) below the limit depth we have reached. If the value of *counter* is non-positive we may assume that splaying is not required. On the other hand, a positive value suggests that the operation is needed.

Taking discounted history into account ensures that isolated accesses outside the working set do not restructure the tree needlessly. On the other hand, giving too much weight to earlier accesses makes the data structure slow to react to changes in the working set.

We could as well let the value of $limit_w$ change during the execution of algorithm. In fact, in our empirical evaluation we use a dynamically changing $limit_w$. However, for a more straightforward analysis we assume for the time being that the value is constant. We also assume that the accessed item can be found in the tree; if not, the value of *counter* should be left unchanged.

There are alternatives for our approach. We could, for example, get rid of the whole discounting philosophy and use individual access counters for the keys [5]. These counters should be included either in the nodes of the tree or kept in a separate data structure. However, e.g., Lai and Wood [5] have already inspected the first approach with splay tree. Also Seidel and Aragon [18] introduced randomized search trees and Cheetham et al. [19] conditional rotating based on this approach. Because keeping the nodes free from additional information is an essential part of splay trees, we would nevertheless like to find another way.

On the other hand, if the counters were kept in a separate data structure, updating this information would be problematic. We are, anyway, mostly interested in only the last access operations. Thus, too static counters would not react quickly enough to the altering working set. Hence, we should have a technique to decrease the significance of old access operations. We are not aware of a solution that would not raise the running time too much. Lee and Martel [10] solve the problem by counting only the amount of deep accesses. Thus their solution loses information about the access depths.

procedure WSPLAY(x)

(1) **if** *counter* > 0 **then**
(2)   *depth* $\leftarrow$ SPLAY(x)
(3)   *counter* $\leftarrow$ *depth* $-$ *limit$_w$*
    else
(4)   *depth* $\leftarrow$ BSTACCESS(x)
(5)   *counter* $\leftarrow$ *counter* $\cdot$ *d* $+$ *depth* $-$ *limit$_w$*
(6)   **if** *counter* > 0 **then**
(7)   SPLAY(x)
(8)   *counter* $\leftarrow$ *depth* $-$ *limit$_w$*

Algorithm 1. The WSPLAY algorithm

4   The Algorithm Detecting Working Sets: Wsplay

Let us now introduce an algorithm based on the information collecting approach described above. WSPLAY (Algorithm 1) simply splays whenever the condition counter implies that the working set is changing.

Function BSTACCESS implements the standard BST access and returns the depth of accessed item. As we want to execute the more efficient top-down version of splaying, we splay if *counter* indicates the need of splaying in the beginning of access. This is the case when the previous access operation—which necessarily included splaying—was deep. Thus we avoid doing unnecessary BST access before every top-down splay. Otherwise we access the item, update the condition counter, and splay if the updated counter value indicates a need for it. Observe that after each execution of the SPLAY function, the history of access depths is erased.

We assume that the function SPLAY also returns the original depth of the accessed item. Note that setting $d = 0$ makes WSPLAY very similar to the algorithm introduced by Sleator and Tarjan in the Long Splay Theorem [2, Theorem 7].

We now prove a logarithmic bound for the running time of WSPLAY. Recall that m is the length of the access sequence and n the number of nodes in the splay tree. The sequence of access operations H is divided in two disjoint categories. Let H_s consist of access operations including splaying and H_n of those without a splay operation. Because the tree structure does not change as a result of the accesses in H_n, the time consumption of these two sequences can be analyzed separately.

The sequence H_s essentially consists of splay operations. Only constant time overhead is caused by counter updating. Thus, the time consumption on H_s is bounded by Theorem 1.

To derive a bound for the time consumption on the sequence H_n, we analyze the values of variable *counter* during the access operations of a single continuous sequence $H_c \subseteq H_n$, $H_c = \langle h_1, h_2, \ldots, h_c \rangle$. Either the first access operation h_1 is the first access in the whole sequence H or its predecessor includes splaying. During H_c splay operations are not executed. Let us denote the values of the variable *counter* right after the first update in the algorithm on line 5 with subindices, respectively. Note that the updates on lines 3 and 8 are not executed because no splay operations are executed in H_n. In particular, $counter_0$ is the value of the variable in the beginning of the access h_1. By the definition of H_c $counter_0 \leq 0$. Let $depth(h_i)$ be the depth of access operation h_i.

With these definitions we give a bound for the average access depth in the sequence H_c. To achieve this, we need to prove that if splay operations are not done the value of $counter_i$ gives a sort of a bound for the depth of access operation h_i. Intuitively the idea is to show that if the *counter* is never above 0, the average depth of accesses cannot be too much larger than $limit_w$.

Lemma 1. *If for all l, $0 \leq l < c$, $counter_l \leq 0$, then for all i, $1 \leq i \leq c$,*

$$\sum_{j=1}^{i}(depth(h_j) - limit_w) + counter_0 \leq counter_i.$$

Proof. During the access operations h_i, $1 \leq i \leq c$, no splay operation is executed and, thus, it holds that

$$counter_i = \sum_{j=1}^{i}(depth(h_j) - limit_w)\, d^{i-j} + d^i\, counter_0.$$

We prove the claim by induction over the index i.

Let $i = 1$. Because $0 \leq d < 1$, it is clear that $depth(h_1) - limit_w + counter_0 \leq depth(h_1) - limit_w + d \cdot counter_0 = counter_1$. Hence, the claim holds in this case.

Let us then assume that the claim holds when $1 \leq i = k < c$. We focus on the situation $i = k + 1$. By assumption we know that $counter_k \leq 0$. Thus,

$$
\begin{aligned}
counter_{k+1} &= counter_k \cdot d + depth(h_{k+1}) - limit_w \\
&\geq counter_k + depth(h_{k+1}) - limit_w \\
&\geq \sum_{j=1}^{k+1}(depth(h_j) - limit_w) + counter_0.
\end{aligned}
$$

Hence, the lemma is valid.

Because the value of $counter_0$ is assigned during the last splayed access, we know that $counter_0 \geq -limit_w$. We also know that $counter_c \leq 0$ and, thus, by Lemma 1 we have that

$$\sum_{i=1}^{c} depth(h_i) - limit_w \cdot (c+1)$$

$$\leq \sum_{i=1}^{c} depth(h_i) - limit_w \cdot c + counter_0 \leq counter_c \leq 0$$

$$\Leftrightarrow \frac{1}{c}\sum_{i=1}^{c} depth(h_i) \leq \left(1 + \frac{1}{c}\right) limit_w \leq 2\, limit_w.$$

The bound is strict: it happens, e.g., when $counter_0 = -limit_w$, $depth(h_1) = (1+d)limit_w$, and $c = 1$.

The above bound for average depth holds for all of $H_c \subseteq H_n$. Thus writing $|H_n| = m_n$ and $|H_s| = m_s$, we have the following theorem:

Theorem 2. *Let $C = n \log_2 n$ and $m_n + m_s = m$. The total access time incurred by* WSPLAY *is at most $2m_n limit_w + m_s(3 \log_2 n + 1) + C$.*

5 Test Setting

As reference algorithms in our empirical evaluation we use splay trees, RBT, and a randomized version of splay trees. All splaying is implemented in a simple top-down fashion. The programming environment is Microsoft Visual C++ 2005 and we use full optimization for speed. The test environment is a PC with a 3.00 GHz Intel Pentium 4 CPU and 1 GB RAM. The cache sizes are 16 kB for L1 and 2 MB for L2.

5.1 The Evaluated Version of Wsplay

The problem in evaluating WSPLAY is to control its two parameters d and $limit_w$. Examining all value combinations would be a massive task. However, moderate changes in the value of parameter d do not affect the results much in practice. Hence, we use a constant value $d = 0.9$.

The parameter $limit_w$ is more problematic because with different values the results vary. However, we are mostly interested in how efficient is splay reduction by monitoring. Thus, a very natural way is to compare its efficiency against data structures that reduce splaying without such monitoring. Randomized splay trees [3,4] do it at random. The data structure of Albers and Karpinski [4], referred here as RSPLAY, matches WSPLAY perfectly in the sense that it contains no other modifications to basic splay trees than reduced splaying.

RSPLAY needs as a parameter the probability p. With the probability $1 - p$ standard BST access is executed instead of splaying. We modify WSPLAY to execute the same amount of splaying by including simple adaptation for the variable $limit_w$. We adjust it in the beginning of every access by adding a value proportional to difference between p and amount of executed splay operations. To prevent algorithm from executing all the available splaying in the beginning, we start the sequence with a $1/p$ length margin where no splaying is allowed. Values of $0, 1, 1/2, 1/4, \ldots, 1/512$ for p are examined. This works well in practice: the difference between p and amount of WSPLAY splaying was at most $\min(0.1\%, 0.05p)$. Note that this modification does not affect the bound of Theorem 2. We can easily keep everything in $O(\log_2 n)$ time given a maximum value for $limit_w$. With this modification it is possible to relate RSPLAY and WSPLAY with different values of p.

5.2 Description of the Data

The keys are inserted in trees in random order. For BSTs this usually leads to a well-balanced tree [20]. We examine only accesses with integer keys in nodes.

The locality of reference is often modeled with *Zipf's distribution* (ZD) [21, p. 400]. In it the ith most commonly accessed item is accessed with a probability p_i inversely proportional to i. We want to experiment with different kinds of ratios of locality or *skewness*. Therefore, we use the modification in which we have an additional parameter $\alpha \geq 0$ [21] so that the ith item has access probability

$$p_i = \frac{1}{i^\alpha C},$$

where $C = \sum_{j=1}^{n}(1/j^\alpha)$ and n is the number of nodes in tree. This distribution is uniform when $\alpha = 0$ and the pure ZD when $\alpha = 1$. The parameter α alone does not give very good control over the skewness of the data. Therefore, Bell and Gupta [7] defined another parameter, the *skew factor*:

$$\beta = \sum_{j=1}^{n/100} p_j.$$

Hence, β is the probability of accessing the 1% of items that is most frequently accessed. With $\beta = 0.01$ the distribution, obviously, is uniform.

To achieve dynamic locality we change the access distribution after every $t = 128$ accesses. When the value of t was very low the results were more like with uniform distribution, but otherwise changing t moderately did not affect the results substantially.

We report the evaluation for tree size $n = 2^{17}$ and $m = 2^{20}$ accesses. We also evaluated the tests at least partially for tree sizes ranging up to 2^{25}. The results were similar with different sizes of trees. However, splay tree benefited from larger tree sizes compared to WSPLAY and RSPLAY. For $n = 2^{25}$ splay tree was at least as efficient as the other two with all the skewness values. For the parameter β we used values $0.01, 0.1, 0.2, \ldots, 0.9$. The values of α in the same order are $0, 0.504, 0.664, 0.764, 0.844, 0.914, 0.981, 1.052, 1.135$, and 1.258.

6   Empirical Evaluation

6.1   Average Depth of an Access

The average access depths for the data structures are depicted in Fig. 1. For WSPLAY and RSPLAY the averages over all values of the parameter p are shown. An interesting observation is that the average access depth for RBT is essentially optimal (recall that $n = 2^{17}$). This may be a reason for the practical efficiency of RBT. Our results resemble those of Bell and Gupta [7]: RBT is superior with low skewness values and with high ones splay trees excel. With low values of β there is no real locality in referencing and, thus, distribution sensitivity is of little use. On the other hand, with high values of β it is useful to raise all the items near the root as soon as possible.

Detailed results for WSPLAY and RSPLAY are presented in Fig. 2. Only results for the best values of p are reported. As expected, WSPLAY was better than R-SPLAY with all value combinations of β and p. For both algorithms the same

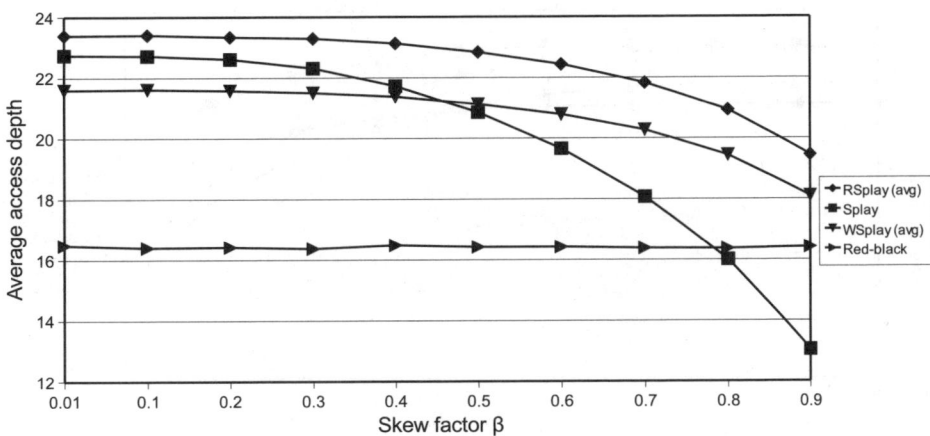

Fig. 1. The average access depths for all data structures

trend for changing the value of p occurs. The best performance for WSPLAY is obtained with $p = 1/16$ and for RSPLAY the best value is $p = 1/4$. The difference between the performance of the algorithms increases with low values of p: With $p = 1/2$ the difference is only 0–2% depending on skewness, with $p = 1/4$ it is already 4–5%, and raises to 7–10% when $p = 1/16$.

The difference of the algorithms on uniform data is probably due to balancing. While RSPLAY does the balances at random, $limit_w$ of WSPLAY settles to a depth in which the amount p of splaying is executed. Thus, the whole tree is treated as a working set. Only roughly the deepest portion p of all accesses are splayed. Splaying the deepest node in a tree tends to halve the depth of a very unbalanced tree [2]. Hence, splaying the deepest nodes in all situations seems to be a better strategy than random splaying. The greater difference in performance between the algorithms for low values of p could also be based on the same reason. For low values of p the only reason to splay is to keep tree balanced and WSPLAY does this better.

As expected, it seems that with high skewness factor values higher values of p were more suitable. It is clearly useful to splay the new working set near the root as soon as possible. Also splaying items near the root does not seem to make the tree more unbalanced. E.g., splaying two different items to the root in turns does not affect the overall balance at all.

6.2   Access Times

The access times for the data structures shown in Fig. 3 are all averages over five evaluations. On the whole the results resemble those of average access depths. RBT is even more superior in these results and splay trees take the most benefit from locality of reference. The overall decrease in access time for splay tree from $\beta = 0.01$ to $\beta = 0.9$ is nearly 50%. An unexpected result is that also RBT decreases its access time by 35% without restructuring the tree.

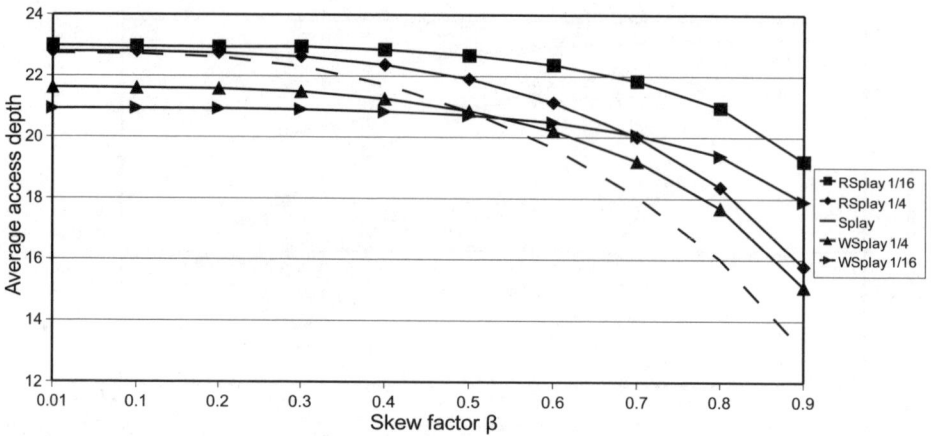

Fig. 2. The average access depths of WSPLAY and RSPLAY with different values of p. The curve for splay tree is marked with the dashed line.

Also the most efficient WSPLAY and RSPLAY are presented in Fig. 3. Understandably the running times decrease as the skewness increases. Also the relative success of the algorithms with different values of p and β is similar to average depths. However, the costly monitoring in WSPLAY reduces its absolute performance. Nevertheless, the most efficient WSPLAY ($p = 1/32$) is usually 5% faster than the most efficient RSPLAY ($p = 1/16$). With $\beta = 0.9$ the difference is only 2%. With low values of β WSPLAY is 7–12% faster than the original splay tree.

With values $p < 1/4$ WSPLAY is 3–5% more efficient than RSPLAY, but with values 0, 1, 1/2, and 1/4 of p the latter prevails. This seems to indicate that in WSPLAY the cost of monitoring the input is compensated only when the amount of splay operations is small. With larger values of p RSPLAY reacts to the change of distribution soon enough. Also RSPLAY seems to need a little more splaying for similar effect, because random splay operations are often not useful. However, absolutely both data structures are most efficient with values $1/4 \geq p \geq 1/64$.

Also with uniform distribution large values of p do not have as good balancing effect as with lower values. In other words, also WSPLAY does too much splaying on accesses that are not very deep in the tree. Thus, restructuring may set the tree out of balance. However with lower values of p WSPLAY is able to splay only the deepest accesses and thus keep the tree balanced.

However, an interesting discovery raises if we examine the efficiency of WSPLAY and RSPLAY for parameter value $p = 1$ (always splay). By comparing these to the original splay tree we get a picture of the implementation specific overhead for the monitoring in WSPLAY and decision-making in RSPLAY. WSPLAY uses 4–7% and RSPLAY 1–3% more time than a splay tree. The access times for RSPLAY do not include the generation of random numbers. This raises the question whether to compare WSPLAY to splay or WSPLAY with $p = 1$.

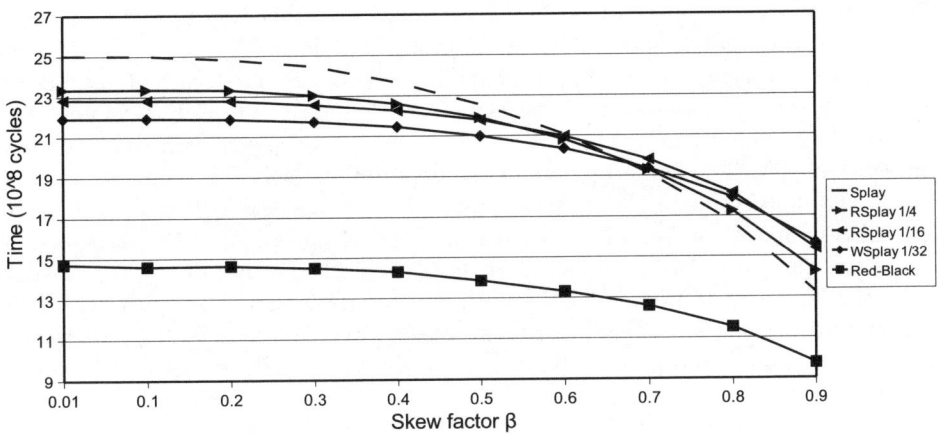

Fig. 3. The average access times for all data structures. For WSPLAY and RSPLAY the best values of p are shown. The curve for splay tree is marked with the dashed line.

WSPLAY increases its efficiency at most more than 15% when compared to the version where WSPLAY always splays. Similar value for RSPLAY is 10%.

6.3   On the Suitability of Zipf's Distribution

Let us briefly analyze the observation that also RBTs seem to be more efficient with higher locality of reference when ZD is used as input (Fig. 3). To the best of our knowledge this has not been reported before.

In Fig. 1 the average depth of RBT does not change. Hence, that is not the reason for the observation. We made sure that setting off compiler optimizations and altering the interval of changing the distribution (value t) did not matter. However, decreasing the size of tree n reduced the advantage of locality of reference down to 25%, but even with very small trees (e.g., $n = 8$) the phenomenon occurred. This contradicts the observation of Bell and Gupta [7] who had a tree of size 4 095. A natural explanation for this is that both operating system and hardware efficiency have progressed significantly since their evaluation. In particular, cache management has progressed in recent years.

In fact, caching could be the reason for this phenomenon. ZD weights the most accessed items very much and thus, with a high probability, only few items are accessed. Let us assume a cache so small that only one path to a node in the whole tree fits it. With high locality even this is useful because a single item is accessed most of the time and, thus, very often the access path is already in cache. Modern memory hierarchy generally consists of many cache layers and for every one of them there is an amount of nodes or paths that fit in. Thus, ZD makes static BSTs use caching very efficiently [22]. Paging complicates the analysis in practice, but the basic idea is the same. A splay tree does not gain as much benefit from caching. In it the access times are more related to the average depth of an accesses in the tree. A natural cause for this could be the amount of restructuring and memory writing the splay tree does.

We also tried the same evaluation with high skewness but with only one level of working sets. The items in the working set were accessed multiple times uniformly. With a changing probability items were also accessed uniformly outside the working set. The same phenomenon did present itself only slightly. In this case fitting only some of the items in the cache is not at all as useful as it is for ZD. This issue, of course, ought to be studied more thoroughly. Nevertheless, it raises the question of how cautious we should be when generalizing the results with ZD. However, there are studies with real-life data that rank splay tree very efficient in certain situations [9]. Maybe other alternatives (e.g. Lévy distribution [23] or actual web page requests) for modeling locality of reference should be used.

7   Conclusion

This work studied whether it is possible to gain advantage for splay trees by taking the properties of the access sequence into account. On one hand, both the average access depth and time were reduced when compared to randomized splay. This was also the case when compared to splay with high skewness. On the other hand, splay trees excelled with high locality of reference. Also red-black trees were still usually more efficient in these experiments.

Further variations for our algorithms can easily be designed. For example the method can be applied to most of the splay tree versions introduced in [2]. These versions include bottom-up splaying and semi-splaying. Also executing a window of splay operations during change of working set is possible.

Acknowledgments

We would like to thank the anonymous reviewers for insightful and helpful comments. This work was supported by Academy of Finland projects INTENTS (206280), ALEA (210795), and "Machine learning and online data structures" (119699). Moreover, the work of T. Aho and J. Kujala is financially supported by Tampere Graduate School in Information Science and Engineering.

References

1. Cormen, T.H., Leiserson, C.E., Rivest, R.L., Stein, C.: Introduction to Algorithms, 2nd edn. The MIT Press, Cambridge, MA (2001)
2. Sleator, D.D., Tarjan, R.E.: Self-adjusting binary search trees. Journal of the ACM 32(3), 652–686 (1985)
3. Fürer, M.: Randomized splay trees. In: Proceedings of the Tenth Annual ACM-SIAM Symposium on Discrete Algorithms, Baltimore, MD, pp. 903–904. SIAM, Philadelphia, PA (1999)

4. Albers, S., Karpinski, M.: Randomized splay trees: Theoretical and experimental results. Information Processing Letters 81(4), 213–221 (2002)
5. Lai, T.W., Wood, D.: Adaptive heuristics for binary search trees and constant linkage cost. In: Proceedings of the Second Annual ACM-SIAM Symposium on Discrete Algorithms, San Francisco, CA, pp. 72–77. SIAM, Philadelphia (1991)
6. Williams, H.E., Zobel, J., Heinz, S.: Self-adjusting trees in practice for large text collections. Software: Practice and Experience 31(10), 925–939 (2001)
7. Bell, J., Gupta, G.: An evaluation of self-adjusting binary search tree techniques. Software: Practice and Experience 23(4), 369–382 (1993)
8. Heinz, S., Zobel, J.: Performance of data structures for small sets of strings. Australian Computer Science Communications 24(1), 87–94 (2002)
9. Pfaff, B.: Performance analysis of BSTs in system software. ACM SIGMETRICS Performance Evaluation Review 32(1), 410–411 (2004)
10. Lee, E.K., Martel, C.U.: When to use splay trees. Software: Practice and Experience 37(15), 1559–1575 (2007)
11. Denning, P.J.: Working sets past and present. IEEE Transactions on Software Engineering 6(1), 64–84 (1980)
12. Subramanian, A.: An explanation of splaying. Journal of Algorithms 20(3), 512–525 (1996)
13. Badr, G.H., Oommen, B.J.: Self-adjusting of ternary search tries using conditional rotations and randomized heuristics. The Computer Journal 48(2), 200–219 (2005)
14. Iacono, J.: Alternatives to splay trees with $O(\log n)$ worst-case access times. In: Proceedings of the Twelfth Annual ACM-SIAM Symposium on Discrete Algorithms, Washington, DC, pp. 516–522. SIAM, Philadelphia (2001)
15. Sleator, D.D., Tarjan, R.E.: Amortized efficiency of list update and paging rules. Communications of the ACM 28(2), 202–208 (1985)
16. Amer, A., Oommen, B.J.: Lists on lists: A framework for self-organizing lists in environments with locality of reference. In: Àlvarez, C., Serna, M.J. (eds.) WEA 2006. LNCS, vol. 4007, pp. 109–120. Springer, Heidelberg (2006)
17. Dhodapkar, A.S., Smith, J.E.: Managing multi-configuration hardware via dynamic working set analysis. In: Proceedings of the 29th Annual International Symposium on Computer Architecture, Anchorage, AK, pp. 233–244. IEEE Computer Society, Los Alamitos (2002)
18. Seidel, R., Aragon, C.: Randomized search trees. Algorithmica 16(4), 464–497 (1996)
19. Cheetham, R.P., Oommen, B.J., Ng, D.T.: Adaptive structuring of binary search trees using conditional rotations. IEEE Transactions on Knowledge and Data Engineering 5(4), 695–704 (1993)
20. Martínez, C., Roura, S.: Randomized binary search trees. Journal of the ACM 45(2), 288–323 (1998)
21. Knuth, D.E.: The Art of Computer Programming. Sorting and Searching, 2nd edn., vol. 3. Addison-Wesley, Boston (1998)
22. Brodal, G.S., Fagerberg, R., Jacob, R.: Cache oblivious search trees via binary trees of small height. In: Proceedings of the Thirteenth Annual ACM-SIAM Symposium on Discrete Algorithms, San Francisco, CA, pp. 39–48. SIAM, Philadelphia, PA (2002)
23. Applebaum, D.: Lévy processes — from probability to finance and quantum groups. Notices of the American Mathematical Society 51(11), 1336–1347 (2004)

Engineering Burstsort: Towards Fast In-Place String Sorting

Ranjan Sinha and Anthony Wirth

Department of Computer Science and Software Engineering,
The University of Melbourne, Australia
rsinha@csse.unimelb.edu.au, awirth@csse.unimelb.edu.au

Abstract. Burstsort is a trie-based string sorting algorithm that distributes strings into small buckets whose contents are then sorted in cache. This approach has earlier been demonstrated to be efficient on modern cache-based processors [Sinha & Zobel, JEA 2004]. In this paper, we introduce improvements that reduce by a significant margin the memory requirements of burstsort. Excess memory has been reduced by an order of magnitude so that it is now less than 1% greater than an in-place algorithm. These techniques can be applied to existing variants of burstsort, as well as other string algorithms.

We redesigned the buckets, introducing sub-buckets and an index structure for them, which resulted in an order-of-magnitude space reduction. We also show the practicality of moving some fields from the trie nodes to the insertion point (for the next string pointer) in the bucket; this technique reduces memory usage of the trie nodes by one-third. Significantly, the overall impact on the speed of burstsort by combining these memory usage improvements is not unfavourable on real-world string collections. In addition, during the bucket-sorting phase, the string suffixes are copied to a small buffer to improve their spatial locality, lowering the running time of burstsort by up to 30%.

1   Introduction

This paper revisits the issue of sorting strings efficiently. String sorting remains a key step in solving contemporary data management problems. Arge *et al.* [3] note that "string sorting is the most general formulation of sorting because it comprises integer sorting (i.e., strings of length one), multikey sorting (i.e., equal-length strings) and variable-length key sorting (i.e., arbitrarily long strings)". Compared to sorting fixed-length keys (such as integers), efficient sorting of variable-length string keys is more challenging. First, string lengths are variable, and swapping strings is not as simple as swapping integers. Second, strings are compared one character at a time, instead of the entire key being compared, and thus require more instructions. Third, strings are traditionally accessed using pointers; the strings are not moved from their original locations due to string copying costs.

C.C. McGeoch (Ed.): WEA 2008, LNCS 5038, pp. 14–27, 2008.

1.1   Traditional Approaches to String Sorting

Standard string sorting algorithms, as taught in a typical undergraduate education, start by creating an array of pointers to strings; they then permute these pointers so that their order reflects the lexicographic order of the strings. Although comparison of long string keys may be time-consuming, with this method at least the cost of moving the keys is avoided. The best existing algorithms for string sorting that honour this rule include multikey quicksort [6] and variants of radix sort [2], including the so-called MBM algorithm [16].

Traditionally, we are taught to measure algorithm speed by the number of instructions, such as *comparisons* or *moves*. However, in recent years the cost of retrieving items from main memory (when they are not in cache), or of translating virtual addresses that are not in the *translation lookaside buffer* (TLB) have come to dominate algorithm running times. The principal principle is *locality of reference*: if data is physically near data that was recently processed, or was itself processed not long ago, then it is likely to be in the cache and may be accessed quickly. We leave further details to Section 1.3. Note that stability is ignored in this paper as any sorting algorithm can be made stable by appending the rank of each key in the input [11,20].

1.2   Burstsort

Burstsort is a technique that combines the burst trie [13] with standard (string) sorting algorithms [6,16]. It was introduced by the first author, and its variants are amongst the fastest algorithms for sorting strings on current hardware [21]. The standard burstsort algorithm is known as P-burstsort, P referring to *pointer*. In P-burstsort, sorting takes place in two stages: (i) the strings are inserted into the trie structure, effectively partitioned by their prefixes into *buckets*, (ii) the trie is traversed in-order and the contents of each bucket (strings with a common prefix) are sorted and pointers to the strings are output in lexicographic order.

The trie. As shown in Figure 1, each trie node contains a collection of pointers to other nodes or to buckets. There is one pointer for each symbol in the alphabet so a node effectively represents a string prefix. Note that each node contains an additional pointer (labelled ⊥) to a special bucket for strings that are identical to the prefix that the node represents.

The trie starts off with one node, but grows when the distribution of strings causes a bucket to become too large. Whenever a bucket does become *too large*, it is *burst*: the trie expands in depth at that point so that there is now a child node for each symbol of the alphabet, each node having a full collection of pointers to buckets.

Cache efficiency. Although the use of a trie is reminiscent of radixsort, in burstsort each string is only accessed a small number of times: when inserted (cache-efficient), whenever its bucket is burst, or when its bucket is sorted. In practice this low dereferencing overhead makes the algorithm faster than radixsort or quicksort.

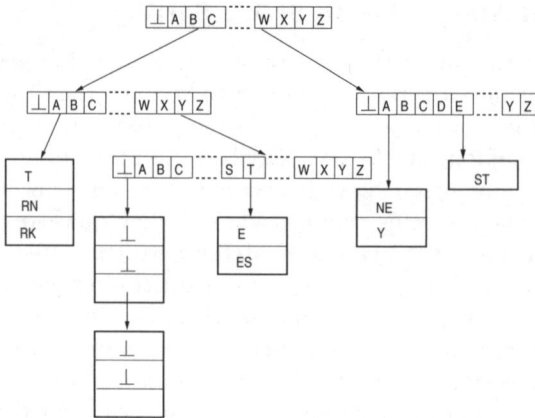

Fig. 1. Index structure of P-Burstsort. Strings inserted into the burst trie are "bat", "barn", "bark", "by", "by", "by", "by", "byte", "bytes", "wane", "way", and "west". The figure has been adapted from [21].

Sorting of buckets. The standard P-burstsort [21] uses a fast multikey quicksort for sorting the buckets, and on real-world collections has been observed to be almost twice as fast as previous algorithms. The original burstsort paper also proposes that buckets be grown *exponentially*, starting with a small bucket, then growing by a certain factor whenever the bucket is full, until a maximum size, when the bucket is burst. This wastes less memory than using a fixed size for the bucket, but increases the memory management load.

Figure 2 reminds the reader of the significant gains in running time of P-burstsort compared to three of the best string sorting algorithms: adaptive radixsort, multikey quicksort and MBM radixsort.

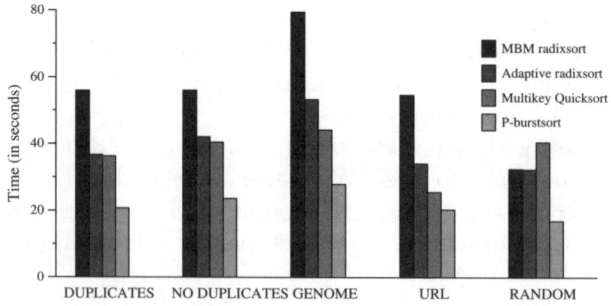

Fig. 2. Time (in seconds) to sort with adaptive radixsort, multikey quicksort, MBM radixsort, and P-Burstsort for all five collections (see Section 4) on a Pentium IV machine with a small 512 KB L2 cache. The bucket threshold used for P-burstsort is 32768.

Previous improvements. A reduction in the number of bursts results in a reduction in string accesses. To that end, in the sampling-based burstsort variants we used a small sample of strings (selected uniformly at random) to create an initial approximate trie, prior to inserting all the strings in the trie [22]. Although this approach reduced the number of cache misses, we believe there remains scope for investigation into more sophisticated schemes.

In P-burstsort, strings are accessed by pointers to them; only the pointers are moved, as is the case in traditional string sorting methods. It is vital that the locality of string accesses is improved, especially during bursts and when the buckets are being sorted. Hence, in the copy-based approach [20], strings were actually copied into the buckets from the start to improve string locality, admittedly at the cost of higher instruction counts. However, we found that the performance improves significantly, largely due to reduced cache and TLB misses.

Memory use. The priority in the earlier versions of burstsort was to increase the speed of sorting. Analysing the memory demand has so far been largely ignored, but is a major focus in this paper; we outline the contributions below.

1.3   Related Work

String sorting. There have been several advances in fast sorting techniques designed for strings. These have primarily focused on reducing instruction count, assuming a unit-cost RAM model [1,14]. For example, three-way partitioning is an important quicksort innovation [7]. Splaysort, an adaptive sorting algorithm, introduced by Moffat *et al* [17], is a combination of the splaytree data structure and insertionsort. Improvements to radixsort for strings were proposed by McIlroy *et al* [16], and by Andersson and Nilsson [2]. Bentley and Sedgewick [6] introduced a hybrid of quicksort and radixsort named three-way radix quicksort [18]; they then extended this to produce multikey quicksort [6].

In this paper, we compare our algorithms with adaptive radixsort [2], multikey quicksort [6] and MBM radixsort [16], as they have been observed to be amongst the fastest [21]. The performance of other algorithms can be obtained from the first author's earlier papers [20,21].

Cache-aware algorithms. While the radix sorts have a low instruction count—the traditional measure of computation speed—they do not necessarily use the cache efficiently for sorting variable-length strings. In earlier experiments [21], on the larger datasets there were typically 5 to 10 cache misses per string during radix-based sorting on a machine with 1 MB L2 cache. Accesses to the strings account for a large proportion of cache misses. Approaches that can make string sorting algorithms more cache-friendly include: (a) using data structures that reduce the number of string accesses; (b) improving the spatial locality of strings, so that strings that are likely to be compared are kept nearer each other; and (c) reducing or eliminating inefficient pointer-based string references.

Cache-oblivious algorithms. Frigo *et al.* [10] introduced *cache-oblivious* algorithms, a novel design approach that respects memory hierarchies. While (previously-mentioned) cache-aware algorithms need to be aware of machine parameters and may extract the best performance from a particular machine, they may not be portable. In contrast, the notion of cache-oblivious design suggests a highly portable algorithm. Though the cache-oblivious model makes several simplifying assumptions [9], it is nevertheless an attractive and simple model for analyzing data structures in hierarchical memory. Recent results [8,4] indicate that algorithms developed in this model can be competitive with cache-aware implementations. Finally, while there has been related work in the external memory domain, the techniques do not necessarily transfer well to in-memory algorithms.

1.4   Our Contributions

The goal of practical string sorting is to produce a fast and, ideally, an in-place algorithm that is efficient on real-world collections and on real-world machines. In this paper, we investigate the memory usage of burstsort and improve the cache efficiency of the bucket sorting phase.

First, we redesign the buckets of P-burstsort so that the memory requirement of the index structure is drastically reduced. Second, we also introduce a moving field approach whereby a field from the trie node is moved to the point in the bucket where a string is about to be inserted—and is thus shifted with each string insertion—resulting in a further reduction in memory use. These memory reduction techniques show negligible adverse impact on speed and are also applicable to the copy-based [20] variants of burstsort, though we do not evaluate the effects here. As a consequence, memory usage is just 1% greater than an in-place algorithm.

Third, the cache efficiency of bucket sorting is further improved by first copying the string suffixes to a small string buffer before they are sorted. This ensures spatial locality for a process that may well access strings frequently.

1.5   Paper Organization

The remainder of the paper is organized as follows. In Section 2 we show how a substantial redesign of the buckets results in significant reduction in memory waste, so that the algorithm is barely more expensive than an in-place approach. This work is enhanced in Section 2 with a brief discussion of the benefits of moving various bookkeeping fields from the trie nodes to the buckets. In Section 3 we show how buffering a bucket's strings, before sorting them, can lower the running time by a further 30%. In Section 4 we outline the experiments that we performed, and then analyze the results in Section 5. We conclude the paper and set out future tasks in Section 6.

2   Bucket Redesign

The primary aim in this paper is to reduce the memory used by P-burstsort, especially by the buckets. The bucket design used in the original P-burstsort [21], is an array that grows dynamically, by a factor of 2, from a minimum size of

2 up to a maximum size of 8192. While this design proved to be fast, it may lead to significant amounts of *unused* memory. For example, theoretically, if the number of pointers in the buckets were distributed uniformly in the range 1 to L, on average about $L/6$ spaces in each bucket would be unused. Of course, on most collections the distribution of the number of string pointers in a bucket would not be uniform, but there is still waste incurred by barely-filled buckets.

In the BR variant of P-burstsort, we replace each bucket with an array of pointers to sub-buckets. Each sub-bucket is allocated as needed: starting at size 2, it grows exponentially, on demand, until it reached size L/k, at which point the next sub-bucket is allocated. When the total amount of sub-bucket space exceeds L, the node and buckets are burst (as in P-burstsort). Naturally, there is a trade-off between the time spent (re)allocating memory, and the space wasted by the buckets.

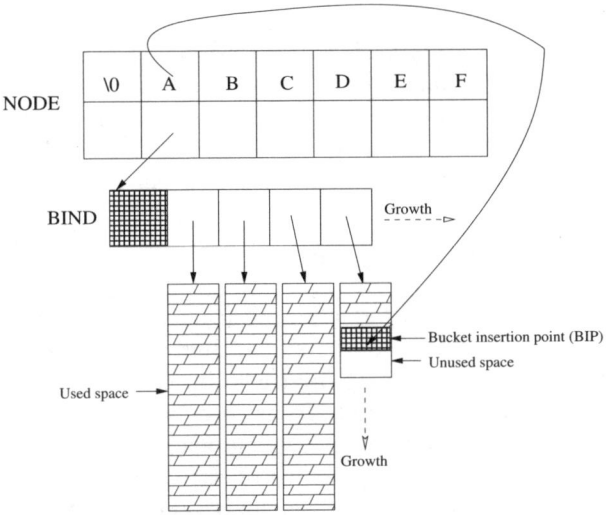

Fig. 3. P_{+BR+MF}-Burstsort: This figure shows the index structure of P-burstsort after incorporating the bucket redesign (BR) and moving field (MF) optimizations

In addition, an index structure is created to manage the sub-buckets; this auxiliary structure must be small so its own creation does not outweigh the benefit of smaller (sub-)buckets. The first component of the index structure, the *bucket index* (BIND), is the dynamically-growing array of pointers to the sub-buckets. It has some auxiliary fields that maintain information about the current state of the sub-bucket structure. The BIND array grows cell-by-cell, only on demand, therefore not wasting space. The other component is at the *bucket insertion point* (BIP), where the next string suffix pointer would be inserted, and therefore moves ahead one word with each insertion (see Figure 3).

In this P_{+BR}-Burstsort variant, only pointers to the strings are copied to the buckets, and the trie node contains two fields: a pointer to the BIND and a pointer to the BIP. The only bucket that does not grow in this BR manner is the special bucket ($\perp$) for short (consumed) strings, whose BIND index can grow arbitrarily large.

Note that the BIND structure is only accessed during the creation of the bucket, adding a new sub-bucket, and during bursting. These occurrences are relatively rare, compared with the number of times strings are accessed.

$O(\sqrt{L})$ *bucket growth.* From a theoretical point of view, exponentially-growing buckets make sense principally when the maximum bucket size is unknown. Given that we have a bound on the bucket size, from a worst-case, or uniform distribution, point of view, buckets that grow following the sequence $\sqrt{L}, 2\sqrt{L}, \ldots, L$ seem to make more sense (see for example Exercise 3.2.3 in Levitin's text [15]). With typical values of L and k being 8192 and 32, our sub-bucket data structure does not quite match this, but it is close to a practical application of this principle.

Moving fields from trie nodes to bucket. It is advantageous to keep the trie nodes as compact as possible so that they are mostly cache-resident. In the Moving Field (MF) approach, we copy the latter field fields to the unused space in the bucket, just at the bucket insertion point (BIP). This approach makes burstsort more scalable (due to compact nodes), while simultaneously saving memory.

3   Buffer-Based String Sorting

In P-burstsort, string suffixes are not moved from their original locations: the aim is to sort the pointers to the strings, rather than the strings themselves. Nevertheless, a significant proportion of the cache misses of burstsort occur when the contents of the buckets are sorted. To be compared, the string suffixes must be fetched into cache as required: on average each string must be compared $\Theta(\log(L/k))$ times. In practice, the fetching of string suffixes observes poor spatial locality, especially for strings that have large prefixes in common. Moreover, with large lines in cache, those imported may not contain only bucket-string content, but other data that is not useful.

It was observed, by Sinha *et al.* [20], that actually copying the strings into buckets improves spatial locality, and cache efficiency. In that spirit, we introduce a small buffer to copy the string suffixes into during the bucket-sorting phase to improve their spatial locality and effectively use the cache lines. In a single pass of the bucket of pointers, we fetch all the string suffixes. We then create pointers to the new locations of the string suffixes, while keeping track of the pointers to the actual strings. The extra buffer storage is reused for each bucket and thus its effect on total memory used is negligible. Once the strings are sorted, a sequential traversal of the pointer buffer copies the original pointers to the source array in sorted order.

In Section 5 we show that this String Buffer (SB) modification increases the sort speed, especially for collections with large distinguishing prefixes, such as

URLs. Such collections require the most accesses to the strings and thus stand to benefit most.

The only disadvantage is the amount of work involved in copying suffixes and pointers to these buffers and then back to their source arrays. The SB approach may perform poorly for collections with small distinguishing prefix, such as the random collection, in which the strings may only be accessed once during bucket sorting anyway. But even for such collections, this approach is expected to scale better and be less dependent on the cache line size.

4 Experimental Design

Our experiments measure the time and memory usage of string sorting. In addition, we use cache simulators such as `cachegrind` [19] to measure the instruction count and L2 cache misses.

Data Collections. We use four real-world text collections: *duplicates, no duplicates, genome,* and *URL.* In addition, we also create a *random* collection in which the characters are generated uniformly at random. These collections, whose details are provided in Table 1, are similar to those used in previous works [21,22,20].

Table 1. Statistics of the data collections used in the experiments

	Size (Mbytes)	Distinct Words ($\times 10^5$)	Word Occurrences ($\times 10^5$)
Duplicates	304	70	316
No duplicates	382	316	316
Genome	302	2.6	316
Random	317	260	316
URL	304	13	100

The *duplicates* and *no duplicates* collections were obtained from the Wall Street Journal subcollection of TREC web data [12]. The *duplicates* collection contains words in occurrence order and includes duplicates, while the *no duplicates* collection contains only unique strings that are word pairs in occurrence order. The genome collection consists of fixed-length strings (of length 9 characters), extracted from nucleotide strings from the Genbank collection [5]. The URL collection is obtained in order of occurrence from the TREC web data.

Algorithms compared with. The performance of the new burstsort enhancements is compared to the original P-burstsort [21], adaptive radixsort [2], a fast multi-key quicksort [6], and MBM radixsort [16].

Algorithm parameters. The buckets in burstsort are grown exponentially by a factor-of-2 starting from a size of 2 to L, where L is the bucket threshold. In our experiments we varied L from a minimum of 8192 to a maximum of $131,072$.

Note that for each individual sub-bucket, L/k is 256, where k is the number of sub-buckets. The alphabet size was restricted to 128 symbols with the random collection having 95 symbols.

Table 2. Architectural parameters of the machine used for experiments

Workstation	Pentium	Pentium	PowerPC
Processor type	Pentium IV	Core 2	PowerPC 970
Clock rate	2800 MHz	2400 MHz	1800 MHz
L1 data cache (KB)	8	32	32
L1 line size (bytes)	64	64	128
L2 cache (KB)	512	4096	512
L2 block size (bytes)	64	64	128
Memory size (MB)	2048	2048	512

Machines. The experiments were conducted on a 2800 MHz Pentium IV Machine with a relatively small 512 KB L2 cache and 2048 MB memory. The operating system was Linux with kernel 2.6.7 under light load. The highest compiler optimization O3 has been used in all the experiments. We also used a more recent dual-core machine with relatively large 4096 KB L2 cache as well as a PowerPC 970 architecture. Further details of the machines are shown in Table 2.

All reported times are measured using the `clock` function, and are the average of 10 runs. As these runs were performed on a machine under light load and on 300 megabyte data sets, the standard deviation is small. On the PowerPC, owing to the smaller memory, we used a smaller data set with 10 million strings [21].

5   Discussion

Bucket redesign and moving fields. All pointer-based string sorting algorithms must create space for the pointer array. The key memory overhead of P-Burstsort is its burst trie-style index structure. Table 3 shows this extra memory usage, including unused space in buckets, memory allocation bytes and associated index structures of the previous algorithm, and the variants introduced here.

The BR and MF modifications cause a large reduction in memory use. For the three real-world collections (Duplicates, No Duplicates, and Genome) there is at least a factor 35 reduction. Table 3 also confirms that increasing bucket sizes result in smaller indexes in P_{+BR+MF}-Burstsort, unlike in P-Burstsort (except for the Random collection).

The BIND and BIP structures require additional maintenance and dynamic memory allocation. Table 4 shows that the number of dynamic memory allocations, in the new variants, increased by over an order-of-magnitude. The good news is that although the memory demand drops significantly, running times increase by only 10% (see Table 5).

We also observe that the MF technique speeds up the sorting of the URL collection in Table 5, due to the relatively large number of trie nodes in that

Table 3. Memory use (in megabytes) incurred by the index structure of P-Burstsort and (the new) P_{+BR+MF}-Burstsort for different bucket thresholds and collections

		Collections				
Threshold	P-Burstsort	Duplicates	No duplicates	Genome	Random	URL
8192	None	94.37	109.37	53.37	47.37	23.85
	+BR+MF	10.37	12.37	4.37	3.37	3.85
16384	None	94.37	100.37	57.37	50.37	24.85
	+BR+MF	5.37	6.37	2.37	3.37	1.85
32768	None	94.37	95.37	48.37	50.37	24.85
	+BR+MF	3.37	3.37	1.37	3.37	0.85
65536	None	90.37	92.37	55.37	61.37	25.85
	+BR+MF	2.37	2.37	1.37	3.37	0.85

Table 4. Number of dynamic memory allocations of P-burstsort and P_{+BR+MF}-Burstsort. The bucket threshold is 32768.

Algorithm	Duplicates	No duplicates	Genome	Random	URL
P-Burstsort	271,281	307,572	76,625	109,922	79,451
P_{+BR+MF}-Burstsort	1,936,175	1,828,805	1,948,281	1,205,117	1,293,854
Factor increase	7.13	5.94	25.42	10.96	16.28

collection. Making these trie nodes compact with the MF enhancement makes up for the cost of shifting the fields. Thus, these enhancements not only significantly reduce the memory usage but also aids in speeding up sorting.

String buffer. The SB modifications, described in Section 3, are a successful enhancement overall. Table 5 shows that for all real-world collections, which have reasonable distinguishing prefix, this approach is beneficial. Only for the Random collection, whose strings may need be fetched only once during bucket sorting, does the string buffer approach result in a slight slowdown. The average number of strings (for the Random collection) in each bucket is less than half the number of cache lines, thus the strings are mostly cache-resident as they are sorted.

Table 5. Sorting time (in seconds) as a function of algorithm modification for all five collections on the Pentium IV

	Collections				
P-Burstsort	Duplicates	No duplicates	Genome	Random	URL
None	20.76	23.64	27.92	16.98	20.36
+SB	18.44	21.16	20.25	18.31	13.33
+BR	22.59	25.25	29.88	20.80	21.03
+BR+MF	23.11	26.02	30.00	22.40	20.65
+SB+BR+MF	22.04	24.71	23.08	29.39	13.87

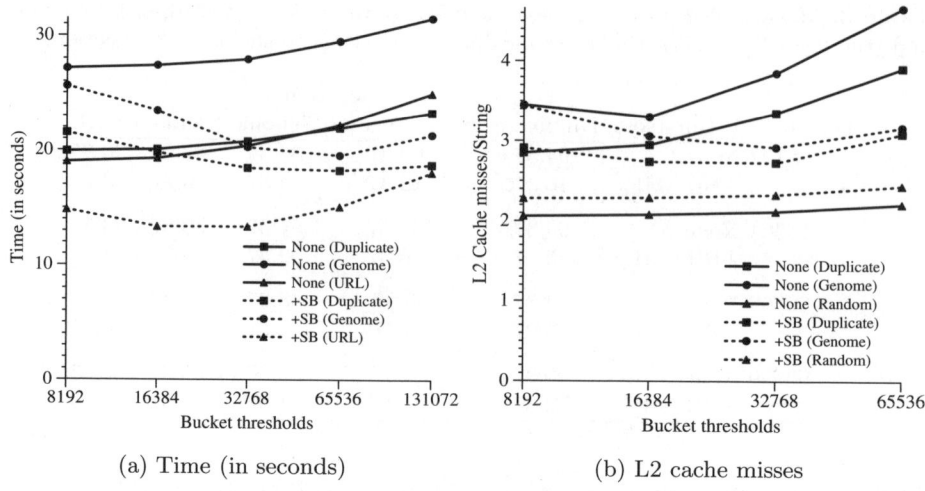

(a) Time (in seconds)   (b) L2 cache misses

Fig. 4. Sorting time (in seconds) and L2 cache misses per string incurred by P-Burstsort and P_{+SB}-Burstsort as a function of bucket thresholds on the Pentium IV

Figure 4 and Table 5 confirms that for collections with larger distinguishing prefix, such as genome and URL, the approach is indeed the most successful and reduces running time by about 30%.

The increase in instruction count (using cachegrind), by up to 80% is more than compensated for by small reductions in the number of L2 cache misses, shown in Figure 4 (b). Moreover, on machines such as PowerPC (discussed below), where the TLB misses are expensive, such an approach is beneficial. The SB approach adapts better to the cache capacity and enhances scalability.

We observe that the BR and MF modifications lead to lower instruction counts (by 7%) due to the reduced copying costs from using small sub-buckets (even for the Random collection). The small increase in cache misses (of 8%) by $P_{+SB+BR+MF}$-Burstsort over P_{+SB}-Burstsort for the real-world collections are due to BIND accesses and moving fields during string insertion and bursts. Combining all three modifications results in a lowering of running time in the Genome and URL collections, a small increase in the Duplicates and No Duplicates collections (of 6% and 4.5% respectively), but a poor performance in the (unrealistic) Random collection. Below, we show that the performance of these approaches on other machine architectures can simultaneously reduce memory usage and indeed improve performance.

Other machine architectures. On the Pentium Core 2 machine, the running time of P_{+BR}-burstsort is faster than that of P-burstsort for *all* real-world collections (shown in Table 6). Similarly, on the PowerPC, P_{+BR}-burstsort was up to 10% faster than that of P-burstsort (shown in Table 7). On another small cache machine (PowerPC), the SB modification reduced the running time by up to 40% (see Table 7). Thus, using a small buffer to copy string suffixes prior to bucket

Table 6. Sorting time (in seconds) for all five collections on the Pentium Core 2 machine. The bucket threshold is 32768.

Algorithm	Duplicates	No duplicates	Genome	Random	URL
Adaptive radixsort	13.63	15.04	17.75	9.72	9.01
MBM radixsort	15.57	16.01	22.38	10.61	13.53
Multikey	12.19	14.04	13.09	12.95	6.39
P-Burstsort	7.45	8.81	8.63	5.80	4.92
P_{+BR}-Burstsort	7.25	8.64	8.31	6.10	4.60
P_{+BR+MF}-Burstsort	7.26	8.75	8.23	6.28	4.54

Table 7. Sorting time (in seconds) for all five collections on the PowerPC 970 machine. The collections contain 10 million strings. The bucket threshold is 32768.

Algorithm	Duplicates	No duplicates	Genome	Random	URL
Adaptive radixsort	15.48	17.74	23.43	13.64	55.33
MBM radixsort	15.24	16.20	24.59	9.05	74.33
Multikey	14.91	17.16	21.94	18.68	58.75
P-Burstsort	10.22	11.71	16.69	7.19	48.26
P_{+SB}-Burstsort	8.24	9.24	10.48	8.35	23.53
P_{+BR}-Burstsort	9.26	10.79	14.90	7.66	45.60
P_{+BR+SB}-Burstsort	7.68	8.74	9.25	9.50	22.61

sorting is beneficial to using the cache capacity productively while reducing the TLB misses. The BR and SB techniques combine to produce the fastest times while simultaneously reducing the memory usage significantly.

These results demonstrate that the modifications work well across different machine architectures and aids in improving the speed while simultaneously enhancing scalability.

6   Conclusions and Further Work

String sorting remains a fundamental problem in computer science. It needs to be revisited because changes in computer architecture have not so much changed the objective functions, but have changed the estimates we have of them. Burstsort was already known to be fast: in this paper, its demands on main memory have been significantly reduced, without running time being compromised.

The BR enhancement enables large reductions in the bucket size, with negligible impact on sorting time, even though it requires an order-of-magnitude more dynamic allocations. The MF technique reduces the trie node memory usage by moving fields to the unused space in the bucket and shifting them with each string insertion. The success of the SB enhancement is further evidence that accessing strings in arbitrary locations (using pointers) is inefficient and there are benefits in improved spatial locality.

Now that the index structure can be reduced to around 1% of the size of the input arrays, we have produced an almost in-place string sorting algorithm that is fast in practice. Briefly, burstsort with these optimizations, is a fast and an almost in-place string sorting algorithm that is demonstrably efficient on real-world string collections, including those with large distinguishing prefixes.

Further work. With large caches now available in multicore processors, it would be interesting to see if our sampling approaches [22] can be developed further: larger caches are expected to be more tolerant of sampling errors. Can burstsort make significant speed increases by using multiple cores for sorting the buckets? In future implementations, we intend to explore the effect of trie layouts such as using an approximate van Emde Boas layout in a dynamic environment.

Acknowledgments

This work was supported by the Australian Research Council.

References

1. Aho, A., Hopcroft, J.E., Ullman, J.D.: The Design and Analysis of Computer Algorithms. Addison-Wesley, Reading (1974)
2. Andersson, A., Nilsson, S.: Implementing radixsort. ACM Jour. of Experimental Algorithmics 3(7) (1998)
3. Arge, L., Ferragina, P., Grossi, R., Vitter, J.S.: On sorting strings in external memory. In: Leighton, F.T., Shor, P. (eds.) Proc. ACM Symp. on Theory of Computation, El Paso, pp. 540–548. ACM Press, New York (1997)
4. Bender, M.A., Colton, M.F., Kuszmaul, B.C.: Cache-oblivious string b-trees. In: PODS 2006: Proceedings of the twenty-fifth ACM SIGMOD-SIGACT-SIGART symposium on Principles of database systems, New York, NY, USA, pp. 233–242. ACM Press, New York (2006)
5. Benson, D.A., Karsch-Mizrachi, I., Lipman, D.J., Ostell, J., Wheeler, D.L.: Genbank. Nucleic Acids Research 31(1), 23–27 (2003)
6. Bentley, J., Sedgewick, R.: Fast algorithms for sorting and searching strings. In: Saks, M. (ed.) Proc. Annual ACM-SIAM Symp. on Discrete Algorithms, New Orleans, LA, USA. Society for Industrial and Applied Mathematics, pp. 360–369 (1997)
7. Bentley, J.L., McIlroy, M.D.: Engineering a sort function. Software—Practice and Experience 23(11), 1249–1265 (1993)
8. Brodal, G.S., Fagerberg, R., Vinther, K.: Engineering a cache-oblivious sorting algorithm. ACM Jour. of Experimental Algorithmics 12(2.2), 23 (2007)
9. Demaine, E.D.: Cache-oblivious algorithms and data structures. In: Lecture Notes from the EEF Summer School on Massive Data Sets, BRICS, University of Aarhus, Denmark, June 2002. LNCS (2002)
10. Frigo, M., Leiserson, C.E., Prokop, H., Ramachandran, S.: Cache-oblivious algorithms. In: Beame, P. (ed.) FOCS 1999: Proceedings of the 40th Annual Symposium on Foundations of Computer Science, Washington, DC, USA, pp. 285–298. IEEE Computer Society Press, Los Alamitos (1999)

11. Graefe, G.: Implementing sorting in database systems. Computing Surveys 38(3), 1–37 (2006)
12. Harman, D.: Overview of the second text retrieval conference (TREC-2). Information Processing and Management 31(3), 271–289 (1995)
13. Heinz, S., Zobel, J., Williams, H.E.: Burst tries: A fast, efficient data structure for string keys. ACM Transactions on Information Systems 20(2), 192–223 (2002)
14. Knuth, D.E.: The Art of Computer Programming: Sorting and Searching, 2nd edn., vol. 3. Addison-Wesley, Reading (1998)
15. Levitin, A.V.: Introduction to the Design and Analysis of Algorithms, 2nd edn. Pearson, London (2007)
16. McIlroy, P.M., Bostic, K., McIlroy, M.D.: Engineering radix sort. Computing Systems 6(1), 5–27 (1993)
17. Moffat, A., Eddy, G., Petersson, O.: Splaysort: Fast, versatile, practical. Software—Practice and Experience 26(7), 781–797 (1996)
18. Sedgewick, R.: Algorithms in C, 3rd edn. Addison-Wesley Longman Publishing Co., Inc., Boston (1998)
19. Seward, J.: Valgrind—memory and cache profiler (2001),
    http://developer.kde.org/~sewardj/docs-1.9.5/cg_techdocs.html
20. Sinha, R., Ring, D., Zobel, J.: Cache-efficient string sorting using copying. ACM Jour. of Experimental Algorithmics 11(1.2) (2006)
21. Sinha, R., Zobel, J.: Cache-conscious sorting of large sets of strings with dynamic tries. ACM Jour. of Experimental Algorithmics 9(1.5) (2004)
22. Sinha, R., Zobel, J.: Using random sampling to build approximate tries for efficient string sorting. ACM Jour. of Experimental Algorithmics 10 (2005)

Comparing Integer Data Structures for 32 and 64 Bit Keys

Nicholas Nash* and David Gregg

Dept. of Computer Science, Trinity College Dublin, Ireland
{nashn, dgregg}@cs.tcd.ie

Abstract. In this paper we experimentally compare a number of data structures operating over keys that are 32 and 64-bit integers. We examine traditional comparison-based search trees as well as data structures that take advantage of the fact that the keys are integers, such as van Emde Boas trees and various trie-based data structures. We propose a variant of a *burst trie* that performs better in both time and space than all the alternative data structures. Burst tries have previously been shown to provide a very efficient base for implementing cache efficient string sorting algorithms. We find that with suitable engineering they also perform excellently as a dynamic ordered data structure operating over integer keys. We provide experimental results when the data structures operate over uniform random data. We also provide a motivating example for our study in *Valgrind*, a widely used suite of tools for the dynamic binary instrumentation of programs, and present experimental results over data sets derived from Valgrind.

1 Introduction

1.1 Background and Motivation

Maintaining a dynamic ordered data structure over a set of ordered keys is a classic problem, and a variety of data structures can be used to achieve $O(\log n)$ worst-case time for insert, delete, successor, predecessor and search operations, when maintaining a set of n keys. Examples of such data structures include AVL trees [10], B-trees [2,10] and red-black trees [4]. Red-black trees in particular see widespread use via their GNU *C++* STL implementation [19].

Where the keys are known to be integers, better asymptotic results can be obtained by data structures that do not rely solely on pair-wise key comparisons. For example, the stratified trees of van Emde Boas [21] support all operations in $O(\log w)$ worst-case time, when operating on w-bit keys, while Willard's q-fast tries [22] support all operations in $O(\sqrt{w})$ worst-case time.

Such data structures are attractive because of their superior worst-case times compared to comparison-based data structures. However, it is a significant challenge to construct implementations that reveal their better asymptotic

---

* Work supported by the Irish Research Council for Science, Engineering and Technology (IRCSET).

C.C. McGeoch (Ed.): WEA 2008, LNCS 5038, pp. 28–42, 2008.

performance, especially without occupying a large amount of extra space compared to comparison-based data structures.

In this paper we experimentally evaluate the performance of a variety of data structures when their keys are either 32 or 64-bit integers. In particular we find that a carefully engineered variant of a *burst trie* [7] provides the best performance in time, and for moderate to large numbers of keys it requires less space than even a well implemented comparison-based search tree.

A noteworthy application of our data structure occurs in the dynamic binary instrumentation tool *Valgrind* [12]. Valgrind comprises a widely used suite of tools for debugging and profiling programs. Internally, Valgrind frequently queries a dynamic ordered data structure that maps machine words to machine words. On current platforms, these machine words are either 32 or 64 bits in length. At present, Valgrind uses an AVL tree to perform these mappings. A significant performance improvement could be obtained by replacing the AVL tree with a more efficient data structure. We present experimental results for a number of data structures when used to track every data memory access done by a program, as for example some Valgrind-based tools do.

1.2   Related Work and Contributions

In this paper we compare the performance of a carefully engineered variant of a burst trie in both time and space to AVL trees, red-black trees and B-trees. Aside from these commonplace general purpose data structures, we also experimentally examine the performance of two slightly more ad-hoc data structures [5,11] which are tailored for the case of integers keys, and have been shown to perform well in practice. We briefly describe these data structures in the remainder of this section.

Dementiev *et al.* [5] describe the engineering of a data structure based on stratified trees [21] and demonstrate experimentally that it achieves superior performance to comparison-based data structures. We refer to their engineered data structure as an S-tree. Although highly efficient in time, the S-tree is tailored around keys of 32-bits in length and generalizing the data structure to 64-bit keys would not be feasible in practice because of the large amount of space required to maintain efficiency. Indeed, even for 32-bit keys the data structure requires more than twice as much space as a typical balanced search tree.

Korda and Raman [11] describe a data structure similar to a *q*-fast trie [22] and experimentally show that it offers performance superior to comparison-based data structures. Unlike the S-tree data structure engineered by Dementiev *et al.* this data structure is not restricted to 32-bit keys and requires less space in practice. We now briefly describe the features of Korda and Raman's data structure relevant to our discussion. We refer to their data structure as a KR-trie. A KR-trie consists of a path compressed trie containing a set of *representative* keys, $K_1 < K_2 < \cdots < K_m$. Associated with each representative key K_i is a bucket data structure B_i containing the set of keys $\{k \in S : K_i \leq k < K_{i+1}\}$ for $i < m$, and $\{k \in S : k \geq K_m\}$ for $i = m$, where S is the entire set of keys in the data structure.

Each bucket contains between 1 and $b-1$ keys. When a new key is inserted into the data structure the compressed trie is first searched for its predecessor key, giving a representative key K_i. If the associated bucket B_i already contains $b-1$ keys, a new representative key is added to the compressed trie that partitions the bucket into two new buckets containing $b/2$ keys each. Deletions operate in a similar manner to insertions, except that when two adjacent buckets B_i and B_{i+1} contain fewer than $b/2$ keys in total the keys of B_{i+1} are inserted into B_i and K_{i+1} is deleted from the trie. A search in the data structure is accomplished by a predecessor query in the compressed trie, followed by a search in the relevant bucket data structure.

There are many other non-comparison-based data structures in addition to the two just mentioned, both practical and theoretical. Two practical examples are LPC-tries [13] and the cache-friendly tries of Achyra *et al.* [1], however, we believe these data structures are less efficient or less general than the two data structures described above. For example, LPC-tries offer search operations more efficiently than binary search trees, but insertions are slower. In contrast, the data structures described above perform better than binary search trees for all operations. The tries of Achyra *et al.* appear efficient, but require knowledge of cache parameters and focus only on trie search. It is not clear how operations like predecessor and successor could be efficiently implemented, since hash tables are used inside the trie nodes.

The contribution of our experimental study is to show that a carefully engineered data structure based on the burst trie described by Heinz *et al.* [7] performs better than both the S-tree and KR-trie data structures described above, as well as the traditional comparison-based data structures.

The work of Heinz *et al.* focuses on the problem of *vocabulary accumulation*, where the keys are variable length strings. The only operations performed are insert and search, with a final in-order traversal of the burst trie. In contrast, we consider the case of integer keys with all the operations usually associated with a dynamic ordered data structure.

The contributions of our work are as follows:

- We provide a thorough experimental comparison of dynamic data structures over 32 and 64-bit integer keys. We provide time and space measurements over random data as well over data sets that occur in Valgrind, a notable application of such data structures.

- We show that burst tries extend efficiently to a dynamic ordered data structure, showing how the operations usually associated with such data structures can be implemented efficiently through careful engineering.

- We show that the data structure is more efficient in time than the best previous data structures that have been engineered for the case of integer keys. We also show that for large numbers of keys, the data structure requires less space than even space efficient implementations of comparison-based search trees.

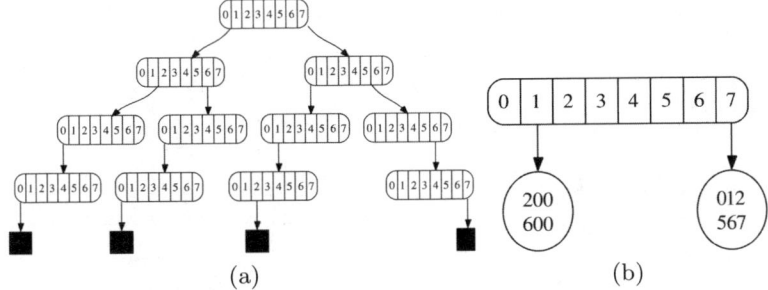

Fig. 1. (a) Shows a trie holding the keys 1200, 1600, 7012 and 7567. The leaves of the trie (black squares) hold the satellite data associated with the keys. A corresponding burst trie is shown in (b).

2   Background

In this section we provide the definition of a burst trie and some basic background information regarding the data structure.

We assume the burst trie contains fixed length keys, each of length l. A key is a sequence $k_1 \cdots k_l$ where each k_i, $1 \leq i \leq l$ is drawn from a set of digits $\{0, \ldots, U - 1\}$. In practice, our keys are 32 and 64-bit integers and we often choose $U = 256$, for implementation reasons. Given a trie T over a set of keys, we call a node *small* only if its parent has more than c descendant leaves, but the node itself has at most c descendant leaves. A burst trie with bucket size c is obtained from a trie T by replacing every small node x in T with a bucket data structure containing the keys corresponding to the leaves descendant from x and discarding all descendants of small nodes. It follows that if two keys $k_1 \cdots k_m$ and $k_1' \cdots k_m'$ reside in the same bucket of a burst trie at depth d, then $k_i = k_i'$ for $1 \leq i < d$, and only their suffixes need be stored in the bucket data structure. Figure 1(a) shows an example of a trie while Figure 1(b) shows a burst trie corresponding to it.

Although we refer to what has just been described as a burst trie, using some kind of bucketing in a trie is an old technique. Sussenguth [20] provides an early suggestion of the technique, while Knuth analyses bucketed tries [10]. In addition, Knessl and Szpankowski [8,9] analyse what they refer to as b-tries — tries in which leaf nodes hold up to b keys.

We use the term burst trie of Heinz *et al.* [7] because their work was the first to provide a large scale investigation of alternative bucket data structures, the time and space trade-offs in practice resulting from bucketing, and the *bursting* of bucket data structures during insertions, which we describe below.

Searching in a burst trie is similar to searching in a conventional trie. The digits of the key are used to determine a path in the trie that either terminates with a NIL pointer, in which case the search terminates unsuccessfully, or a bucket is found. In the latter case, the search finishes by searching the bucket data structure for the key suffix.

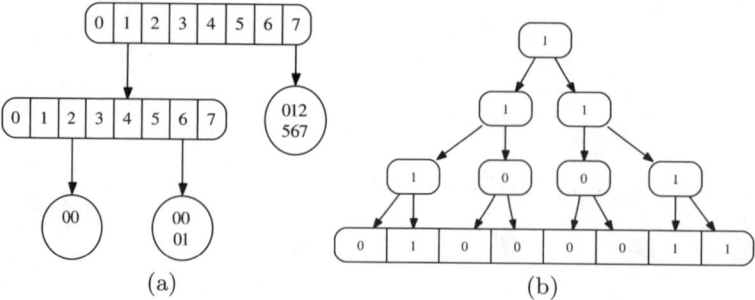

Fig. 2. (a) Shows the burst trie of Figure 1(b) after inserting the key 1601. Assuming the buckets can hold at most two key suffixes, inserting the key 1601 causes the left bucket shown in Figure 1(b) to burst. In (b) an OR-tree is shown, a possible in-node data structure for implementing a burst trie.

Insertion of a key into a burst trie is also straightforward. The digits of the key are used to locate a bucket where the key suffix should be stored. If no such bucket exists, one is created. On the other hand, if a bucket is found and it contains fewer than c keys it need not be burst and the key suffix is simply added to that bucket. Otherwise, if the bucket already contains c keys, it is burst. This involves replacing the bucket with a trie node and distributing the keys suffixes of this bucket into new buckets descending from this new trie node. Figure 2(a) shows an example of a burst operation occuring on the burst trie of Figure 1(b). It is possible that all keys from the burst bucket belong in the same bucket in the newly created node. In this case, the bursting process is repeated.

Deleting a key k from a burst trie is performed by first searching for the bucket where k is stored, as described above. If there is no such bucket, no deletion need occur. Otherwise, k is deleted from some bucket b at a node x. If b is then empty, it is deleted from x. If x then has only NIL child and bucket pointers x is deleted from the trie. This step is repeated, traversing the path from x to the root of the trie deleting ancestors encountered with only NIL child or bucket pointers. The traversal terminates when either a node with a non-nil pointer is encountered, or the root of the trie is reached.

3   Engineering Burst Tries

Although the burst trie data structure described in the preceding section leads to a highly efficient data structure, especially for strings, as shown by Heinz *et al.* [7], a little care must be taken when engineering it for the case of an ordered data structure for integer keys. Our variant of a burst trie makes use of two data structures that have a significant influence on its performance: (1) The bucket data structures at the leaves of the trie, and (2) the data structures inside the nodes of the burst trie. We describe the alternatives for this latter data structure in the next section.

3.1   In-Node Data Structures

Given a node x in a trie-based data structure with branching factor b, and an index i, $0 \leq i < b$, it is often necessary to find $\text{SUCC}(i)$, that is, the smallest $j > i$ such that $x[j] \neq \text{NIL}$. This is the bucket or child node pointer directly following $x[i]$. It is also often required to find $\text{PRED}(i)$, the largest $j < i$ such that $x[j] \neq \text{NIL}$. These operations upon nodes are required, for example to support queries on the trie for the smallest key greater than or equal to some given key. We elaborate on the precise use of these operations in Section 3.3.

The simplest data structure supporting these predecessor and successor operations is just a linear search over a bit-vector. This data structure requires only $O(1)$ time when a new bucket or child is added or removed from the node, however, PRED and SUCC are inefficient, requiring $O(b)$ time.

An alternative in-node data structure is an OR-tree. Figure 2(b) shows an example of this data structure. A breadth-first traversal of an OR-tree can be laid out in an array inside each node, requiring an additional $O(b)$ space compared to a simple bit-vector approach. However, an OR-tree offers all operations in $O(\lg b)$ time.

As a compromise between these two data structures, PRED and SUCC can be implemented using $\lceil \sqrt{b} \rceil$ counters. Where the ith counter, $0 \leq i < \lceil \sqrt{b} \rceil$ holds a count of the non-zero bits in the range $[i \lceil \sqrt{b} \rceil, i \lceil \sqrt{b} \rceil + \lceil \sqrt{b} \rceil - 1]$ (except perhaps for the last counter, which covers the range $[b - \lceil \sqrt{b} \rceil, b - 1]$). This data structure allows insertions and deletions in $O(1)$ time and supports PRED and SUCC in $O(\sqrt{b})$ time, requiring at most $\lceil \sqrt{b} \rceil$ counters to be examined followed by at most $\lceil \sqrt{b} \rceil$ bits.

To determine the most efficient in-node data structure we conducted a number of experiments, randomly populating a bit-vector and then performing a large number of successor queries. Figure 3(a) shows that the OR-tree is much less efficient than either performing a simple linear scan, or using counters to guide the search. Figures 3(b) and (c) reveal why this is so. Firstly, as Figure 3(b) shows the OR-tree causes by the far most branch mispredictions. Intuitively, one expects that an algorithm with a better time complexity increases the information it extracts from each branch instruction — thus making each branch instruction less predictable. Secondly, as Figure 3(c) shows, the OR-tree has very bad cache performance compared to the linear scan or counter based search. This is to be expected since the the OR-tree's breadth-first layout is cache unfriendly, while the other two algorithms perform linear searches, which make full use of every cache line. Note that the bad performance of the OR-tree is despite the fact that it executes the fewest instructions of any of the algorithms.

We selected the counter based search for our burst trie implementation because it has very similar performance to the linear scan in practice and better performance in the worst-case (i.e. when a trie node is very sparse). Other in-node data structures could be used. For example, recursively applying the $O(\sqrt{b})$ approach essentially leads to a stratified tree which would provide all operations in $O(\lg \lg b)$ time. However, since the branching factors of our trie nodes are never greater than 2^{16} this approach is unlikely to yield a performance benefit.

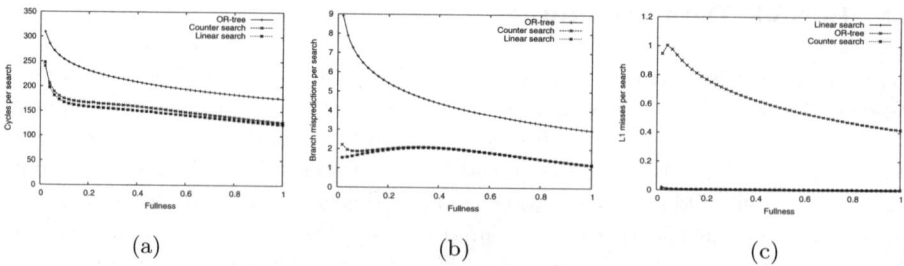

(a)   (b)   (c)

Fig. 3. (a) Shows the cycles per successor operation (i.e. time) on randomly populated 2^{14} entry bit vector. "Fullness" denotes the number of randomly inserted bits as a fraction of the total size of the bit vector. The OR-tree performs much worse than either a simple linear scan or counter based search. This is due to the large number of branch mispredictions, shown in (b) and cache misses, shown in (c), incurred by the OR-tree compared to the other data structures. The cache misses shown are level 1 data cache misses, since the entire bit-vector fits in the level 2 cache. These results are averaged over several thousand repetitions, and were gathered using PAPI [6].

In the next section we describe the second important data structure used by burst tries — the bucket data structure.

3.2   Bucket Data Structures

The choice of data structure used for the buckets of a burst trie is critical in achieving good performance. Heinz *et al.* [7] concluded that unbalanced binary search trees holding at most 35 strings offered the best performance as a bucket data structure. They also experimented with linked lists and splay trees [18]. Since the maximum number of keys stored in each bucket is modest (at most 35), a simple bucket data structure, even with bad asymptotic behaviour, may perform well. We experimented with balanced binary trees as well as with sorted arrays as bucket data structures, and found that sorted arrays are far more efficient in practice than the search trees. We also found that a bucket size of about 256 keys gave best performance. It is likely that the sorted arrays incur far fewer cache misses than the search trees. In fact, unsorted arrays of strings have been used as bucket data structures for burst tries as a basis for the *burstsort* algorithm [14,15,16,17], a cache-efficient radix sorting algorithm.

In contrast to the array buckets of the burstsort algorithm, our buckets are sorted and much smaller (in burstsort the buckets are allowed to grow until they reach the size of the processor's 2nd level cache, which can be several megabytes in size) holding at most 256 keys. The buckets are implemented as growable sorted arrays, and an insertion involves possibly doubling the size of the bucket followed by a linear scan to find the correct position for the key to be inserted.

Often the most frequent operation executed on a data structure is a search, and so searching buckets in particular should be efficient. We use a binary search that switches to a linear search when the number of keys which remain to be searched falls below a certain threshold. We found a threshold of between 16 and

32 keys gave a performance improvement over a simple binary search. Our burst trie implementation is designed to provide a mapping from a key to the satellite data associated with that key, which we refer to as the *value* for the key. To improve the spatial locality of searches the keys and values of a bucket should not be interleaved. Rather, all the keys should be stored sequentially, followed by all the values of that bucket. This ensures searching for a key makes better utilization of the processor's cache lines.

If the maximum bucket capacity is c, it takes $O(c)$ time to insert into a bucket and $O(\lg c)$ time to search in a bucket. Finally, since bursting a bucket just involves splitting the sorted sequence of keys it contains into a number of other sorted sequences, bursting a bucket also takes $O(c)$ time.

3.3   Operations

The preceding two sections have described the two main data structures required for extending burst tries to an ordered data structure. We now show how these data structures can be used efficiently to provide a burst trie with all the usual operations associated with a dynamic ordered data structure. Note that in order that predecessor and successor operations are supported efficiently, it is wise to maintain the leaves of the burst trie (i.e. the buckets) in order in a doubly linked list.

Locate. We first describe the *locate* operation, which finds the value associated with the smallest key greater than or equal to a supplied key k (or NIL if there is no such key). Assuming the path in the burst trie determined by k leads to a bucket, then that bucket is searched for the smallest key suffix greater than or equal to k's suffix, and its corresponding value is returned. In this case, the locate operation takes $O(h + \lg c)$ time, where h is the maximum height of the trie, and each bucket holds at most c keys. In the case where k does not lead to a bucket, the in-node data structure is queried to find a bucket requiring $O(\sqrt{b})$ time, thus *locate* requires $O(h + \max(\lg c, \sqrt{b}))$ time.

Insert. If inserting a new key k requires the creation of a new bucket the in-node data structure and doubly linked list of buckets must be updated. This requires finding the two buckets whose keys are the immediate predecessors and successors of k, and can be accomplished in time $O(h + \sqrt{b})$ time. Note that the in-node data structures should be augmented with indices storing the minimum and maximum non-nil pointer at each node, which we refer to as the node's LOW and HIGH fields respectively. The LOW and HIGH fields are used to avoid avoid the process of locating the predecessor and successor buckets requiring $O(h\sqrt{b})$ time. In the case where an existing non-full bucket is found for k, the insertion takes time $O(h + c)$ (recall that the buckets are simply sorted arrays). Finally, in the case where bursting must occur, the insertion can take time $O(hc)$ at worst, since an insertion to a full bucket may repeatedly cause all key suffixes to enter the same new bucket deeper in the trie. A straightforward argument can be used to show insertion requires $O(h + \max(c, \sqrt{b}))$ amortized time.

Other Operations. Deletion is carried out as was described in Section 2, except that when an empty bucket is deleted from a node, the in-node data structure and linked list of buckets should also be updated. Detecting whether a node should be removed from the trie following a deletion is accomplished by examining its LOW and HIGH fields. Note that when a node x is to be removed from the trie, its parent's in-node data structure should only be updated if the parent itself is not also to be removed as a result of the removal of x. This ensures that $O(\sqrt{b})$ time is spent updating in-node data structures, rather than $O(h\sqrt{b})$ time. Since it takes $O(c)$ time to delete a key from a bucket, deletion takes $O(h + \max(\sqrt{b}, c))$ time in total.

Predecessor and successor operations can be implemented with minor modifications to the locate operation described above. Often, predecessor and successor operations on a data structure are supported via iterators. In this case, by using the linked list of buckets, predecessor and successor both operate in constant time.

4 Results

4.1 Experimental Setup

We now describe the experimental comparison of our burst trie variant with a number of other data structures. For our experiments over 32-bit keys we used an Intel Core 2 processor with a clock-speed of 2.13GHz, a second level cache size of 2MB and 4 GB of main memory. For our experiments over 64-bit keys we used an Intel Core 2 processor with a clock speed of 2.0GHz, a second level cache size of 4MB and 4GB of main memory. Note that our experiments investigate the case where the entire data structure fits in main memory. All results we present below are averaged over several thousand runs. Although not presented below, to verify the robustness of our results we have also conducted experiments on Sun SPARC as well as PowerPC architectures, and observed results very similar to the ones we describe below.

We compare our implementation to the C++ STL map implementation [19], which uses a red-black tree. We also compare to the AVL tree implementation used internally in Valgrind [12], except that we have implemented a custom memory allocator to reduce memory usage and improve performance. We also include a comparison with an optimized B-tree implementation, as well as with the stratified tree based data structure of Dementiev et al. [5], which we refer to below as an S-tree. Finally, we include a comparison with a KR-trie (described in Section 1.2). For the KR-trie we use the same in-node data structures and bucket data structures as we used for the burst trie[1].

We used uniform random data as well as data generated internally by Valgrind to assess the relative performance of the data structures. We used Brent's

---

[1] This differs slightly from the implementation of the KR-trie described by Korda and Raman [11], since they use fixed size rather than growable arrays as buckets. However, using growable arrays improves performance and reduces memory consumption. Moreover, Korda and Raman do not precisely specify the in-node data structure they use.

[3] pseudorandom number generator implementation for generating both 32 and 64-bit random numbers. The data sets generated using Valgrind consist of the memory addresses of all the data memory accesses performed during the execution of a program. This reflects the use of the data structure to track every memory access performed by a program, as is done by some Valgrind-based tools. We generated data sets for the Linux program Top (a task viewer) as well as three applications from the K Desktop Environment: Amarok (a music player), Konqueror (a web browser and file manager) and KPDF (a PDF viewer). Each of these data sets contain between 10^7 and 10^9 operations in total, and 70-80% of the operations are loads. A load generates a search operation on the appropriate data structure while a store generates an insert operation. Currently, Valgrind uses an AVL tree to perform these operations. Note that iteration over the data structure is also required at certain times, removing the possibility of using a hash table.

4.2   Random Data

Figure 4(a) shows the time per insertion for the data structures for uniform random 32-bit keys. Note that the S-tree and red-black tree use all available memory at 2^{25} and 2^{26} keys respectively. The burst trie performs best of all the data structures, although the S-tree and KR-trie are also competitive. The comparison-based data structures are not as competitive, although the B-tree performs quite well.

Figure 4(b) shows the time per search operation, the vast majority of the searches are for keys that are not in the data structure. Before it runs out of memory, the S-tree data structure performs slightly better than the burst trie. Note that this is the only operation for which the S-tree out-performs the burst trie. The KR-trie performs worse than the burst trie and S-tree, followed by the B-tree. The binary search trees perform quite badly in comparison to these data structures.

Figure 4(c) shows the memory consumption of the data structures. The S-tree is clearly a very memory hungry data structure. The comparison-based data structures are attractive because of their uniform memory overhead. It is noteworthy that the use of a custom memory allocator for the AVL tree allows it to occupy significantly less memory than the red-black tree. The B-tree, which uses nodes consisting of growable arrays of keys of up to 4KB in size, uses the least memory of any of the comparison-based data structures. The KR-trie and burst trie are highly inefficient in their memory usage until the number of keys becomes large. However, when the number of keys becomes large, beyond about 2^{18} (262,144) keys the burst trie in particular has a very modest memory overhead. The rapid increase and then decrease in memory consumption of the KR-trie beginning at 2^{20} keys is a result of the fact that the KR-trie uses a compressed trie to control access to its buckets. Beginning at 2^{20} keys, many compressed branches of the trie are expanded, resulting in poor space utilization of their links and buckets. Subsequently, the buckets and links become filled and used, and space utilization improves again.

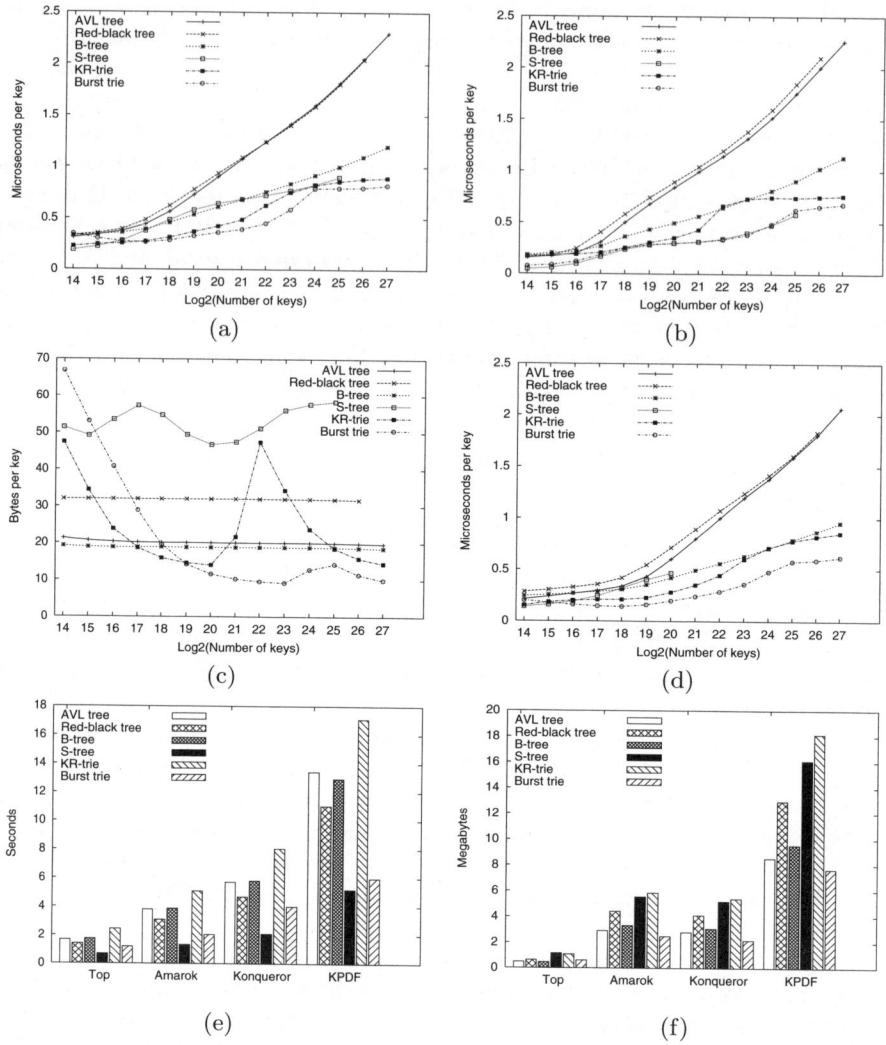

Fig. 4. These figures show a comparison of the data structures when operating on 32-bit keys. (a) Shows the time per insertion operation for the data structures. (b) Shows the time per locate operation for the data structures, a locate operation returns the smallest key greater than or equal to a given key. (c) Shows the number of bytes per key of memory consumed by the data structures following a sequence of insertions. (d) Shows the time for a mixed sequence of equiprobable insertion and deletion operations. The results of (a)—(d) are over uniform random keys. The charts in (e) and (f) show, respectively, the time and space required by the data structures required to process various Valgrind data sets. The results are discussed in Section 4.

Figure 4(d) shows the time per operation required for a mixed sequence of insertions and deletions, which occur randomly and are equiprobable. The burst

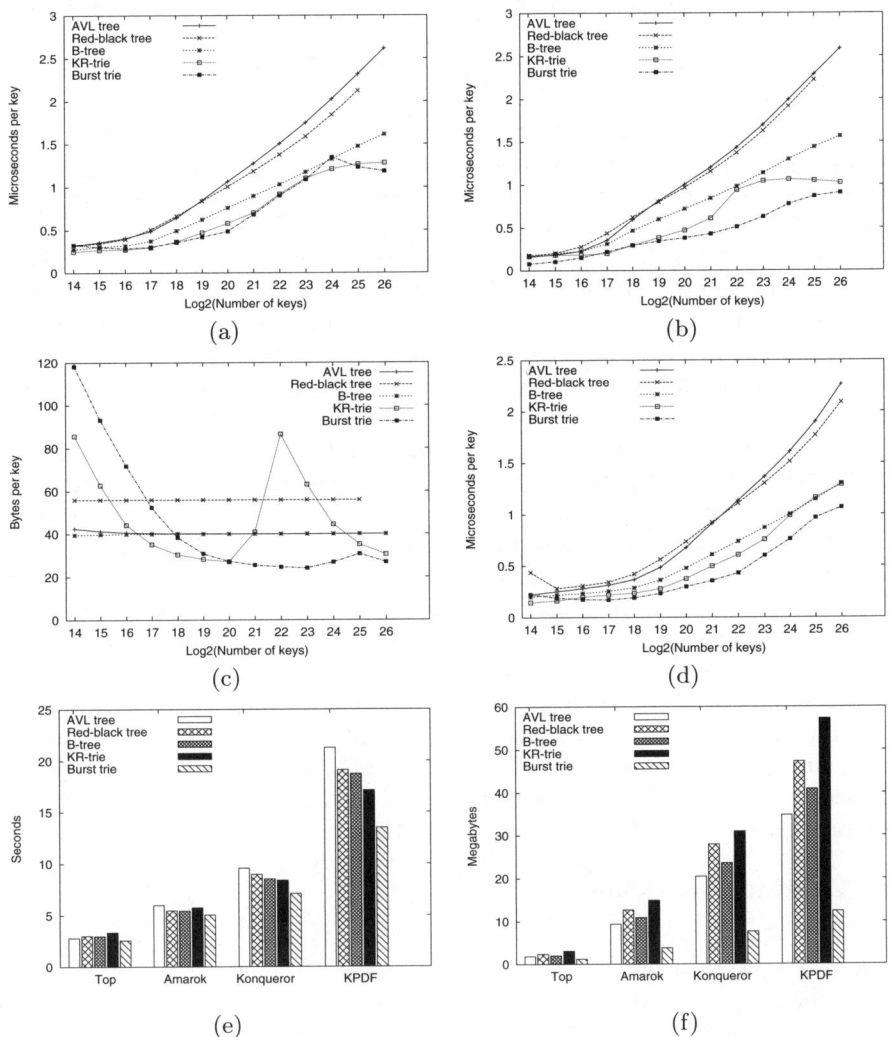

Fig. 5. These figures show a comparison of the data structures when operating on 64-bit keys, and are broadly similar to the results in the 32-bit case, shown in Figure 4. The S-tree is absent because it is restricted to 32-bit keys. (a) Shows the time per insertion operation for the data structures. (b) Shows the time per locate operation for the data structures, a locate operation returns the smallest key greater than or equal to a given key. (c) Shows the number of bytes per key of memory consumed by the data structures following a sequence of insertions. (d) Shows the time for a mixed sequence of equiprobable insertion and deletion operations. The results of (a)—(d) are over uniform random keys. The charts in (e) and (f) show, respectively, the time and space required by the data structures required to process various Valgrind data sets. The results are discussed in Section 4.

trie is also the best performing data structure in this case, although the KR-trie and B-tree also perform quite well. Unfortunately the only S-tree implementation available to us[2] had bugs in its delete operation causing it to fail on inputs larger than 2^{20} keys.

Over the full range of operations, on random 32-bit keys the burst trie's performance is competitive with or superior to that of all the other data structures in time. In addition, when the number of keys is large it also uses the least memory of any of the data structures.

The results for random 64-bit keys are shown in Figure 5(a)—(d). The S-tree is excluded from these results because it is tailored specifically for 32-bit keys, and extending it efficiently to 64-bit keys requires an enormous amount of extra space. For the remaining data structures the results for the 64-bit case are broadly similar to those observed in the 32-bit case. The burst trie and KR-trie perform better than the comparison-based data structures, with the B-tree performing the best of the comparison-based data structures. In addition, it appears the burst trie has the edge over the KR-trie in both time and space.

It is noteworthy that the burst trie achieves its space efficiency in part because it stores a trie over the common prefixes of keys (described fully in Section 2) and as a result only stores key suffixes in buckets. This also improves the cache performance of searches in the buckets. However, the trie of representative keys (described in Section 1.2) stored by the KR-trie does not guarantee that the keys in the same bucket of a KR-trie share a common prefix.

In addition, the space occupied by the burst trie could perhaps be further reduced by ensuring its trie is compressed. However, at least over uniform random data chains of single-children nodes are less probable than in a traditional trie, and so the space saved by compression may be modest. In addition, maintaining a compressed trie can be quite expensive in time.

4.3   Valgrind Workloads

Figure 4(e) shows the time for processing 32-bit Valgrind data sets of various programs (these data sets are described in Section 4.1). The S-tree is the most efficient data structure in time, followed by the burst trie. It is notable that the KR-trie performs the worst on these Valgrind data sets. Figure 4(f) shows the memory consumed by the data structures in processing the data sets. Except for on the smallest trace, the burst trie requires the least memory of any of the data structures. Both the S-tree and KR-trie require much more space than the comparison-based data structures and the burst trie.

Figure 5(e) shows the time for processing the Valgrind data sets in the 64-bit case. The S-tree is excluded because it cannot operate on 64-bit keys. The burst trie is the most efficient data structure, with the KR-trie and B-tree also performing quite well. As Figure 5(f) shows, the burst trie is by far the most space efficient data structure on the data sets.

On the 32-bit data sets, the S-tree performs better than the burst trie, however, it requires almost twice as much memory. On the 64-bit data sets, the burst

---

[2] Obtained from http://www.mpi-inf.mpg.de/~kettner/proj/veb/index.html

trie is the best performing data structure, as well as requiring the least space of any data structure.

5   Conclusion

This paper has provided an experimental comparison of efficient data structures operating over 32 and 64-bit integer keys. In particular we have shown that extending burst tries to an ordered data structure for integer keys provides a data structure that is very efficient in both time and space.

In comparisons using uniform random data with AVL trees, red-black trees and B-trees we have shown that for moderate to large sized inputs, burst tries provide all operations more efficiently in both time and space. We have also compared our extended version of burst tries to Dementiev *et al.*'s S-tree data structure based on stratified trees, and found that while Dementiev *et al.*'s data structure is competitive in time, it requires far more memory than a burst trie and is less general, being restricted to 32-bit keys. We have also compared burst tries to KR-tries, a data structure based on Willard's *q*-fast tries. We carefully engineered an implementation of KR-tries, using the same bucket and node data structures as our burst trie, and found that they are generally slightly less efficient than burst tries. One significant advantage of a burst trie over a KR-trie is that because of a burst trie's organisation, it need only store key suffixes in buckets, improving space usage as well as cache performance.

The data structure presented in this paper has wide applicability, and furthermore our results are robust, having been verified on several different architectures. We have presented results for an application of our data structure in Valgrind where the keys are 32 and 64-bit integers. Our results show that in the 32-bit case only the S-tree data structure operates faster, but the S-tree requires almost twice as much space as the burst trie. In the 64-bit case, the burst trie requires less space and operates more rapidly than any of the alternative data structures. This paper demonstrates, through the application of our data structure in Valgrind together with the results presented over random data, that burst tries should be considered as one of the many alternative data structures for applications requiring a general purpose dynamic ordered data structure over keys such as integers or floating point numbers.

Acknowledgements. The authors are grateful to Julian Seward for all his patient assistance with Valgrind. We also thank the anonymous reviewers for their helpful comments.

References

1. Acharya, A., Zhu, H., Shen, K.: Adaptive algorithms for cache-efficient trie search. In: Goodrich, M.T., McGeoch, C.C. (eds.) ALENEX 1999. LNCS, vol. 1619, pp. 296–311. Springer, Heidelberg (1999)
2. Bayer, R., McCreight, E.M.: Organization and maintenance of large ordered indices. Acta Inf. 1, 173–189 (1972)

3. Brent, R.P.: Note on marsaglia's xorshift random number generators. Journal of Statistical Software 11(5), 1–4 (2004)
4. Cormen, T.H., Leiserson, C.E., Rivest, R.L., Stein, C.: Introduction to Algorithms, 2nd edn., pp. 273–301. MIT Press, Cambridge, MA, USA (2001)
5. Dementiev, R., Kettner, L., Mehnert, J., Sanders, P.: Engineering a sorted list data structure for 32 bit keys. In: Proc. of the Sixth SIAM Workshop on Algorithm Engineering and Experiments, New Orleans, LA, USA, pp. 142–151 (2004)
6. Dongarra, J., London, K., Moore, S., Mucci, P., Terpstra, D., You, H., Zhou, M.: Experiences and lessons learned with a portable interface to hardware performance counters. In: IPDPS 2003: Proc. of the 17th International Symposium on Parallel and Distributed Processing, Washington, DC, USA, p. 289.2. IEEE Computer Society, Los Alamitos (2003)
7. Heinz, S., Zobel, J., Williams, H.E.: Burst tries: a fast, efficient data structure for string keys. ACM Trans. Inf. Syst. 20(2), 192–223 (2002)
8. Knessl, C., Szpankowski, W.: Heights in generalized tries and patricia tries. In: Gonnet, G.H., Viola, A. (eds.) LATIN 2000. LNCS, vol. 1776, pp. 298–307. Springer, Heidelberg (2000)
9. Knessl, C., Szpankowski, W.: A note on the asymptotic behavior of the heights in b-tries for b large. Electr. J. Comb. 7 (2000)
10. Knuth, D.E.: The Art Of Computer Programming. Sorting And Searching, 2nd edn., vol. 3, pp. 458–478, 482–491, 506. Addison Wesley Longman Publishing Co., Inc., Redwood City, CA, USA (1998)
11. Korda, M., Raman, R.: An experimental evaluation of hybrid data structures for searching. In: Proc. of the 3rd International Workshop on Algorithm Engineering (WAE), London, UK, pp. 213–227 (1999)
12. Nethercote, N., Seward, J.: Valgrind: A framework for heavyweight dynamic binary instrumentation. SIGPLAN Not. 42(6), 89–100 (2007)
13. Nilsson, S., Tikkanen, M.: An experimental study of compression methods for dynamic tries. Algorithmica 33(1), 19–33 (2002)
14. Sinha, R.: Using compact tries for cache-efficient sorting of integers. In: Ribeiro, C.C., Martins, S.L. (eds.) WEA 2004. LNCS, vol. 3059, pp. 513–528. Springer, Heidelberg (2004)
15. Sinha, R., Ring, D., Zobel, J.: Cache-efficient string sorting using copying. J. Exp. Algorithmics 11, 1.2 (2006)
16. Sinha, R., Zobel, J.: Cache-conscious sorting of large sets of strings with dynamic tries. J. Exp. Algorithmics 9, 1.5 (2004)
17. Sinha, R., Zobel, J.: Using random sampling to build approximate tries for efficient string sorting. J. Exp. Algorithmics 10, 2.10 (2005)
18. Sleator, D.D., Tarjan, R.E.: Self-adjusting binary search trees. J. ACM 32(3), 652–686 (1985)
19. Stroustrup, B.: The C++ Programming Language, 3rd edn. Addison-Wesley Longman Publishing Co., Inc., Boston, MA, USA (1997)
20. Sussenguth, E.H.: Use of tree structures for processing files. Commun. ACM 6(5), 272–279 (1963)
21. van Emde Boas, P.: Preserving order in a forest in less than logarithmic time and linear space. Inf. Process. Lett. 6(3), 80–82 (1977)
22. Willard, D.E.: New trie data structures which support very fast search operations. J. Comput. Syst. Sci. 28(3), 379–394 (1984)

A New Graph-Theoretical Model for k-Dimensional Guillotine-Cutting Problems

François Clautiaux[1], Antoine Jouglet[2], and Aziz Moukrim[2]

[1] Université des Sciences et Technologies de Lille, LIFL UMR CNRS 8022, Parc de la Haute Borne, Bâtiment INRIA, 59655 Villeneuve d'Ascq
francois.clautiaux@univ-lille1.fr
[2] HeuDiaSyC UMR CNRS 6599 UTC, BP 20529 60205, Compiègne
{antoine.jouglet, aziz.moukrim}@hds.utc.fr

Abstract. We consider the problem of determining if a given set of rectangular items can be cut in a large rectangle, using guillotine cuts only. We introduce a new class of arc-colored and oriented graphs, named *guillotine graphs*, which model guillotine patterns. Then we show that an uncolored and non-oriented multigraph is sufficient to obtain any guillotine pattern. We propose linear algorithms for recognizing these graphs, and computing the corresponding patterns. Finally we explain how the model can be used in a constraint programming approach.

1 Introduction

The two-dimensional orthogonal guillotine-cutting problem consists in determining if a given set of small rectangles (items) can be cut from a larger rectangle (bin), using *guillotine cuts* only. A guillotine cut is straight and has to be performed from one edge to the opposite edge of a current available rectangle. As it is a decision problem, it can be seen either as an *open-dimension packing* problem, or as a *knapsack* problem [12]. It occurs in industry if pieces of steel, wood, or paper have to cut from larger pieces.

Two ways are used in the literature for seeking an exact solution for this problem (see [8]). The first method [3] consists in iteratively cutting the bin into two rectangles, following an horizontal or a vertical cut, until all rectangles are obtained. The second method [10] recursively fusions items into larger rectangles, using so-called horizontal or vertical *builds* [11].

In this paper we propose a new graph-theoretical model for the two-dimensional guillotine-cutting problem. For this purpose, we introduce the notion of *guillotine-cutting class*, which is similar to the concept of *packing class* [6,7], but leads to less redundancies for the guillotine-case.

First we describe a model that is based on a unique oriented arc-colored graph, where circuits are related to horizontal or vertical builds. Then we show that a unique undirected uncolored multigraph is sufficient to represent exactly two guillotine-cutting classes. We describe several algorithms of linear complexity for recognizing these graphs and for constructing the guillotine pattern related to a given graph. All proofs can be found in [5].

C.C. McGeoch (Ed.): WEA 2008, LNCS 5038, pp. 43–54, 2008.
© Springer-Verlag Berlin Heidelberg 2008

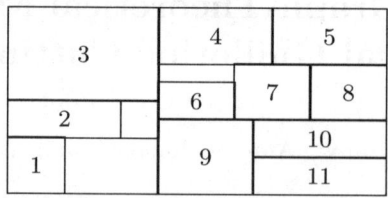

Fig. 1. A guillotine pattern

In Section 2, we describe our notation. Section 3 is devoted to our new model, based on colored and oriented graphs. In Section 4, we deal with the undirected uncolored multigraph model. In Section 5 we propose an algorithm to recover a guillotine pattern from such a graph. In Section 6, we describe a constraint programming approach to solve the decision problem using the model proposed.

2   Notation and Graph-Theoretical Concepts

2.1   Problem Formulation and Notation

A guillotine-cutting instance D is a pair (I, B). I is the set of n items i to cut. An item i has a width w_i and an height h_i ($w_i, h_i \in \mathbb{N}$). The bin B is of width W and height H. All items have to be cut, items cannot be rotated, all sizes are discrete, and only guillotine cuts are allowed (see Figure 1). A *cutting pattern* is a set of coordinates for the items to cut. A pattern is *guillotine* if it can be obtained using guillotine cuts only. It can be checked in $O(n^2)$ time [2].

A build [11] consists in creating a new item by combining two items. The result of an horizontal build of two items i and j, denoted $build(i, j, horizontal)$, is an item of label $min\{i, j\}$, of width $w_i + w_j$, and of height $max\{h_i, h_j\}$. A vertical build can be defined similarly. A valid sequence of builds is an ordered list $b_1, b_2, \ldots, b_{n-1}$, such that for $k = 1 \ldots, n-1$, $b_k = build(i_k, j_k, o_k)$, i_k and j_k two valid labels for step k, and o_k an orientation (horizontal or vertical). A valid sequence of builds is said to be *normal* if for any b_k, $i_k < j_k$. Any guillotine pattern can be obtained by a normal sequence of builds.

2.2   Graph-Theoretical Concepts

An *undirected graph* G is a pair (V, E), and a *directed graph* G is a pair (X, A). The notation $[u, v]$ is used for an edge in an undirected graph, and (u, v) for an arc in a directed graph. For a given vertex v, its *neighborhood* $N(v)$ is the set of vertices u such that $[v, u] \in E$. When an edge may appear more than once, we have a multigraph. An *Hamiltonian chain* (resp. cycle) μ is a chain (resp. cycle) that visits each vertex exactly once.

We give the definition of the classical concept of *edge-contraction* that plays an important role in this paper. Contracting an edge $e = (u, v)$ in the graph $G = (V, E)$ consists in deleting u, v and all edges incident to u or incident to v

and introducing a new vertex v_e and new edges, such that v_e is incident to all vertices that were incident to u or incident to v.

3   A New Graph-Theoretical Model for the Guillotine-Cutting Problem

In this section, we propose a new graph-theoretical model for the guillotine-cutting problem. For this purpose, we introduce the concept of *guillotine-cutting classes*, which are similar to the *packing classes* of [6]. We study a new class of oriented and arc-colored graphs, and show that such graphs can be associated with a guillotine-cutting class. We name these graphs *guillotine graphs*.

3.1   Guillotine-Cutting Classes

In order to avoid equivalent patterns in the non-guillotine orthogonal-packing problem ($2OPP$), Fekete and Schepers [6] proposed the concept of *packing class*. Packing classes are general and model any pattern, guillotine or not. When only guillotine patterns are sought, packing classes may not be suited to the problem, as two different packing classes may lead to patterns with the same combinatorial structure. In order to gather such guillotine patterns, we introduce the concept of *guillotine-cutting class*. It takes into account the fact that exchanging the positions of two rectangular blocks of items does not change the combinatorial structure of the solution. The definition uses the notion of builds reminded in Section 2. Note that from a given normal sequence of builds, one may obtain several patterns: for each vertical build, one may choose to cut the first item above the second, or the opposite (the same applies to horizontal cuts).

Definition 1. *Two solutions belong to the same* guillotine-cutting *class if they can be obtained from the same normal sequence of horizontal and vertical builds.*

Clearly if a member of a guillotine-cutting class is feasible, also are the other members. In this case, we say that the guillotine-cutting class is feasible. This concept takes into account the specificity of the guillotine cuts and reduces dramatically the number of equivalent patterns compared to a direct application of the model of [6]. This is not surprising, since the concept of packing classes is not designed for this specific problem. On the contrary, the model of guillotine-cutting class cannot model any non-guillotine orthogonal-packing patterns.

3.2   A New Graph-Theoretical Model

In the sequel we propose a new class of graphs that represent guillotine-cutting classes. In order to give a definition of this new class, we introduce the concept of *circuit contraction*, similarly to the classical concept of arc contraction used in graph theory (see Section 2).

Definition 2. *Let $G = (V, E)$ be a graph, and $\mu = [v_{i_1}, v_{i_2}, \ldots, v_{i_k}, v_{i_1}]$ a cycle of G. Contracting μ is equivalent to iteratively contracting each edge of μ.*

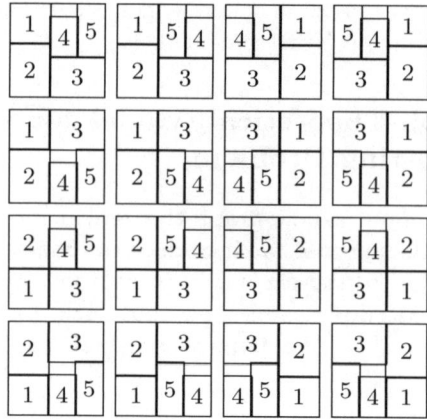

Fig. 2. A guillotine-cutting class

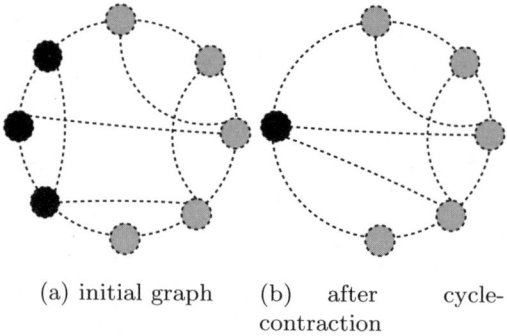

(a) initial graph   (b) after cycle-contraction

Fig. 3. Cycle-contraction

For undirected graphs, we use the term cycle contraction. The same concept can be applied to directed graphs, in which case we will use the term *circuit contraction*. In Figure 3, contracting the black cycle of the left-hand graph leads to the right-hand graph. Note that the order in which the arcs/edges are contracted does not change the resulting graph. The index of the vertex obtained by contracting a circuit μ is the smallest index of an item in μ.

In our new model, a vertex x_i is associated with each item i, and a circuit is associated with a list of horizontal or vertical builds. Let $G = (X, A)$ be a directed graph. In order to make a difference between horizontal and vertical builds, we associate the color red to horizontal builds and the color green to vertical builds. Let $\xi : A \rightarrow \{red, green\}$ be a coloration of the arcs of G. We say that a circuit is monochromatic if all arcs of the circuit have the same color. In the graph, circuit-contracting a red (resp. green) circuit corresponds with a list of horizontal (resp. vertical) builds. When a circuit μ is contracted, the size associated with the residual vertex is the size of the item built, and its label is

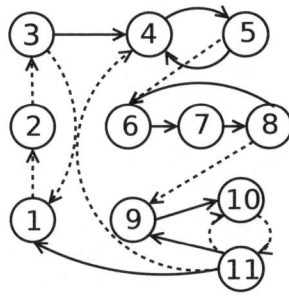

Fig. 4. Modeling the pattern of Figure 1 with a guillotine graph

the smallest label of a vertex of μ. We now give a definition of *guillotine graphs*, which model guillotine patterns.

Definition 3. *Let G be an arc-colored oriented graph. G is a* guillotine graph *if the two following conditions hold:*

1. *G can be reduced to a single vertex by iterative contractions of monochromatic circuits*
2. *no steps are encountered where a vertex belongs to two different monochromatic circuits*

In Figure 4, we depict the dominant graph that models the configuration of Figure 1. Many equivalent graphs can be associated with a given guillotine-cutting class, depending on the order of the vertices in the circuits. To avoid equivalent graphs, we only consider *dominant guillotine graphs*.

Definition 4. *Let G be a valid guillotine graph. G is a* dominant guillotine graph *if in all graphs obtained by applying circuit-contractions to G, vertices in a monochromatic circuit are ordered by increasing index and when a circuit is contracted, only two vertices are of degree greater than two.*

Theorem 1. *If G is a dominant guillotine graph, G can be associated with a unique guillotine-cutting class. Moreover for each normal sequence of builds, there is exactly one dominant guillotine graph.*

If several normal sequences of builds lead to the same guillotine pattern there will be several graphs associated with the same pattern. This may occur when items of the same size are cut such that the order of a vertical and an horizontal cut can be exchanged. Handling these symmetries remains an issue for our model.

4   Cycle-Contractable Graphs

Now we show that colors and orientations are not mandatory in our model. For this purpose, we introduce a new class of undirected multi-graphs, which are

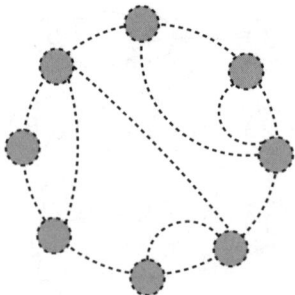

Fig. 5. A cycle-contractable graph

named *cycle-contractable graphs* and we show that these graphs are undirected and uncolored guillotine graphs.

A cycle-contractable graph is associated with two guillotine patterns, depending on the chosen coloring. These graphs have many interesting properties, which are taken into account to design algorithms of linear complexity for recognizing them and computing the corresponding guillotine patterns.

4.1 Cycle-Contractable Graphs and Guillotine Graphs

Let $G = (V, E)$ be an undirected multigraph. If there is an Hamiltonian cycle $\mu = [v_{i_1}, v_{i_2}, \ldots, v_{i_n}, v_{i_1}]$, a corresponding ordering σ can be associated with the vertices of V: $\sigma(i_1) = 1$, and $\sigma(i_{j+1}) = \sigma(i_j) + 1$ for $j = 1, \ldots, n-1$. In the sequel, when a graph G has an Hamiltonian cycle, any edge that is not included in the cycle is named a *backward edge*.

Definition 5. *Let* $G = (V, E)$ *be an undirected multigraph.* G *is a cycle-contractable graph if* G *contains an Hamiltonian cycle* μ *with a corresponding ordering* σ *such that*

1. G *does not include two backward edges* $[v_i, v_j]$ *and* $[v_i, v_j]$ *connecting the same pair of vertices*
2. G *does not include two backward edges* $[v_i, v_j]$ *and* $[v_k, v_l]$ *such that* $\sigma(i) < \sigma(k) \le \sigma(j) < \sigma(l)$

The graph of Figure 5 is a cycle-contractable graph. It can be pictured as a circle of vertices and non-crossing chords.

We now show that dominant guillotine graphs have the structure of cycle-contractable graphs. For this purpose we consider the graph obtained by removing the color and the orientation of the considered guillotine graph.

Theorem 2. *An uncolored non-oriented dominant guillotine graph is a cycle-contractable multigraph.*

The next proposition states that the number of edges in a cycle-contractable graph is in $O(n)$. This result allows us to propose $O(n)$ algorithms in the sequel.

Algorithm 1: Finding the Hamiltonian cycle in a cycle-contractable graph

---

Data: $G = (V, E)$: multigraph;

$\mu \leftarrow \emptyset$;

$L \leftarrow \emptyset$;

forall *edge that appears twice in E* **do** delete one of the two edges $[v_i, v_j]$;

forall *i such that $|N(v_i)| = 2$* **do** $L \leftarrow L \cup \{v_i\}$;

repeat

    Let v_i be a vertex in L and let v_j and v_k be its two neighbors;

    $L \leftarrow L \setminus \{v_i\}$;

    if *$[v_i, v_j]$ is not backward* **then** $\mu \leftarrow \mu \cup \{[v_i, v_j]\}$;

    if *$[v_i, v_k]$ is not backward* **then** $\mu \leftarrow \mu \cup \{[v_i, v_k]\}$;

    $G \leftarrow G \setminus \{v_i\}$;

    if *$[v_j, v_k] \notin G$* **then** $G \leftarrow G \cup \{[v_j, v_k]\}$;

    mark $[v_j, v_k]$ as *backward*;

    if *$|N(v_j)| = 2$* **then** $L \leftarrow L \cup \{v_j\}$;

    if *$|N(v_k)| = 2$* **then** $L \leftarrow L \cup \{v_k\}$;

until *$n = 3$ or L is empty*;

if *$n > 3$* **then** exit with the **FAIL** status;

add each remaining edge in μ if it is not backward;

return μ;

Proposition 1. *In a guillotine graph G with at least two vertices, the number m of arcs in G is in $[n, 2n - 2]$, and the bounds are tight.*

In the remainder, any graph will be connected and with at most $2n - 2$ edges.

4.2   Finding the Hamiltonian Cycle in a Cycle-Contractable Graph

Cycle-contractable graphs can be recognized in linear time. When such a graph is considered, the first step for computing the corresponding guillotine graphs is to determine which edges belong to the Hamiltonian cycle μ, and which edges are backward. Algorithm 1 finds the cycle μ in linear time, by using the fact that: if there are two edges $[v_i, v_j]$ then one of them is in μ; if a vertex v_i has two neighbors, the two edges incident to v_i belong to μ. We also use the fact that removing these edges leads to a graph that is still cycle contractable.

Proposition 2. *If G is cycle-contractable, Algorithm 1 finds the Hamiltonian cycle of G.*

5   From Cycle-Contractable Graphs to Guillotine Patterns

We have shown that a given dominant guillotine graph leads to a unique cycle-contractable graph. In this section we show that a given cycle-contractable graph leads to at most two guillotine-cutting classes. The first step consists in deducing the unique suitable orientation of the edges. Then a choice remains for the coloring of the arcs. The two possible colorings lead to two possible guillotine graphs.

5.1   Finding Suitable Orientation and Coloring for the Edges

Not all cycle-contractable graphs lead to dominant guillotine graphs, depending on the possible ordering of the vertices in the cycles. In order to avoid non-dominant solutions, we introduce the *dominant cycle-contractable graphs*, which lead to dominant guillotine graphs.

Proposition 3. *A cycle-contractable graph leads to a dominant guillotine-graph if and only if, for one of the two possible orientations, for all obtained arcs* (x_i, x_j) *of the main circuit* $(j \neq 1)$, *either* $i < j$, *or there is a backward arc* (x_l, x_i) *such that* $l < j$.
  We say that such a graph is a dominant cycle-contractable graph.

Algorithm 2 returns true if and only if the input cycle-contractable graph is dominant. This property is checked using the result of Proposition 3. In this case, the algorithm finds a suitable orientation (*i.e.* an orientation of the arcs such that the obtained graph is a dominant circuit-contractable graph) and coloring for determining the corresponding guillotine graph. The algorithm visits the vertices v of the obtained directed graph following the Hamiltonian circuit. Each time a backward arc has v as final extremity, it means that a new circuit is included in the current circuit, so the current color is changed. This color is changed as many times as there are such backward edges. Similarly, when backward edges have v as initial extremity, each backward edge is considered by decreasing value of index and is colored with the current color, and then the color is changed.

Proposition 4. *Algorithm 2 colors the arcs of a dominant cycle-contractable graph in such a way that this graph is a dominant guillotine graph.*

Corollary 1. *Given the color of one arc, there is only one valid coloring for a cycle-contractable graph.*

The results above can be summarized by Theorem 3, which is the main result of this section. As we have shown that each guillotine pattern can be obtained from a cycle-contractable graph, we can state the guillotine-cutting problem in a new way: finding a dominant cycle-contractable graph leading to a guillotine pattern fitting the input bin.

Theorem 3. *Each dominant cycle-contractable graph is related to two guillotine-cutting classes and every dominant sequence of builds is related to one dominant cycle-contractable graph.*

5.2   Computing the Size of the Guillotine Pattern

Algorithm 3 computes the width and the height of the guillotine pattern associated with the guillotine graph G. First the ordering σ is computed using Algorithm 1. Then the vertices are considered following σ. Initially, a dummy build b is created, with the current item only. When there is a backward arc, the new build associated with the corresponding circuit is computed and stored in b, and then pushed on the top of S. At the end of the algorithm, S only contains one element, which corresponds with the guillotine pattern.

Algorithm 2: Orienting and coloring a cycle-contractable graph

---

Data: $G = (V, E)$: a cycle-contractable graph;

Use Algorithm 1 to determine the backward edges;

Choose an orientation for the edges, which is consistent with the hamiltonian cycle ;

$test \leftarrow true$;

forall arc (v_i, v_j) *of the Hamiltonian circuit* **do**

    | **if** $j < i$ *and* $\nexists k < j$ *s.t.* (v_k, v_i) *is a backward edge* **then** $test \leftarrow fail$;

if $test = fail$ **then**

    | choose the other orientation for the edges;

    | **forall** arc (v_i, v_j) *of the Hamiltonian circuit* **do**

        | **if** $j < i$ *and* $\nexists k < j$ *s.t.* (v_k, v_i) *is a backward edge* **then** return **false**;

compute the corresponding ordering σ;

choose a color;

for $i : 1 \rightarrow n$ **do**

    | $v \leftarrow \sigma(i)$;

    | Let S^+ be the set of backward arcs a such that $a = (u, v)$;

    | **forall** $a \in S^+$ **do** change the current color;

    | Let S^- be the set of backward arcs a such that $a = (v, u)$;

    | **foreach** *backward arc* a *of* S^- *by decreasing value of label* **do**

        | color a with the current color;

        | change the current color;

    | $u = \sigma(i + 1)$;

    | color the arc (v, u) with the current color;

return **true**;

Theorem 4. *Recognizing a cycle-contractable graph, and computing the two guillotine-cutting classes related to this graph takes $O(n)$ time and space.*

6   A Constraint-Programming Approach

In this section, we explain briefly how our new graph-theoretical model can be used in a constraint programming approach. We chose this approach since it was shown to be efficient for the non-guillotine rectangle packing problem (see [1, 4] for example).

Constraint programming is a programming paradigm aimed at solving combinatorial optimization problems that can be described by a set of variables, a set of possible values for each variable, and a set of constraints between the variables. The set of possible values of a variable is called the *variable domain*. A constraint between variables expresses which combination of values for the variables are allowed. The question to be answered is whether there exists an assignment of values to variables, such that all constraints are satisfied. The power of constraint programming method lies mainly in the fact that constraints can be used in an active process termed "constraint propagation" where certain deductions are performed, in order to reduce computational effort. Constraint propagation removes values from the domains, deduces new constraints, and detects inconsistencies.

Algorithm 3: Computing the size of the guillotine pattern related to a guillotine graph

---

Data: G: a valid guillotine graph;
σ: the corresponding ordering on the items ($\sigma(1) = 1$);
Let S be an initially empty stack of builds b_k;
for $i : 1 \rightarrow n$ **do**
    $v_j \leftarrow \sigma(i)$;
    Let b_j be a new build of size $w_j \times h_j$ and of label j;
    foreach *backward arc* (v_j, v_k) *of color* c *by decreasing value of* $\sigma^{-1}(v_k)$ **do**
        repeat
            remove from S its top element b_t;
            $b_j \leftarrow build(b_j, b_t, c)$;
        until b_j *has for label* v_k;
    push b_j on the top of S;
return *the unique element of* S;

The problem is modeled using two sets of variables and constraints. The first set of variables is used to model a graph with adjacency matrices that specify which arcs belong to the guillotine-graph and what are their *state*. The state of an arc determines its orientation and if it is a backward arc or not. A state is also associated with each vertex, specifying if it is contracted in another vertex, its position in the Hamiltonian circuit, and its current dimensions. Its dimensions are the results of the contractions that have been performed from this vertex. Constraints and propagation techniques are used to ensure that a (partial) graph can always lead to a dominant guillotine graph.

A valid dominant guillotine graph may lead to a guillotine-cutting class that does not fit the bin. Consequently we use a second set of variables, which represent the coordinates of a member of the guillotine-cutting class in construction. Constraints and propagation techniques are then used to ensure that this solution is valid according to the dimensions of the bin.

All along the search, constraint-propagation techniques are used to reduce the search space by eliminating non relevant values from the domain of the variables. These techniques perform different deductions:

- they eliminate potential arcs that cannot lead to a dominant guillotine graph or to a valid solution;
- they eliminate potential coordinates that cannot lead to a valid solution;
- they add some arcs that are mandatory to obtain a dominant guillotine graph and a valid solution;
- they update the possible orientations or the backward status of arcs.

Note that these techniques are used to adjust the domains of variables of the graph according to the domains of coordinate variables and vice-versa.

This approach has been implemented in C++ using ILOG Solver [9] and run on a Genuine Intel CPU T2600 2,16 GHz. We used instances [8] derived from

Table 1. Experimental results

instance	IMVB			IGG		
	nodes	cpu (s)	cpu x MIPS	nodes	cpu (s)	cpu x MIPS
SCP1	69	0.1	25	17	0.1	1651
SCP2	1797	3.4	857	80	0.2	2683
SCP3	3427	6.8	1714	61	0.3	3714
SCP4	6356	78.6	19807	759	0.7	9699
SCP5	5	0.1	25	16	0.1	1032
SCP6	8012	54.6	13759	22	0.1	1651
SCP7	1195	1.8	454	46	0.2	2683
SCP8	484	0.4	101	54	0.3	3302
SCP9	60	0.7	176	37	0.2	2270
SCP10	4	0.1	25	12	0.2	2270
SCP11	11036	221.5	55818	25	0.2	2889
SCP12	908	1.3	328	65	1.0	13001
SCP13	4359	39.5	9954	73	0.1	1651
SCP14	4782	41.7	10508	63	0.4	4746
SCP15	673	0.7	176	110	0.7	9080
SCP16	85627	654.8	165010	2253	2.7	35081
SCP17	13668	227.3	57280	1361	1.7	22493
SCP18	22087	321.5	81018	3419	4.2	54892
SCP19	39550	1794.3	452164	2733	2.3	30954
SCP20	36577	874.3	220324	1909	1.7	22493
SCP21	26748	1757.6	442915	8624	8.0	105450
SCP22	40909	606.0	152712	640	2.6	34875
SCP23	29512	691.9	174359	1754	1.4	18779
SCP24	117013	6265.0	1578780	12402	10.7	140738
SCP25	69434	3735.8	941422	6485	10.7	141563
Average	20972	695.2	175188	1721	2.0	26786

strip cutting-packing problems. In strip cutting problem, the width of the bin is fixed and the minimal feasible height for the bin is sought. Consequently this problem leads to several decision problems.

In Table 1, we compare the incremental modified version of Viswanathan and Bagchi's algorithm (IMVB) of [8] with our incremental guillotine-graph based algorithm (IGG). Let $[H_{min}, H_{max}]$ respectively be the lower and upper bounds of the height of the bin provided by [8]. Both algorithms solve several decision problems beginning with $H = H_{min}$ and by incrementing H by one until there is a solution for the problem. For each method we provide the number of nodes ("nodes") and the computing time in seconds ("cpu(s)").

It is tricky to compare the computing times between the methods since the results of the IMVB method are the ones provided by [8] and run in 1998 on a Sparc-Server 20/712. In order to obtain a fair comparison, we also provide a third column ("cpu x MIPS") for both methods, which represents the CPU time multiplied by the number of Million Instructions per Seconds (MIPS) of the machine (252 MIPS for the Sparc-Server 20/712 and 13207 MIPS for our machine). This indicator can be used as an approximated comparison.

Note that these are preliminary results, which will be improved in the future. However, these results hints that our model is able to lead to interesting results, since our method is able to solve decision problems in reasonable time and search space specially for hard instances such as SCP24 and SCP25 (see [8] for example).

7 Conclusion

We have proposed a new graph-theoretical model for the two-dimensional guillotine-cutting problem. It uses an arc-colored directed graph, which can be replaced by an uncolored undirected multigraph. We show that such a graph is related to two classes of guillotine patterns. We also propose linear algorithms for recognizing these graphs and for computing the guillotine patterns. Our model can be used to design heuristic and exact methods for any $k-$dimensional case.

References

1. Beldiceanu, N., Carlsson, M., Poder, E., Sadek, R., Truchet, C.: A generic geometrical constraint kernel in space and time for handling polymorphic k -dimensional objects. In: Bessière, C. (ed.) CP 2007. LNCS, vol. 4741, pp. 180–194. Springer, Heidelberg (2007)
2. Messaoud, S.B.: Caractérisation, modélisation et algorithmes pour des problèmes de découpe guillotine. PhD thesis, Université de Technologie de Troyes (2004)
3. Christofides, N., Hadjiconstantinou, E.: An exact algorithm for orthogonal 2-d cutting problems using guillotine cuts. European Journal of Operational Research 83, 21–38 (1995)
4. Clautiaux, F., Jouglet, A., Carlier, J., Moukrim, A.: A new constraint programming approach for the orthogonal packing problem. Computers and Operations Research 35(3), 944–959 (2008)
5. Clautiaux, F., Jouglet, A., Moukrim, A.: A new graph-theoretical model for k-dimensional guillotine-cutting problems. Technical report, Université des Sciences et Technologies de Lille (2008),
   http://www.iut-info.univ-lille1.fr/~clautiau/graph.pdf
6. Fekete, S., Schepers, J.: A combinatorial characterization of higher-dimensional orthogonal packing. Mathematics of Operations Research 29, 353–368 (2004)
7. Fekete, S., Schepers, J., Van Der Ween, J.: An exact algorithm for higher-dimensional orthogonal packing. Operations Research 55(3), 569–587 (2007)
8. Hifi, M.: Exact algorithms for the guillotine strip cutting/packing problem. Computers and Operations Research 25(11), 925–940 (1998)
9. Ilog: Ilog Solver Reference Manual, Gentilly, France (2004)
10. Viswanathan, K.V., Bagchi, A.: Best-first search methods for constrained two-dimensional cutting stock problem. Operations Research 41, 768–776 (1993)
11. Wang, P.Y.: Two algorithms for constrained two-dimensional cutting stock problems. Operations Research 31, 573–586 (1983)
12. Wäscher, G., Haussner, H., Schumann, H.: An improved typology of cutting and packing problems. European Journal of Operational Research 183, 1109–1130 (2007)

Layer-Free Upward Crossing Minimization

Markus Chimani, Carsten Gutwenger, Petra Mutzel, and Hoi-Ming Wong

Department of Computer Science, Technical University of Dortmund, Germany
{markus.chimani,carsten.gutwenger,petra.mutzel,
hoi-ming.wong}@cs.uni-dortmund.de

Abstract. An upward drawing of a DAG G is a drawing of G in which all edges are drawn as curves increasing monotonically in the vertical direction. In this paper, we present a new approach for upward crossing minimization, i.e., finding an upward drawing of a DAG G with as few crossings as possible. Our algorithm is based on a two-stage upward planarization approach, which computes a feasible upward planar subgraph in the first step, and re-inserts the remaining edges by computing constraint-feasible upward insertion paths. An experimental study shows that the new algorithm leads to much better results than existing algorithms for upward crossing minimization, including the classical Sugiyama approach.

1 Introduction

Drawing hierarchical graphs is one of the fundamental issues in graph drawing, having received a lot of attention in the past. Given a directed acyclic graph (DAG) G, we are interested in an *upward drawing* of G, i.e., a drawing of G in which all edges are drawn as curves, monotonically increasing in the vertical direction. This problem has numerous applications and occurs whenever a natural flow of information shall be visualized.

Surprisingly, the state-of-the-art in drawing DAGs still facilitates the general framework by Sugiyama et al. in 1981 [12], which consists of three steps:

1. **Layer Assignment:** Assign the nodes to layers such that edges point from lower to higher layers. Split long edges spanning several layers such that edges connect only nodes on neighboring layers.
2. **Crossing Reduction:** Reorder the nodes on each layer with the objective to reduce the number of edge crossings.
3. **Coordinate Assignment:** Assign final node (and bend-point) coordinates.

The individual steps are usually solved heuristically. Several refinements and improvements to this approach have been published; most notable are the work by Gansner et al. [8] and the fast Sugiyama implementation by Eiglsperger et al. [7]. Nevertheless, one inherent drawback of the framework is not overcome by any of these modifications: assigning nodes to fixed layers in the first step can severely affect the subsequent crossing minimization step, requiring edge crossings that would be unnecessary if a "better" layer assignment had been chosen, cf. Figure 1.

C.C. McGeoch (Ed.): WEA 2008, LNCS 5038, pp. 55–68, 2008.

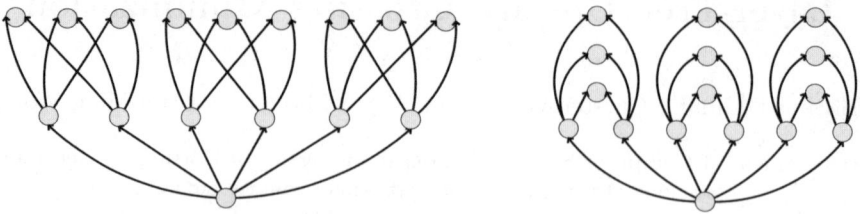

Fig. 1. An unfortunate layering (left) can force unnecessary crossings. In this example the graph is even upward planar (right).

In this paper, we propose a replacement for the first two steps of the Sugiyama framework, thus combining layer assignment and crossing minimization in order to obtain drawings with fewer crossings. We borrow ideas from the *planarization approach* [1,10], which is the most successful heuristic for minimizing crossings in undirected graphs. This approach computes a large planar subgraph in the first step, and then tries to re-insert the remaining edges one-by-one, each time replacing the required crossings with dummy vertices, so that the graph remains planar. The sequence of edges crossed while inserting an edge is also called an *insertion path*. The final outcome is a *planarized representation* of the input graph in which edge crossings are represented as dummy vertices of degree four.

However, adapting this approach to upward drawings is by far not straightforward. First of all—while planarity can be tested efficiently—testing upward planarity is NP-complete [9]; on the other hand, upward planarity can efficiently be tested for graphs with a single source [2], so-called *sT-graphs*. Secondly, while any planar subgraph is suitable as a starting point in the undirected case, further constraints are necessary for upward drawings. And finally, it is not sufficient to find an upward insertion path for an edge independent of the remaining edges. Details and examples on these challenges are given in Section 2.

First successful results on applying the idea of upward planarization are given by Eiglsperger et al. [6] in the context of mixed-upward graphs (graphs in which only a subset of the edges needs to be drawn upward). However, their proposed heuristic for upward edge insertion still needs to layer the graph. Our aim is to completely forgo layering the graph in the crossing minimization step.

The paper is organized as follows. In Sections 2–4, we describe our new approach to upward planarization, where Section 2 gives an informal overview on the algorithm. Section 3 provides necessary formal definitions and describes our basic tool for modeling the dependencies between the edges to be inserted. Section 4 details the upward edge insertion procedure. Finally, an experimental study in Section 6 proves the superiority of the new approach with respect to solution quality compared to existing algorithms.

2 Upward Planarization Approach

In the following, we are given an sT-graph $G = (V, A)$ that we want to planarize with as few crossings as possible. If G would have multiple sources, we can add

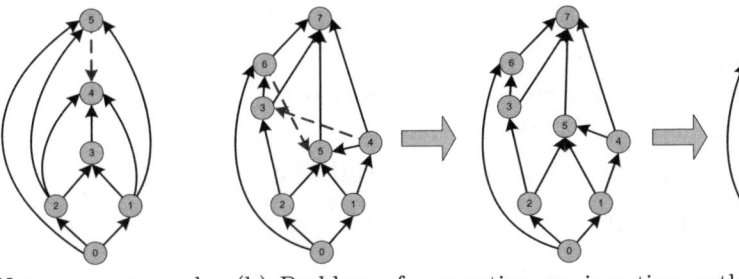

(a) Not every upward planar subgraph is feasible: the edge $(5,4)$ cannot be inserted.

(b) Problem of computing an insertion path: The given graph (left) and a FUPS (middle) obtained by removing the edges $(4,3)$ and $(6,5)$. A crossing minimal insertion path for $(4,3)$ is infeasible as $(6,5)$ can no longer be inserted (right).

Fig. 2. Problems with simple edge insertion approaches

a super source node $\hat{s}$, connect it to all original sources, and set the costs for crossing these additional edges to 0. Like the traditional planarization approach for undirected graphs, our algorithm consists of two stages: in the first we will construct a *feasible* upward planar subgraph $U = (V, A')$, formally defined in Section 3. In the second step, we will iteratively insert the edges not yet in U so that few crossings arise; these crossings are replaced by dummy nodes so that the graph in which edges are inserted can always be considered planar. Inserting an edge e into a planar graph thereby means that all arising crossings will lie on e; we do not introduce additional crossings purely on the planar graph itself.

The main challenge with this approach is that—in contrast to the approach for undirected graphs—the edge insertion steps are not independent of each other. In particular:

– Assume we construct the upward planar subgraph straight-forwardly by adding edges to an initially empty subgraph; after each edge we test for upward planarity. The process stops when no more edges can be inserted without loosing upward planarity. As shown in Figure 2(a), we may not be able to insert the remaining edges *at all*, no matter how many crossings we may use and which upward embedding we choose for the subgraph. Hence we need to identify a *feasible* upward planar subgraph (FUPS).
– Similarly, even if we have a FUPS, we cannot easily insert edges iteratively into it in a crossing minimal fashion, without taking the not-yet inserted edges into account. Figure 2(b) shows that, even though it is possible to insert both edges into U, inserting one edge inconsiderately may make inserting the other one impossible.

We resolve these problems by introducing a *merge graph* $\mathcal{M}$, i.e., a graph representing not only the current (probably planarized) subgraph U, but also modeling all edges that are yet to be inserted. The special properties of $\mathcal{M}$, cf. Section 3, allow us to use a simple acyclicity test on it to determine whether all

---

Algorithm 1. Upward planarization algorithm
___
Require: sT-Graph $G = (V, A)$
Ensure: Upward-planar planarized representation of $G = (V, A)$

 1: Identify spanning tree $U = (V, A')$ of G, directed from s outwards
 2: ▷ Compute feasible upward planar subgraph (FUPS)
 3: $B := \emptyset$
 4: **for** each $e \in A \setminus A'$ **do**
 5:     $U' := U + e$
 6:     **if** $\exists$ upward planar embedding Γ' of U' **then**
 7:        **if** $\mathcal{M}(\Gamma')$ is acyclic **then**
 8:          $U := U', \Gamma := \Gamma'$
 9:          **continue**
10:     $B := B \cup \{e\}$

11: ▷ Non-planar edge insertion
12: **for** each $e \in B$ **do**
13:     Compute insertion path p for e into Γ that will ensure property **(M)**
14:     Insert e along p, replacing crossings by dummy nodes → new U, Γ.
15:     **Property (M):** $\mathcal{M}(\Gamma)$ is acyclic

---

remaining edges will be insertable. Algorithm 1 gives an overview on our upward planarization algorithm. The next section will investigate the merge graph, its applicability and its computation further; Section 4 will center on the crossing-minimal edge insertion (lines 15–19 in Algorithm 1).

3   Feasible Graphs and Paths

Let $G = (V, A)$ be an sT-graph and $U = (V, A')$ an upward planar subgraph of G. We define:

Definition 1 (Upward Insertion Path). *An insertion path p_1 w.r.t. some edge $e_1 = (x_1, y_1) \in A \setminus A'$ is an ordered list of edges $a_1, \ldots, a_\kappa \in U$ such that the graph U_1 obtained from realizing p is upward planar. The realization works as follows: we split the edges $a_1, \ldots, a_\kappa$ obtaining the dummy nodes $d_1, \ldots, d_\kappa$, and add the edges $(x_1, d_1), (d_1, d_2), \ldots, (d_\kappa, y_1)$ representing e_1.*

Let Γ_1 be an upward planar embedding of U_1. We say Γ_1 induces an upward planar embedding Γ of U, which is obtained by reversing the realization procedure while maintaining the embedding.

Definition 2 (Upward Insertion Sequence). *An upward insertion sequence is a sequence of k upward insertion paths for all edges $A \setminus A'$. Thereby the first edge in the sequence is inserted into U —introducing dummy nodes—which results in an upward-planar graph U_1. The second edge is then inserted into U_1, etc. After realizing all insertion paths, we hence obtain a final upward planar graph U_k, which is a planarized representation of G.*

In the following, let Γ be a fixed upward planar embedding of U.

Definition 3 ((Constraint) Feasible Upward Insertion Path). *The upward insertion path p_1 is* feasible *w.r.t. Γ, if there exists an upward planar embedding Γ_1 of the realizing graph U_1 which induces Γ. It is* constraint feasible *w.r.t. Γ, if it furthermore allows an upward insertion sequence for the edges $A \setminus \{A' \cup \{e_1\}\}$ into U_1 such that there exists an upward planar embedding Γ_k of the final graph U_k which induces Γ. We say the path p_1 is* (constraint) minimal *if it requires the fewest crossings over all (constraint) feasible insertion paths.*

Definition 4 (Feasible Upward-Planar Subgraph (FUPS) and Embedding). *An upward planar subgraph $U = (V, A')$ of $G = (V, A)$ is* feasible *if there exists an insertion sequence. An upward planar embedding Γ of a FUPS U is* feasible *if there exists an insertion sequence such that Γ_k induces Γ.*

In an upward planar embedding Γ we can define a *sink-switch* [3] w.r.t. some face f, as a node v incident to f for which there exists no edge incident to f and starting at v. Among all sink-switches of an inner face f, we can identify exactly one *top* sink-switch, i.e., the sink-switch which has to be drawn above all other sink-switches. Analogously we can define a *source-switch* as a node v incident to f for which there exists no edge incident to f and ending at v.

Definition 5 (Merge Graph). *The* Merge Graph $\mathcal{M}(\Gamma)$ *of U with respect to Γ is constructed as follows:*

1. *We start with $\mathcal{M}(\Gamma)$ being a copy of G.*
2. *For each internal face f of Γ, we add an arc from each non-top sink-switch of f to the top sink-switch of f. We call these edges* sink arcs.

Let v_1 and v_2 be two nodes of G. If there exists a non-empty path from v_1 to v_2, we say v_1 *dominates* v_2 and denote this by $v_1 \rightsquigarrow v_2$. If there does not exist such a path we write $v_1 \not\rightsquigarrow v_2$. Considering some specific upward drawing of G, we denote by $v_1 \prec v_2$ that v_1 is drawn lower than v_2. An upward planar drawing clearly requires $v_1 \prec v_2$ if $v_1 \rightsquigarrow v_2$. On the other hand, even for any fixed upward planar embedding Γ, there always exists an upward planar drawing with respect to Γ with $v_1 \prec v_2$ if $v_2 \not\rightsquigarrow v_1$.

Lemma 1 (Feasibility Lemma). *The merge graph $\mathcal{M}(\Gamma)$ is acyclic if and only if there exists an insertion sequence such that the resulting graph is upward planar.*

Proof. $\exists$ Seq. $\Longrightarrow \mathcal{M}(\Gamma)$ acyclic: Consider an upward planar drawing $\mathcal{D}_k$ of U_k with respect to Γ_k. We replace all dummy nodes with crossings and delete the edges $A \setminus A'$ in $\mathcal{D}_k$. The graph associated with the new drawing $\mathcal{D}$ is the upward planar subgraph U and the embedding induced by it is Γ. For each face f, we draw an edge from each non-top sink-switch to its according top sink-switch. As Γ is upward planar, these edges are clearly oriented upwards in the drawing. Finally, we reintroduce the edges $A \setminus A'$ into $\mathcal{D}$, drawing them exactly as in $\mathcal{D}_k$.

By these operations we obtained a drawing of the merge graph, and as all edges in the resulting drawing are oriented upwards, the merge graph is acyclic.

$\mathcal{M}(\Gamma)$ acyclic $\Longrightarrow \exists$ Seq.: Let $\mathcal{N}$ be the graph $\mathcal{M}(\Gamma)$ without the edges $A \setminus A'$. $\mathcal{N}$ can be embedded like Γ, embedding the sink arcs planarly within their corresponding faces. Let $\Gamma_\mathcal{N}$ be the resulting embedding. Let $e_i = (x_i, y_i)$, $1 \leq i \leq k$, be the edges not in U. Since $\mathcal{M}(\Gamma)$ contains these edges and is acyclic we know that $y_i \not\rightsquigarrow x_i$ in $\mathcal{N}$, for all $1 \leq i \leq k$. Hence, considering any edge e_i individually, we can find a drawing with respect to Γ where $x_i \prec y_i$.

We show that this holds for all edges together by induction over the number of edges to be inserted. We start with any upward drawing of $\mathcal{N}$ which is stepwise modified. Consider the drawing $\mathcal{D}_j$ of $\mathcal{N}$ in step j where we have $x_i \prec y_i$ for all $1 \leq i < j$. If $x_j \prec y_j$ there is nothing to do in this step, so assume $y_j \prec x_j$. Since $\mathcal{M}(\Gamma)$ is acyclic we know that $y_j \not\rightsquigarrow x_j$ and hence there exists a drawing with $x_j \prec y_j$. We show that we can realize the latter condition without violating the respective order for $e_1, \ldots, e_{j-1}$.

Considering $\mathcal{M}(\Gamma)$, let Y_j be the nodes dominated by y_j, i.e., $y_j \rightsquigarrow z$ in $\mathcal{M}(\Gamma)$ for all $z \in Y_j$. Clearly, $x_j \notin Y_j$ as $\mathcal{M}(\Gamma)$ is acyclic. In $\mathcal{D}_j$, we will move y_j and all nodes Y_j such that y_j is above x_j. Due to the sink arcs, $\mathcal{N}$ does not have any maximal two-connected component C that is within another such component. Hence the upward shift will not result in any crossings.

Assume that the shift would invalidate $x_i \prec y_i$ for some i. This would mean that x_i would be in Y_j but y_i would not. But since (x_i, y_i) is an edge in the merge graph this cannot be the case. Therefore the shift in step j ensures an upward planar drawing of $\mathcal{N}$ where all edges $e_1, \ldots, e_j$ can be drawn into in an upward fashion using straight lines. We can simply extract insertion paths from the final drawing $\mathcal{D}_k$ to generate a valid insertion sequence. $\square$

Corollary 1. *An upward planar embedding Γ of a FUPS U is feasible if and only if the corresponding merge graph $\mathcal{M}(\Gamma)$ is acyclic.*

4   Upward Edge Insertion

We now consider the problem of inserting edges into a FUPS with few crossings, by iteratively adding the edges not in the FUPS. In the following we are again given an sT-graph $G = (V, A)$, a FUPS $U = (V, A')$ and an embedding Γ of U. Let $e_i = (x_i, y_i)$, for $1 \leq i \leq k = |A \setminus A'|$ be the edges not in U, and let p_1 be a feasible upward insertion path for e_1 into U with respect to Γ. Realizing p_1 results in a graph U_1 with embedding Γ_1. Formally we can define the arising problem per edge:

Definition 6 (Constraint Upward Edge Insertion Problem with Fixed Embedding). *Given an sT-graph $G = (V, A)$, a FUPS $U = (V, A')$, a feasible upward embedding Γ of U and an edge $e_1 \in A \setminus A'$. The constraint upward edge*

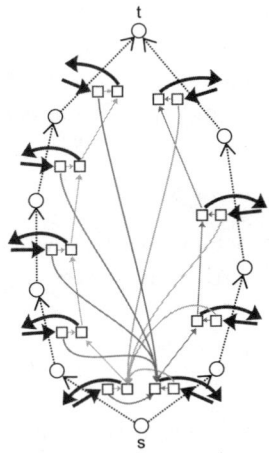 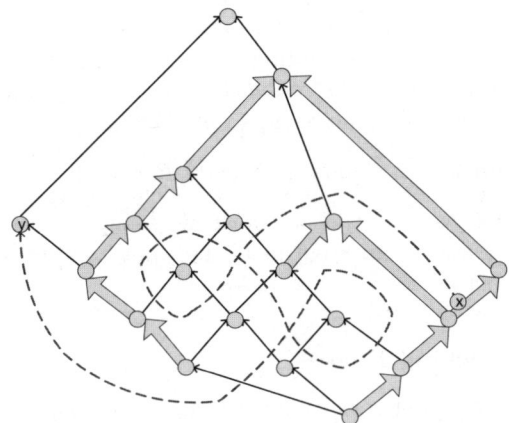

(a) Routing sub-network of a single internal face. The dotted lines are the edges of the underlying graph $\hat{U}$. The bold edges are *crossing arcs*.

(b) Routing in $\hat{U}$ without locking can result in infeasible insertion paths. The thick gray edges denote sub-structures which are expensive to cross. The dashed line denotes the shortest path in the underlying routing network which makes loops and is thus infeasible as an upward insertion path.

Fig. 3. Finding feasible upward insertion paths using the routing network

insertion problem with fixed embedding *is to find a constraint minimal upward insertion path p_1 for e_1 into U with respect to Γ and the edges $A \setminus \{A' \cup \{e_1\}\}$.*

4.1   Routing Network

In order to compute an upward insertion path for $e_1 = (x_1, y_1)$ we use a *routing network R*. For the traditional edge insertion problem, i.e., without considering upward planarity, we simply use the bidirected dual graph of U with respect to Γ, augmented with the start and end nodes of e_1. Due to the required upward planarity we have to use a more heavily augmented routing network.

Firstly, we augment U and Γ with the sink arcs as we did for the merge graph. Furthermore, we add a super sink node $\hat{t}$ and *super sink arcs* $(t, \hat{t})$, for each sink on the outerface of Γ. We call the resulting graph $\hat{U}$ with its upward planar embedding $\hat{\Gamma}$ (inducing Γ). The augmentation guarantees that all faces in $\hat{\Gamma}$ are *simple*, i.e., they have exactly one source- and one sink-switch.

We do not represent single faces as single nodes in R but as well-structured sub-networks, as shown in Figure 3(a) for an internal face; the sub-network for the external face is analogous. Such a sub-network guarantees that when we enter a face f over some edge, we can only leave f either above that edge or on the other side of f. We call the edges that are the dual of some edge of $\hat{U}$ the *crossing arcs*, as a path over these arcs crosses an edge of $\hat{U}$.

Finally, we add nodes x^* and y^* to R which will be the start and end node of the insertion path, corresponding to x_1 and y_1, respectively. Let A_x and A_y be the crossing arcs corresponding to edges starting at x_1 or ending at y_1, respectively. We add edges from x^* to each target node of the arcs A_x, and edges from each source node of the arcs A_y to y^*. We have:

Lemma 2. *The routing network R has $\mathcal{O}(|V|)$ nodes and edges.*

To use R as a routing network, we assign a cost of 1 to each crossing arc which corresponds to an edge in U, i.e., not to sink arcs in $\hat{U}$. All other edges in R have cost 0.

4.2   Locking Edges

The routing network by itself is not strong enough to guarantee that the shortest path between x^* and y^* corresponds to a feasible upward insertion path, cf. Figure 3(b): the shortest path may contain "loops" which clearly violate the upward property of our drawing.

Therefore we will introduce static and dynamic *locks*, i.e., we prohibit edges to be considered during the shortest path computation. While the static locking by itself does not directly ensures feasibility for the upward insertion paths, it is necessary to make the dynamic locking strategy valid. The next lemma follows immediately:

Lemma 3 (Static Locking). *Considering the merge graph $\mathcal{M}(\Gamma)$, let Y_1 be the nodes that are dominated by y_1, including y_1 itself; let X_1 be the nodes that dominate x_1, including x_1 itself. A constraint feasible upward insertion path will not cross edges of $\hat{U}$ that connect two nodes of X_1 or two nodes of Y_1.*

Proof. Assume a constraint feasible upward insertion path p' would cross an edge that connects two nodes $n_1, n_2 \in Y_1$. Let d be the dummy node created by this crossing; it is dominated by either n_1 or n_2. Then the merge graph $\mathcal{M}(\Gamma_1)$ would have a cycle as $y_1 \rightsquigarrow d$ and $d \rightsquigarrow y_1$ through the inserted edge e_1. This contradicts the constraint feasibility of p'. The analogous holds if $n_1, n_2 \in X_1$. □

For any face f in $\hat{\Gamma}$ let $a(f)$ be an edge corresponding to a crossing arc in R with shortest distance $\delta(f)$ to x^*. By construction, each face f consists of two directed paths—one on the left- and one on the right-hand side—from the source-switch to the sink-switch of f. We denote all edges of f between the source-switch and $a(f)$ as the *face-lock* $F(f)$.

Lemma 4 (Dynamic Locking). *For each $a(f)$, there exists a path q in R from x^* to a crossing arc corresponding to $a(f)$, where q has length $\delta(f)$ and uses no edge of $F(f')$ for all faces f'.*

Proof. Let f be a face with minimal $\delta(f)$ for which each $(x^* \rightarrow a(f))$-path in R of length $\delta(f)$ contains at least one face-lock edge. Among all those shortest paths for this f, let q be the path which uses the fewest face-lock edges.

Traversing q from x^*, let f' be the first face for which q uses the first face-lock edge $b \in F(f')$. Let q' be q's segment from x^* to b. By definition, we know that there exists a path $q'' := x^* \rightsquigarrow a(f')$ which is as long as q'. Since q' is shorter than q, we know that q'' does not use any face-lock edges. Hence we can create a path q^* from q by replacing q' with q'': this replacement is straight-forwardly possible as q leaves f' over a non-face-lock edge. Note that q cannot leave f' over another face-lock edge, as this would contradict the minimality of q.

The existence of q^* —which is as long as q, but crosses less face-lock edges— constitutes a contradiction and thus the lemma holds.   □

4.3   Upward Edge Insertion Algorithm

Based on the above routing graph R and Lemmata 3 and 4 we can compute a feasible upward insertion path using a BFS algorithm adapted for 0/1 weights.

We start at x^* but instead of generating the whole routing graph we generate the successor nodes dynamically on the fly. Thereby we ignore all crossing arcs corresponding to the edges specified by static locking (Lemma 3). Furthermore we use Lemma 4 to forbid additional edges: whenever we visit a crossing arc for the first time leading into a face f, the corresponding edge in $\hat{U}$ can be seen as $a(f)$, and hence we will block all face-lock edges $F(f)$ induced by this $a(f)$.

We call this algorithm MINIMALFEASIBLEINSERTIONPATH.

Lemma 5. *The algorithm* MINIMALFEASIBLEINSERTIONPATH *computes a minimal feasible upward insertion path* p_1 *for* x_1 *and* y_1.

Proof. Assume p_1 would be infeasible. There would be a face f which cannot be drawn in an upward fashion. The routing network guarantees that a single segment of p_1 crossing through f cannot lead to such a situation. Hence there have to be at least two segments of p_1 going through f, which together contradict the upwardness of the drawing and therefore cross each other. W.l.o.g. assume that the first segment s_1 goes from some point z_l on the left of f to some point z_r on the right-hand side of f. In order to conflict with the direction of s_1, the second crossing segment s_2 can be only one of the following:

- It starts on the left-hand side above z_l and ends on the right side below z_r. But then we could find a shorter path which goes directly from z_l to the exit point of s_2, removing the part of p_1 between s_1 and s_2 and all the crossings induced by it.
- It starts on the right-hand side above z_r and ends on the left side below z_l. Due to our dynamic locking, we forbade all crossings over edges below z_l, hence s_2 cannot exit the face there.

The minimality of p_1 follows from the validity of the 0/1-BFS algorithm and Lemmata 3 and 4.   □

Corollary 2. *Let* p_1 *be a minimal feasible upward insertion path obtained by* MINIMALFEASIBLEINSERTIONPATH, *and* Γ_1 *the upward planar embedding arising from realizing* p_1. *If the merge graph* $\mathcal{M}(\Gamma_1)$ *is acyclic,* p_1 *is a constraint minimal upward insertion path.*

There may be the situation that the computed minimal insertion path is not constraint feasible, i.e., the resulting merge graph contains a cycle. In such cases we have to resort to a heuristic for finding a constraint feasible upward insertion path which though may not be minimal:

The algorithm CONSTRAINTFEASIBLEINSERTIONPATH works similar to MIN-IMALFEASIBLEINSERTIONPATH but uses no dynamic locking and instead computes an *intermediate merge graph* $\mathcal{I}$ whenever the BFS algorithm relaxes some crossing arc r: we insert "part of" e_1 along the shortest path up to r, ending at some new dummy node ξ. We then build the merge graph for this graph, adding the edge (ξ, y_1) instead of (x_1, y_1) to $\mathcal{I}$. If $\mathcal{I}$ contains a cycle we know that selecting r in our current path would lead to an infeasible path, and we can forbid to use it in the BFS enumeration. Clearly, this algorithm—though always terminating—will in general not give the optimal solution, as an alternative path up to the rejected edge r might have allowed us to use r and find an overall shorter path:

Lemma 6. *The algorithm* CONSTRAINTFEASIBLEINSERTIONPATH *computes a constraint feasible upward insertion path* p_1 *for* x_1 *and* y_1.

Algorithm 2 gives an overview on the overall edge insertion strategy: we try to add all edges using the optimal path of MINIMALFEASIBLEINSERTIONPATH strategy. Only if there is no more edge insertable by it, we insert a not-yet inserted edge using CONSTRAINTFEASIBLEINSERTIONPATH. Afterwards we again try to use the former algorithm for the remaining edges. We iterate that process unless all edges are inserted. — As we will see in the next section, the heuristic procedure CONSTRAINTFEASIBLEINSERTIONPATH is only used very rarely; hence in most cases all edges are inserted optimally.

Algorithm 2. Reinsert all edges

Require: sT-graph $G = (V, E)$, FUPS $U = (V, A')$ with feas. upward embedding Γ
Ensure: upward planarized graph U^* of G with embedding Γ^*, inducing Γ
1: List $L := A \setminus A'$
2: $U^* := U$, $\Gamma^* := \Gamma$
3: **while** L not empty **do**
4:     boolean $success :=$ **false**
5:     **for** each $e \in L$ **do**
6:         $p :=$ MINIMALFEASIBLEINSERTIONPATH(e, U^*, Γ^*)
7:         $U^\circ, \Gamma^\circ :=$ realizePath(p, U^*, Γ^*)
8:         **if** $\mathcal{M}(\Gamma^\circ)$ acyclic **then** $\triangleright$ p was constraint feasible
9:             $U^* := U^\circ$, $\Gamma^* := \Gamma^\circ$
10:            $success :=$ **true**
11:            L.remove(e)
12:     **if** not $success$ **then**
13:         $e := L$.extractRandomElement()
14:         $p :=$ CONSTRAINTFEASIBLEINSERTIONPATH(e, U^*, Γ^*)
15:         $U^*, \Gamma^* :=$ realizePath(p, U^*, Γ^*)

---

5   Runtime Analysis

We conclude the theoretic description by analyzing the algorithms' runtimes.

Lemma 7. *Let $G = (V, A)$ be an sT-graph. Algorithm 1 computes a FUPS U with a feasible embedding Γ in $\mathcal{O}(|A|^2)$ time.*

Proof. The algorithm 1 performs $|A \setminus A'| = \mathcal{O}(|A|)$ upward planarity and cycle tests. Since upward planarity testing of sT-graphs requires $\mathcal{O}(|V|)$ time [2] and cycle testing can be done in $\mathcal{O}(|A|)$, we have the above lemma.   □

Lemma 8. *Let $\bar{U} = (\bar{V}, \bar{A})$—with embedding $\bar{\Gamma}$—be the planarization obtained after inserting some arcs into the original FUPS U of G. Let (x, y) be the next edge to insert, and let r be the number of edges to insert afterwards.*

(a) Computing a minimal feasible upward insertion path for (x, y) via MINI-MALFEASIBLEINSERTIONPATH and checking its constraint feasibility requires $\mathcal{O}(|\bar{V}| + r)$ time.
(b) CONSTRAINTFEASIBLEINSERTIONPATH computes a constraint feasible upward insertion path for (x, y) in $\mathcal{O}(|\bar{V}|^2 + r|\bar{V}|)$ time.

Proof. By Lemma 2, the routing network for $\bar{U}$ with respect to $\bar{\Gamma}$ can be constructed in $\mathcal{O}(|\bar{V}|)$ time. To compute the static locks we have to construct the merge graph $\mathcal{M}(\bar{\Gamma})$ which requires $\mathcal{O}(|\bar{V}| + r)$ time. The runtime of the 0/1-BFS algorithm, including the computation of the dynamic locks, is bounded by $\mathcal{O}(|\bar{V}|)$. For the constraint feasibility checking, we have to (temporarily) insert the insertion path into $\bar{U}$, construct the corresponding merge graph and test for acyclicity. This can be done in $\mathcal{O}(|\bar{V}| + r)$. Thus the total runtime of *(a)* is dominated by $\mathcal{O}(|V_j| + r)$.

The runtime analysis of CONSTRAINTFEASIBLEINSERTIONPATH is similar. Instead of computing the dynamic locks, we temporarily insert the current insertion path into $\bar{U}$ after each edge relaxation, and check for acyclicity of the corresponding merge graph. Hence the runtime is $\mathcal{O}\left(|\bar{V}| \cdot (|\bar{V}| + r)\right)$.   □

6   Experiments

We implemented the new level-free upward planarization approach using the open-source C++-library OGDF [11] and compared its performance with published results of state-of-the-art algorithms [4,6].

In our implementation we randomize the order of the edges considered in the for-each loop (line 4 of Alg. 1) of the FUPS computation, and the order in which edges are re-inserted (for-each loop in line 5 of Alg. 2). We denote by LFUPi the best result obtained by after i independent calls of the algorithm. Besides comparing with published results, we also applied OGDF's Sugiyama implementation with optimal node ranking, Barycenter heuristic, and 50 randomized runs. We used two public benchmark sets of graphs, described in the following.

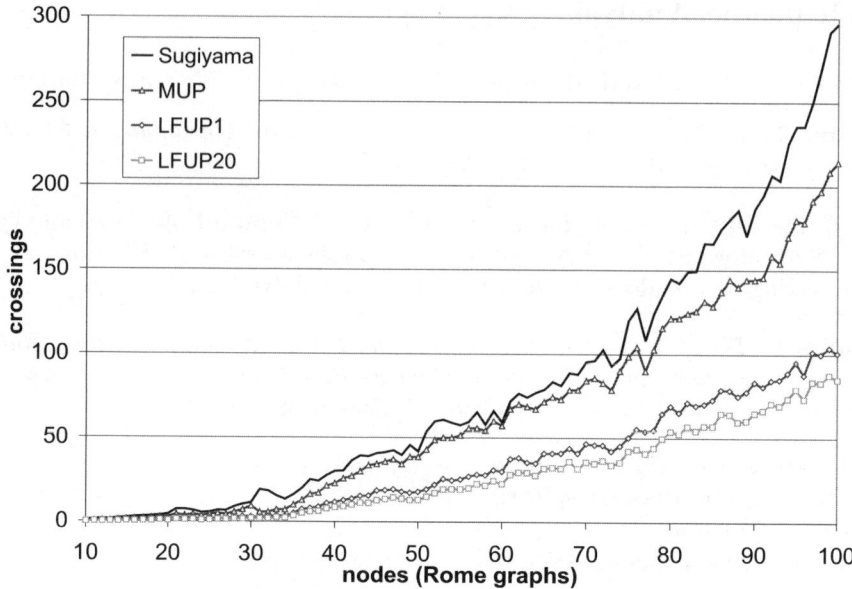

Fig. 4. Rome graphs: average crossings vs. number of nodes

Rome Graphs. The Rome graphs [5] are a widely used benchmark set in graph drawing, obtained from a basic set of 112 real-world graphs. The benchmark contains 11528 instances with 10–100 nodes and 9–158 edges. Although the graphs are originally undirected, they have been used as directed graphs—by artificially directing the edges according to the node order given in the input files—for showing the performance of the mixed-upward planarization (MUP) approach by Eiglsperger et al. [6]. In this case, all edges are directed and the graphs are acyclic; hence their approach turns into an upward planarization method.

Figure 4 shows the results for MUP, OGDF's Sugiyama algorithm, and our new approach for 1 and 20 random calls. Each data point refers to the average number of crossings of all the graphs with the same number of nodes. The new algorithm clearly outperforms the other two approaches. Though MUP is already considerably better than the Sugiyama algorithm, LFUP1 obtains solutions with only half as many crossings as MUP. It can also be seen that the randomization of LFUP can further reduce the number of crossings significantly.

North DAGs. The North DAGs have been introduced in an experimental comparison of algorithms for drawing DAGs [4]. They are 1158 DAGs collected by Stephen North and slightly modified by Di Battista et al. The graphs are grouped into 9 sets, where set i contains graphs with $10i$ to $10i + 9$ edges for $i = 1, \ldots, 9$. From this experimental study, we used the results of the two best algorithms: *Layers* is an implementation of Sugiyama's algorithm according to the original paper and *Dot* is a highly-optimized version of this algorithm

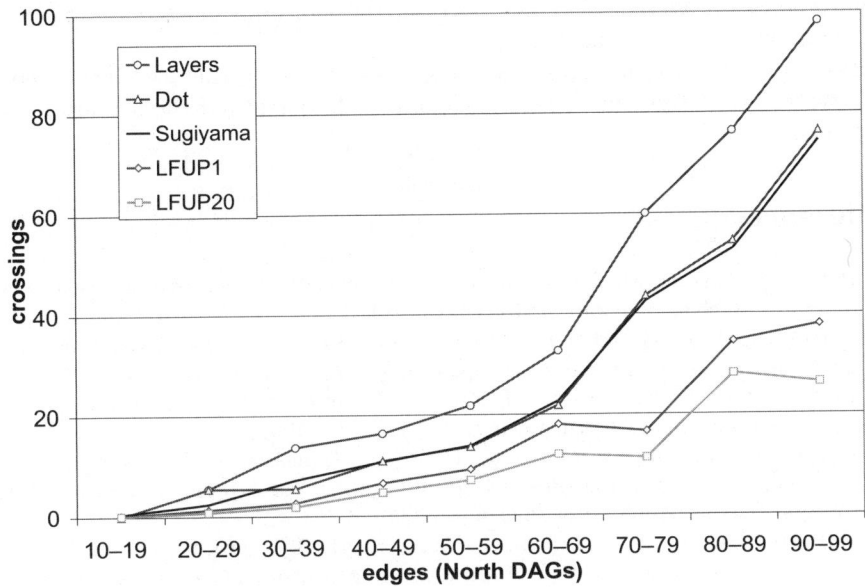

Fig. 5. North DAGs: average crossings vs. number of edges

developed by Koutsofios and North. In our diagrams, we omitted the two algorithms that use a simple method based on planarization of st-graphs as they perform very poorly, achieving roughly 300 crossings on average for the largest graphs.

The results are shown in Figure 5. Again, the new algorithm outperforms the three layer-based algorithms by far. While OGDF's Sugiyama achieves almost the same results as Dot, LFUP1 obtains solutions with roughly half as many crossings; the multi-run version again yields significant improvements, especially for larger graphs.

Constraint feasibility. A very interesting outcome of our studies is that the minimal feasible path obtained by MINIMALFEASIBLEINSERTIONPATH is already constraint feasible in most of the cases and therefore allows us to insert the edge provably optimally.

From the 2,708,474 edge insertion calls performed by FLUP20 in total over all Rome graphs, only 114 (0.004%) require to call the CONSTRAINTFEASIBLEIN-SERTIONPATH heuristic; this corresponds to 0.87% of the instances requiring this heuristic at all. Furthermore, the heuristic was never used for any North DAG.

Runtime. The experiments were conducted on an Intel Core-2 Duo 3.0GHz with 2GB RAM per process under Windows Vista. The maximum computation time of FLUP1 over all instances was 0.19 seconds. The large Rome graphs (90–100 nodes) take under 0.1 second on average, the large North DAGs (90-99 edges) require 60ms on average. For comparison, the runtimes of OGDF's Sugiyama implementation were at most 15ms.

Based on our experimental comparison, we conclude that the layer-free approach achieves large improvements compared to state-of-the-art layer-based methods.

References

1. Batini, C., Talamo, M., Tamassia, R.: Computer aided layout of entity relationship diagrams. J. Syst. Software 4, 163–173 (1984)
2. Bertolazzi, P., Di Battista, G., Mannino, C., Tamassia, R.: Optimal upward planarity testing of single-source digraphs. SIAM J. Comput. 27(1), 132–169 (1998)
3. Bertolazzi, P., Di Battista, G., Liotta, G., Mannino, C.: Upward drawings of triconnected digraphs. Algorithmica 12(6), 476–497 (1994)
4. Di Battista, G., Garg, A., Liotta, G., Parise, A., Tamassia, R., Tassinari, E., Vargiu, F., Vismara, L.: Drawing directed acyclic graphs: An experimental study. Int. J. Comput. Geom. Appl. 10(6), 623–648 (2000)
5. Di Battista, G., Garg, A., Liotta, G., Tamassia, R., Tassinari, E., Vargiu, F.: An experimental comparison of four graph drawing algorithms. Comput. Geom. Theory Appl. 7(5–6), 303–325 (1997)
6. Eiglsperger, M., Kaufmann, M., Eppinger, F.: An approach for mixed upward planarization. J. Graph Algorithms Appl. 7(2), 203–220 (2003)
7. Eiglsperger, M., Siebenhaller, M., Kaufmann, M.: An efficient implementation of Sugiyama's algorithm for layered graph drawing. In: Pach, J. (ed.) GD 2004. LNCS, vol. 3383, pp. 155–166. Springer, Heidelberg (2005)
8. Gansner, E., Koutsofios, E., North, S., Vo, K.-P.: A technique for drawing directed graphs. Software Pract. Exper. 19(3), 214–229 (1993)
9. Garg, A., Tamassia, R.: On the computational complexity of upward and rectilinear planarity testing. SIAM J. Comput. 31(2), 601–625 (2001)
10. Gutwenger, C., Mutzel, P.: An experimental study of crossing minimization heuristics. In: Liotta, G. (ed.) GD 2003. LNCS, vol. 2912, pp. 13–24. Springer, Heidelberg (2004)
11. OGDF – the Open Graph Drawing Framework. Technical University of Dortmund, Chair of Algorithm Engineering, http://www.ogdf.net
12. Sugiyama, K., Tagawa, S., Toda, M.: Methods for visual understanding of hierarchical system structures. IEEE Trans. Sys. Man. Cyb. 11(2), 109–125 (1981)

On the Efficiency of a Local Iterative Algorithm to Compute Delaunay Realizations

Kevin M. Lillis[1,2] and Sriram V. Pemmaraju[1]

[1] Department of Computer Science,
University of Iowa, Iowa City, IA 52242-1419, U.S.A.
{lillis,sriram}@cs.uiowa.edu
[2] Computer and Information Science,
St. Ambrose University, Davenport, IA 52803, U.S.A.
LillisKevinM@sau.edu

Abstract. Greedy routing protocols for wireless sensor networks (WSNs) are fast and efficient but in general cannot guarantee message delivery. Hence researchers are interested in the problem of embedding WSNs in low dimensional space (e.g., $\mathbb{R}^2$) in a way that guarantees message delivery with greedy routing. It is well known that Delaunay triangulations are such embeddings. We present the algorithm FindAngles, which is a fast, simple, local distributed algorithm that computes a Delaunay triangulation from any given combinatorial graph that is Delaunay realizable. Our algorithm is based on a characterization of Delaunay realizability due to Hiroshima et al. (IEICE 2000). When compared to the PowerDiagram algorithm of Chen et al. (SoCG 2007), our algorithm requires on average $1/7^{th}$ the number of iterations, scales better to larger networks, and has a much faster distributed implementation. The PowerDiagram algorithm was proposed as an improvement on another algorithm due to Thurston (unpublished, 1988). Our experiments show that on average the PowerDiagram algorithm uses about 18% fewer iterations than the Thurston algorithm, whereas our algorithm uses about 88% fewer iterations. Experimentally, FindAngles exhibits well behaved convergence. Theoretically, we prove that with certain initial conditions the error term decreases monotonically. Taken together, these suggest our algorithm may have polynomial time convergence for certain classes of graphs. We note that our algorithm runs only on Delaunay realizable triangulations. This is not a significant concern because Hiroshima et al. (IEICE 2000) indicate that most combinatorial triangulations are indeed Delaunay realizable, which we have also observed experimentally.

1 Introduction

A wireless sensor network (WSN) consists of battery-powered nodes that can communicate with one another when within radio broadcast range and can perform local computations. Because there is no centralized control and because each node has only a small amount of memory, WSNs often employ *memoryless*

C.C. McGeoch (Ed.): WEA 2008, LNCS 5038, pp. 69–86, 2008.

routing, in which each node decides to whom a message should be forwarded based solely on the source of the message, its destination, and information gathered from nearby nodes. *Geographic routing protocols* are memoryless routing protocols that use geographic information such as the coordinates of the source, the destination, and nearby nodes. In the geographic routing protocol known as *greedy routing* each node forwards a message by choosing a neighbor that is closest to the destination. While this scheme is simple and efficient, it does not guarantee message delivery. This is because a routed message may become trapped in a cycle [5] or it may reach a *local minimum*; this is a node that is closer to the destination than any of its neighbors [10].

Given a point set $P \subseteq \mathbb{R}^2$, a *Delaunay triangulation* of P is commonly defined as a triangulation of P satisfying the property that the circumcircle of each inner face (triangle) contains no point of P in its interior. This is known as the *empty circle property*. A Delaunay triangulation of P is also well known as the planar dual of the *Voronoi diagram* of P. It is this characterization that has been used to show that greedy geographic routing will always succeed on a Delaunay triangulation [5].

The fact that greedy geographic routing is always successful on a Delaunay triangulation motivates the question of whether a given WSN can be embedded in the plane as a Delaunay triangulation. In cases where the WSN is not a triangulation to start with, standard topology control protocols [16] can be employed to extract a planar spanning subgraph of the WSN which can then easily be converted into a triangulation. Define a *combinatorial triangulation* as a planar graph in which all faces are 3-cycles. Suppose that we are given a WSN $G = (V, E)$ that is a combinatorial triangulation. We seek a one-one mapping $\Phi : V \rightarrow \mathbb{R}^2$ such that if each vertex $v \in V$ is placed on the plane at $\Phi(v)$ and each edge $\{u, v\} \in E$ is represented by a straight line segment with endpoints $\Phi(u)$ and $\Phi(v)$, then the set of points $\Phi(V) = \{\Phi(v) \mid v \in V\}$ and the set of line segments $\Phi(E) = \{\{\Phi(u), \Phi(v)\} \mid \{u, v\} \in E\}$ defines a Delaunay triangulation. The problem of finding the mapping Φ is called the *Delaunay realization problem*, the mapping Φ, if it exists, is called a *Delaunay realization*, and combinatorial triangulations G for which a Delaunay realization exists are called *Delaunay realizable* graphs. The problem of determining whether or not a combinatorial triangulation is Delaunay realizable can be solved in polynomial time; for example by checking if a certain linear system of inequalities defined by Hiroshima et al. [9] has a feasible solution. However, as far as we know, the problem of actually finding a Delaunay realization does not have a polynomial time solution and seems rather difficult. In this paper we present a simple iterative algorithm, called `FindAngles`, that finds a Delaunay realization of a Delaunay realizable graph. We do not prove polynomial time convergence, but we do present substantial experimental evidence indicating that `FindAngles` converges rapidly. The algorithm is inherently local and has an obvious, distributed implementation in which each node updates some local geometric information based on such information at neighboring nodes.

Recently two other approaches have been considered for the problem of finding embeddings that permit successful greedy geographic routing. The first approach, due to Papadimitriou and Ratajczak [15], seeks *greedy embeddings*. Let a *distance decreasing* path in an embedding of a graph be a path $s = v_1, v_2, \ldots, v_k = t$ such that $\|v_i - t\| < \|v_{i-1} - t\|$, $2 \leq i \leq k$. Here $\| \ldots \|$ denotes Euclidean distance. A *greedy embedding* of a graph is an embedding into the Euclidean plane such that there exists a distance decreasing path between every pair of vertices. Papadimitriou and Ratajczak [15] conjecture that every combinatorial triangulation (indeed every 3-connected planar graph) has a greedy embedding. This conjecture was very recently proved by Dhandapani [7]. The proof depends on the Knaster-Kuratowski-Mazurkiewicz Theorem [11] that is known to be equivalent to the Brouwer Fixed Point Theorem, and hence does not seem to immediately lead to a polynomial time algorithm. He does mention an iterative algorithm at the end of his paper, but without any theoretical bounds or experimental results.

A second approach due to Chen at al. [4] uses *power diagrams*. Let $P \subseteq \mathbb{R}^2$ be a planar set of points, with each point $p \in P$ having an associated disk $D(p)$ with center p and radius $r(p) \geq 0$. The *power distance* from any point $q \in \mathbb{R}^2$ to p, denoted $power(q, p)$ is $\|p - q\|^2 - r(p)^2$. The *power diagram* of P is the cell complex in $\mathbb{R}^2$ that associates to each $p \in P$, the convex domain $cell(p) = \{q \in \mathbb{R}^2 \mid power(q, p) < power(q, p') \text{ for all } p' \in P - \{p\}\}$. The Voronoi diagram is a special case of a power diagram, obtained by setting $r(p) = 0$ for all $p \in P$. Just as Delaunay triangulations are duals of Voronoi diagrams, a more general class of triangulations are the planar duals of power diagrams [8]. This motivates the question of whether for a given combinatorial triangulation $G = (V, E)$, we can find an embedding $\Phi : V \to \mathbb{R}^2$ and a radius assignment $r : V \to \mathbb{R}^+$, whose power diagram has, as its planar dual, the graph G. If we can find such an embedding, then greedy geographic routing can be used with power distance in lieu of Euclidean distance. For any combinatorial triangulation G (and in fact for any planar graph), it is known that there exists an embedding $\Phi : V \to \mathbb{R}^2$ and a radius assignment $r : V \to \mathbb{R}^+$ that yields a power diagram whose dual is G. This follows from the celebrated Koebe Representation Theorem [1,2,12,18]. None of the known proofs of Koebe's theorem lead to efficient algorithms [14,18]. Again, we have a situation in which the existence of appropriate embeddings are well known, but the proof of existence does not give a clear indication of how to efficiently compute these embeddings. It should be noted that there are polynomial time algorithms to compute a Koebe representation [13,17], but these use the ellipsoid method and are therefore not practical for obtaining a fast, distributed algorithm. Given this situation, Chen et al. [4] use a simple iterative algorithm to produce a Koebe representation. This algorithm is not guaranteed to run in polynomial time and is obtained in a straightforward way from Thurston's proof of Koebe's theorem. We will call this the `Thurston` algorithm. Chen et al. [4] also present an algorithm that improves on the performance of the `Thurston` algorithm by using fewer iterations to terminate with a power diagram. We call this the `PowerDiagram` algorithm. Both of these algorithms are described in some detail in Section 2.2. In Section 4, we present strong experimental evidence indicating that our FindAngles algorithm is orders of magnitude faster than the PowerDiagram algorithm, which in turn is a bit faster than the Thurston algorithm.

Main results. We present a simple, iterative, local distributed algorithm, called FindAngles, for computing a Delaunay realization of any given combinatorial triangulation that is Delaunay realizable. Our algorithm is based on a characterization of Delaunay realizability due to Hiroshima et al. [9]. Our algorithm uses far fewer iterations as compared to the PowerDiagram algorithm of Chen et al. [4]. In experiments we ran, on average, the number of iterations of our algorithm was about $1/7^{th}$ of the number of iterations of the PowerDiagram algorithm (see Table 1). Using the PowerDiagram algorithm in a distributed setting is problematic because in each iteration, it requires a sweep of the entire graph in order to update coordinates. In a distributed setting, such operations are costly and each iteration of the PowerDiagram essentially corresponds to n communication rounds, where n is the size of the graph. In contrast, each iteration of FindAngles corresponds to exactly one round of communication. In addition, the number of iterations required by our algorithm scales linearly with n. The PowerDiagram algorithm was proposed as an improvement on the Thurston algorithm and in our experimental results, on average, the PowerDiagram algorithm uses about 18% fewer iterations than the Thurston algorithm, whereas our algorithm uses about 88% fewer iterations. Experimentally, FindAngles exhibits very well-behaved convergence; the error term falls rapidly in the beginning and then falls more slowly as the algorithm gets close to a valid Delaunay realization. The error term rarely increases and we are able to prove that it strictly decreases under certain initial conditions. This experimentally observed behavior, along with our analysis, provides some hope that our algorithm may have polynomial time convergence, at least for special classes of combinatorial triangulations. It should be noted that whereas the PowerDiagram algorithm runs on all combinatorial triangulations, FindAngles runs only on Delaunay realizable triangulations. This may not be much of a problem because Hiroshima et al. [9] indicate that most combinatorial triangulations are Delaunay realizable and our experiments strongly support this. We randomly generated 5000 combinatorial triangulations on 100 vertices; of these only one was not Delaunay realizable.

2   Technical Background

Our algorithm is based on a characterization of Delaunay realizable graphs due to Hiroshima, Miyamoto, and Sugihara [9]. We call this the *HMS test* and describe it in Section 2.1. In our experiments, we compare FindAngles with the Thurston algorithm and the PowerDiagram algorithm. Both of these are based on the Koebe Representation Theorem and we describe that and sketch the Thurston algorithm and the PowerDiagram algorithm in Section 2.2.

2.1   The HMS Test for Delaunay Realizability

We describe the HMS test in some detail here because our algorithm (described in Section 3) is based on the proof of correctness of the HMS test.

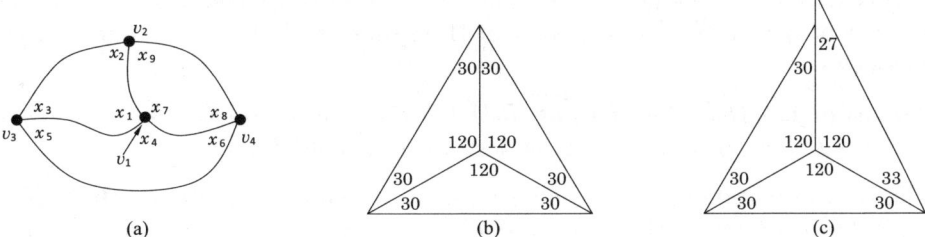

Fig. 1. (a) The connectivity for a complete graph on 4 vertices. (b) A solution to (C1)-(C5) that yields a Delaunay triangulation. (c) A solution to (C1)-(C5) of that does not yield a Delaunay triangulation [9].

Let $G = (V, E)$ be a combinatorial triangulation. For each inner face f_i, three *angle variables* x_{3i+1}, x_{3i+2} and x_{3i+3} are defined, respectively corresponding to the three vertices that bound f_i (see Figure 1(a)). A vertex of G is called an *outer vertex* (*edge*) if it is on the boundary of the outer face of G; otherwise it is called an *inner vertex* (*edge*). Let $\{u, v\}$ be an inner edge and let (u, v, w_1) and (w_2, v, u) be the two faces containing edge $\{u, v\}$. Then the angle variables at w_1 associated with face (u, v, w_1) and the angle variable at w_2 associated with face (w_2, v, u) are the two *facing angle variables* associated with edge $\{u, v\}$. For example, in Figure 1(a) the facing angle variables of edge $\{v_1, v_3\}$ are x_2 and x_6. Hiroshima et al. [9] state 5 simple linear constraints that the angle variables must satisfy in any Delaunay realization:

(C1) The sum of the three angle variables associated with each inner face equals 180.
(C2) The sum of the angle variables associated with each inner vertex equals 360.
(C3) The sum of the angle variables associated with each outer vertex is less than or equal to 180.
(C4) The sum of the facing angle variables associated with each inner edge is less than or equal to 180.
(C5) Each angle variable is positive.

In the above list (C4) is particular to Delaunay triangulations and follows easily from the empty circle property. Hiroshima et al. [9] point out that even though every Delaunay realization satisfies conditions (C1)-(C5), these conditions are not sufficient because they do not guarantee that the triangles incident on a vertex v can be consistently "glued" together around v (see Figure 1(c)). Interestingly, Hiroshima et al. [9] were able to show that for any solution that does satisfy (C1)-(C5), there exists a transformation that modifies this solution into one that additionally satisfies a "consistent gluing" condition around each vertex. Thus Hiroshima et al. [9] were able to reduce the problem of testing if a combinatorial triangulation is Delaunay realizable into a problem of testing the feasibility of a linear system of equations and inequalities. Note that the size of this system (i.e., number of variables, number of constraints) is linear in the number of vertices in G. The result of Hiroshima et al. [9] can be summarized as follows.

Theorem 1. *(Hiroshima et al. [9])* *Whether a given combinatorial triangulation is Delaunay realizable can be tested in polynomial time.*

The transformation mentioned above is only shown existentially by Hiroshima et al. [9]. As a result the above theorem does not lead to a polynomial time algorithm for constructing a Delaunay realization. We now review the proof of Hiroshima et al. [9] which prompted our algorithm, presented in Section 3.

Let v be an inner vertex of G and let $w_0, w_1, \ldots, w_{s-1}$ be its neighbors in counter clockwise order. Then the angles $(v, w_0, w_1), (v, w_1, w_2), \ldots, (v, w_{s-1}, w_0)$ are called *cc-facing angles* about v denoted $\phi_0, \phi_1, \ldots, \phi_{s-1}$. The angles $(v, w_1, w_0), (v, w_2, w_1), \ldots, (v, w_0, w_{s-1})$ are called *c-facing angles* about v, denote $\theta_0, \theta_1, \ldots, \theta_{s-1}$ (see Figure 2). Define

$$F(v) = \frac{\sin(\phi_0)\sin(\phi_1)\cdots\sin(\phi_{s-1})}{\sin(\theta_0)\sin(\theta_1)\cdots\sin(\theta_{s-1})}$$

It is shown in Lemma 3.1 of Hiroshima et al. [9] that in order to "glue" the triangles incident on v consistently around v, $F(v)$ must equal 1. Note that since all ϕ_i's and θ_i's are strictly between 0 and 180, we have $F(v) > 0$. Hiroshima et al. define a condition (C6) as follows:

(C6) $F(v) = 1$ for each inner vertex v.

Conditions (C1)-(C6) exactly capture Delaunay realizability; these conditions are both necessary and sufficient. Unfortunately, (C6) is not "well behaved" like conditions (C1)-(C5) and no efficient algorithm is known for solving the system of equations and inequalities implied by conditions (C1)-(C6).

Consider any real α. Let *change*(v, α) denote the operation in which each cc-facing angle ϕ_i around v is changed to $\phi_i + \alpha$ and each c-facing angle θ_i around v is changed to $\theta_i - \alpha$. Let ϕ_{min} denote $\min_i \phi_i$ and let θ_{min} denote $\min_i \theta_i$. For any α, $-1 \times \phi_{min} < \alpha < \theta_{min}$, applying the operation *change*(v, α) keeps all the angles positive. The operation *change*(v, α), for α, $-1 \times \phi_{min} < \alpha < \theta_{min}$, has the very useful property that if conditions (C1)-(C5) are satisfied before the operation, they will continue to be satisfied after the operation. Hiroshima et al. [9] show that given an assignment of values to the angle variables satisfying (C1)-(C5), there exists a set $\{(v_1, \alpha_1), (v_2, \alpha_2), \ldots\}$, such that performing *change*(v_1, α_1), *change*(v_2, α_2), $\ldots$ leads to condition (C6) also being satisfied. While the existence of $\{(v_1, \alpha_1), (v_2, \alpha_2), \ldots\}$ is shown, there is no known polynomial time algorithm to find it.

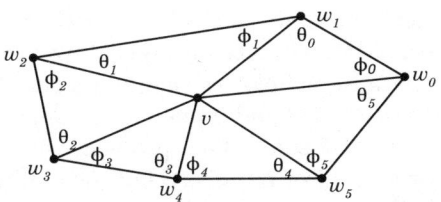

Fig. 2. cc-facing angles about v are $\phi_0, \phi_1, \ldots, \phi_5$. c-facing angles are $\theta_0, \theta_1, \ldots, \theta_5$

2.2   The Koebe Representation Theorem

Koebe [12] in 1936 proved the following remarkable theorem.

Theorem 2. (Koebe [12]) *Given any planar graph* $G = (V, E)$ *with* $V = \{v_1, v_2, \ldots, v_n\}$, *we can find a packing of* n *(not necessarily congruent) disks* $C = \{C_1, C_2, \ldots, C_n\}$ *in the plane with the property that* C_i *and* C_j *touch iff* $\{v_i, v_j\} \in E$ *for* $1 \leq i, j \leq n$.

The fact that the set C is a circle *packing* implies that the interiors of the circles are pairwise mutually disjoint.

From Thurston's proof of Koebe's theorem [18] one can extract an iterative algorithm for (approximately) computing a Koebe representation (i.e., a circle packing promised by Koebe's theorem) for a given planar graph. In the following description we assume, for simplicity, that G is a combinatorial triangulation. Let $\mathbf{r} = (r_1, r_2, \ldots, r_n)$ be arbitrary initial radii assigned to the vertices $v_1, v_2, \ldots, v_n$ respectively. Then $\mathbf{r}$ uniquely defines the three angles in each inner face of G. Let $\sigma_{\mathbf{r}}(v_i)$ denote the sum of the angles at v_i in all of the inner faces to which v_i belongs. The subscript "$\mathbf{r}$" here denotes the fact that the angles depend on the radii assignment $\mathbf{r}$. If $\sigma_{\mathbf{r}}(v_i) = 360$ for all inner vertices v_i, we are done because we can consistently "glue" the triangular faces together. If $\sigma_{\mathbf{r}}(v_i) \neq 360$ for some inner vertex v_i, then we can find a new radius r_i' for v_i such that with respect to $\mathbf{r}' = (r_1, r_2, \ldots, r_i', \ldots, r_n)$, the angle sum $\sigma_{\mathbf{r}'}(v_i) = 360$. Of course this update affects the angle sums at neighbors of v_i. However, it is possible to show that this iterative process converges. Collins and Stephenson [6] suggest several improvements to this basic algorithm that reduce the number of iterations needed for convergence. This algorithm has an obvious distributed implementation in which, in each round, each node v_i for which $\sigma_r(v_i) \neq 360$ updates its radius r_i. New radius values are then exchanged with neighboring nodes in one round of local communication, and the algorithm proceeds to the next round. It is this entire class of algorithms that we refer to as the `Thurston` algorithm.

A Koebe representation $C = \{C_1, C_2, \ldots, C_n\}$ of a combinatorial triangulation $G = (V, E)$ can also be viewed as a power diagram whose planar dual is G. This is the starting point of the work of Chen et al. [3,4], who go on to point out that Koebe representations are special power diagrams in which disks corresponding to adjacent cells are mutually tangent. With this motivation, Chen et al. [3,4] develop a *local power diagram (LPD)* test that takes as input a mapping

$\Phi : V \rightarrow \mathbb{R}^2$ and an assignment of disks $D(v)$ to vertices $v \in V$ and determines if $\{\Phi, D\}$ is a power diagram. It is this algorithm that we call PowerDiagram. This is called a "local" test because it involves checking a condition at each vertex v that is a function of Φ and D at v and its neighbors only. Thus this test can be implemented in a distributed fashion. One problem with LPD is the fact that it depends not just on the radii of the disks, but also on the coordinates of the vertices specified by Φ. This is a problem because after each radii update (as in the description of Thurston's algorithm) we have to recompute the coordinates of all the vertices by using some kind of a global sweep of the graph. Thus, after each round in which all nodes update their radii, we need $\Omega(diameter)$ rounds of communication to update the coordinates. In our view, this problem largely offsets the gains obtained by using LPD.

3   The FindAngles Algorithm

The FindAngles algorithm first obtains an initial angle assignment by finding a feasible solution of the linear program defined by constraints (C1)-(C5). The algorithm terminates if no feasible solution it found. Once an initial angle assignment is found, the algorithm repeatedly adjusts the angle values. It is this angle adjustment process that is the focus of our work. Angles are adjusted through an iterative process that repeatedly applies the $change(v, \alpha)$ operation introduced in Section 2.1. This process consists of a sequence of *rounds*. In each round, we scan through all the inner vertices in an arbitrary order. At each inner vertex v we check if $F(v) = 1$. If this is not the case, we solve for an α^* such that after applying $change(v, \alpha^*)$, $F(v)$ equal 1. We then apply $change(v, \alpha^*)$. To see that it is possible to efficiently solve for such an α^*, define the following function $g : \mathbb{R} \rightarrow \mathbb{R}$:

$$g(\alpha) = \frac{\sin(\phi_0 + \alpha)\sin(\phi_1 + \alpha)\cdots\sin(\phi_{s-1} + \alpha)}{\sin(\theta_0 - \alpha)\sin(\theta_1 - \alpha)\cdots\sin(\theta_{s-1} - \alpha)}. \tag{1}$$

Now suppose that $F(v) < 1$. Then $g(0) = F(v) < 1$ and $\lim_{\alpha \rightarrow \theta_{min}} g(\alpha) = +\infty$ (recall that $\theta_{min} = \min_i \theta_i$). Since $g(\alpha)$ is a continuous function, it is guaranteed that for some $\alpha^* \in (0, \theta_{min})$, $g(\alpha^*) = 1$. In fact $g(\alpha)$ is monotonically increasing as α increases from 0 to θ_{min}. To see this observe that the ratio $\sin(\phi_i + \alpha)/\sin(\theta_i - \alpha)$ is monotonically increasing for $0 < \alpha < \theta_i$ provided $\phi_i + \theta_i \leq 180$. This is clearly so since ϕ_i and θ_i are required to be positive (condition (C5)) and the sum of ϕ_i, θ_i, and a third angle equals 180 (condition (C1)). Thus the equation $g(\alpha) = 1$ can be easily solved using a standard numerical root finding technique such as Newton's method. We have a symmetric situation if $F(v) > 1$. After each iteration, we check if a global termination condition is satisfied. This termination condition depends on a fixed parameter $\varepsilon > 0$ and is defined as either:

Max-termination condition: $|1 - F(v)| < \varepsilon$ for all inner vertices v
or
Sum-termination condition: $\sum_v |1 - F(v)| < \varepsilon$ over all inner vertices v

We have separately implemented both termination conditions and report results for both, however we focus on the first termination condition since it is local, i.e., each node can decide for itself, based on local information, if it needs to participate in the current iteration. Once the iterations terminate we perform a global BFS sweep of the graph to fix planar coordinates of all vertices in G. We Start with an arbitrary pair of adjacent vertices u and v and place them arbitrarily, at distinct points on the plane. The sweep guarantees that for every subsequent vertex w that is processed, there is a triangle of G, (a, b, w) such that a and b are vertices that have already been placed in $\mathbb{R}^2$. Given the fact that the three angles of the triangle (a, b, w) are known, there is a unique location in $\mathbb{R}^2$ for w. Upon termination of our algorithm, each $F(v)$ may differ from 1 by ε. Hence the angle variables themselves yield an *approximate* Delaunay realization (see Figure 3). We therefore geometrically evaluate each approximate Delaunay realization to see if it is indeed a Delaunay triangulation. As expected, as ε becomes smaller, the fraction of approximate Delaunay realizations that are actually Delaunay triangulations, increases and reaches 1 (see Table 3). While `FindAngles` may not run in polynomial time, experimental evidence suggests that it is very fast (see Section 4), leading us to believe that some variant of this algorithm may indeed be shown to have polynomial time convergence. The `FindAngles` algorithm is summarized in the below pseudocode. The algorithm was implemented in *Mathematica 6.0*, using in-built functions for linear programming and root finding.

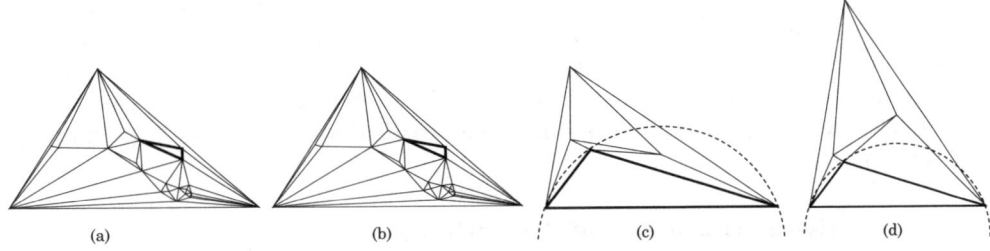

(a)   (b)   (c)   (d)

Fig. 3. (a) and (b) were produced from a single input with $\varepsilon = 10^{-2}$ and $\varepsilon = 10^{-6}$ respectively. (a) is not a Delaunay triangulation (the highlighted triangle violates the empty circle property) while (b) is Delaunay. For illustrative purpose only, (c) and (d) were produced from a single input with $\varepsilon = 20$ and $\varepsilon = 12$ respectively. (c) is not Delaunay while (d) is.

Algorithm `FindAngles`
Input: G: A combinatorial triangulation, $\varepsilon \in \mathbb{R}^+$
Output: Embedding of G that is a Delaunay triangulation, or report none exists.

1. Find initial angle assignment: a feasible solution to (C1) - (C5).
2. **if** (no feasible solution exists)
3.   Report that G is not Delaunay realizable
4.   Stop
5. **end if**

6.   **while** ($\exists$ inner vertex v : $| 1 - F(v) | \geq \varepsilon$)
7.   **for**(each inner vertex u)
8.   **if** ($| 1 - F(u) | \geq \varepsilon$)
9.   Compute α^* such that $F(u) = 1$ following $change(u, \alpha^*)$
10.   Apply $change(u, \alpha^*)$
11.   **end if**
12.   **end for**
13.   **end while**
14.   Arbitrarily select edge $\{u, v\}$ and embed u and v at distinct points in $\mathbb{R}^2$
15.   Perform a BFS traversal starting with u or v. At each vertex w:
16.   Pick a triangle (a, b, w) such that a and b have been embedded
17.   Embed w in $\mathbb{R}^2$ using the angles of triangle (a, b, w)

4   Experimental Results

The basic experiments described in this section were run on a total of 65 input graphs, 13 each of order $n \in \{23, 43, 63, 83, 103\}$. To construct each input graph, $n - 3$ random points were first placed inside 3 boundary vertices. A Delaunay triangulation $\mathcal{T}$ on this set of n points was then constructed. Each edge e of $\mathcal{T}$ was next inspected in an arbitrary order. If e was found to be the diagonal of a convex quadrilateral, then e was removed from $\mathcal{T}$ and the opposite diagonal added. By doing this we have a triangulation that is clearly not a Delaunay triangulation, even though it might be Delaunay realizable via some alternate realization. We then extracted a combinatorial representation of the adjacencies of this triangulation. To determine how well our algorithm scales we used 50 larger graphs up to order $n = 1003$. For a given input graph, we are interested in finding a realization that is the combinatorial dual of a power diagram. For simplicity, when we succeed in finding such a realization, we say we have found a "Power Diagram".

4.1   Relative Performance of Algorithm FindAngles

In this subsection we compare the performance of the FindAngles algorithm with that of algorithms Thurston and PowerDiagram. The comparison is along three dimensions: number of iterations, fraction of realizations that are power diagrams, and running time. Along all three dimensions, FindAngles outperforms both of the other algorithms.

We start by looking at the number of iterations and the processing time required by each algorithm to produce 100% power diagrams (see Table 1 and Figure 4). Recall that the PowerDiagram algorithm always terminates with a valid power diagram. So to perform a meaningful comparison, for each n we pick a largest ε for which all realizations generated by FindAngles are Delaunay triangulations and report the average number of iterations and time required using this ε. Note different ε may be selected for different n. The results for the Thurston algorithm are similarly selected and reported. Chen et al. [4] have proposed the PowerDiagram algorithm as an improvement over the

Thurston algorithm. However, it is clear from Table 1 that the improvement of the PowerDiagram algorithm is marginal, relative to the improvement obtained by using our algorithm. For example, for $n = 103$, the Thurston algorithm needs 1664 iterations, the PowerDiagram algorithm needs 1330 iterations, whereas Algorithm FindAngles needs only 122 iterations! The last three columns of the table show the times (in seconds) for the three algorithms. Again, our algorithm is significantly faster than both the Thurston algorithm and the PowerDiagram algorithm. In our implementation, the PowerDiagram algorithm takes significantly more time than the Thurston algorithm, despite using fewer iterations. This is due to fact that we implemented the local power diagram test by solving a non-linear system of equations using the Mathematica function FindInstance. It is possible that an alternate, purely geometric, implementation of the local power diagram test would speed up the PowerDiagram algorithm.

Table 1. For each n, we report the average number of iterations and time (in seconds) for the PowerDiagram algorithm to terminate. For each of the other two algorithms, for each n, we find the largest ε, for which all realizations produced by the algorithm are power diagrams.

| n | Average Iterations for 100% Power Diagrams | | | Average Time (in sec.) for 100% Power Diagrams | | |
	Thurston	PowerDiagram	FindAngles	Thurston	PowerDiagram	FindAngles
23	256	175	60	9	66	2
43	536	442	49	71	389	4
63	879	758	165	256	1115	23
83	1137	961	134	583	2123	24
103	1664	1330	122	1321	3922	27

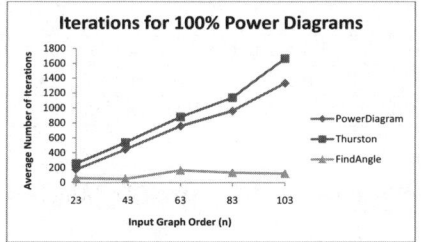
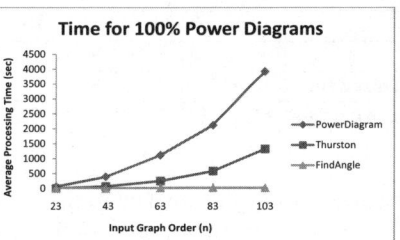

Fig. 4. These charts represent the data from Table 1

Next we compare FindAngles with the Thurston algorithm (see Table 2 and Figure 5). The reported results are for $\varepsilon = 10^{-5}$. Again, the comparison is along three dimensions and again, FindAngles significantly outperforms the Thurston algorithm.

As mentioned in Section 3, our algorithm produces an approximate Delaunay realization, where the approximation depends on ε: as $\varepsilon \to 0$, the realization produced gets "closer" to a Delaunay triangulation. Table 3 shows the increase

Table 2. This summarizes the results of the execution of the `Thurston` algorithm and Algorithm `FindAngles`. The results are for runs with $\varepsilon = 10^{-5}$. Each row represents the average results for 13 different input graphs of the order shown.

Input Graph	Thurston			FindAngles		
Order n	Average Iterations	Average Time (sec)	Percent Power Diag	Average Iterations	Average Time (sec)	Percent Power Diag
23	309	11	100	83	3	100
43	591	78	100	104	9	100
63	879	256	100	165	23	100
83	1101	565	92	171	32	100
103	1341	1063	54	170	40	100

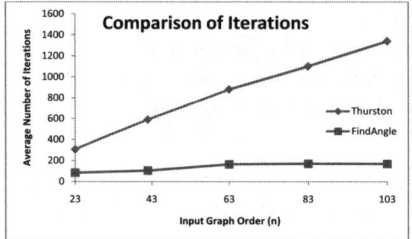 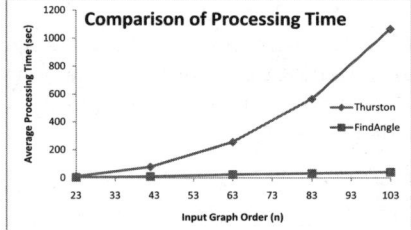

Fig. 5. These charts show the data form Table 2

Table 3. This table shows the increase in the fraction of actual Delaunay triangulations produced `FindAngles` and the `Thurston` algorithm as $\varepsilon \to 0$

ε	$\times 10^{-3}$					$\times 10^{-4}$					$\times 10^{-5}$					$\times 10^{-6}$				
	8	6	4	2	1	8	6	4	2	1	8	6	4	2	1	8	6	4	2	1
Thurston	0	0	0	0	2	3	6	9	17	25	26	32	40	69	89	95	97	97	100	100
FindAngles	11	12	14	20	34	40	45	55	74	86	92	95	98	98	100	100	100	100	100	100

in the fraction of actual Delaunay triangulations produced by `FindAngles` and the `Thurston` algorithm as $\varepsilon \to 0$.

As mentioned in Section 3, we have also used sum-termination condition as an alternate termination condition for `FindAngles`. Being a bit more flexible than the max-termination condition, we expect termination to be reached in fewer iterations using sum-termination. This is confirmed in Table 4. For example, for $n = 103$ and $\varepsilon = 10^{-4}$, it takes on average, 200 iterations for `FindAngles` to terminate with the sum-termination condition. The sum-termination condition requires the sum of the "errors" at all inner vertices to be at most 10^{-4}. We interpret this constraint as being roughly equivalent to the "error" at each vertex being bounded by $10^{-4}/n = 10^{-6}$. We can then compare the 200 iterations needed to terminate with $\varepsilon = 10^{-4}$ with the sum-termination condition with the 243 iterations it takes to terminate with $\varepsilon = 10^{-6}$ under the max-termination condition. Similar comparisons can be made for each of the other values of n and we see that one can ensure roughly similar average error using slightly fewer iterations with the sum-termination condition. The salient point of this comparison is to show that convergence of our algorithm seems fairly robust and is not greatly affected by using alternate termination conditions.

Table 4. The last three entries of each row respectively represent the average number of iterations for $\varepsilon = 10^{-4}$ under sum-termination, $\varepsilon = 10^{-5}$ under max-termination, and $\varepsilon = 10^{-6}$ under max-termination

| | Average Number of Iterations | | |
| | Sum-termination | Max-termination | Max-termination |
n	$\varepsilon = 10^{-4}$	$\varepsilon = 10^{-5}$	$\varepsilon = 10^{-6}$
23	73	83	106
43	103	104	138
63	176	165	220
83	191	171	232
103	200	170	243

4.2 Scaling to Graphs of Order $n = 1000$

Our simulations demonstrate that `FindAngles` is efficient for input graphs with up to 103 vertices. To determine whether it scales to larger graphs, `FindAngles` was run on graphs up to order $n = 1003$ (see Table 5 and in Figure 6). We see that the number of iterations grows linearly with the input size. The processing time grows a bit more rapidly and consists mainly of time spent adjusting angles. The time needed to solve the linear program is just a small fraction of the overall processing time.

Table 5. This table shows how our algorithm scales to inputs of size 1003. These average are taken over five inputs of each size shown.

Scaling to Large Inputs ($\varepsilon = 10^{-5}$)										
n	103	203	303	403	503	603	703	803	903	1003
Average Iterations	157	273	399	340	505	579	637	786	834	951
Average Time to Solve LP	0.02	0.05	0.08	0.11	0.14	0.17	0.20	0.23	0.26	0.28
Average Time to Adjust Angles	38	139	341	396	836	1171	1588	2426	3285	3711

4.3 Most Triangulations Are Delaunay Realizable

One of the main motivations of our work is the observation by Hiroshima et al. [9] that most triangulations are Delaunay realizable. In that work the authors experimentally verified their observation for small size triangulations (with up to 12 vertices). In an effort to corroborate those findings, we randomly constructed 5000 pairwise non-isomorphic combinatorial triangulations of order $n = 103$ and

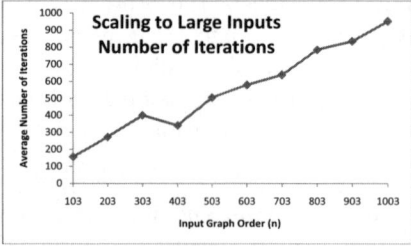 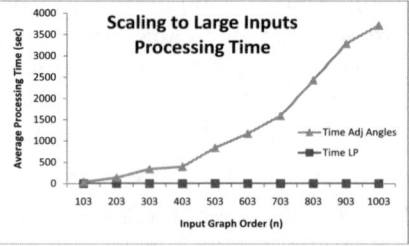

Fig. 6. These charts represent the data from Table 5

used the linear program corresponding to constraints (C1)-(C5) to see how many were Delaunay realizable. The triangulations were generated by first dropping points randomly in the plane, constructing a Delaunay triangulation, "flipping" randomly chosen diagonals repeatedly, and finally extracting a combinatorial representation of the resulting triangulation. The number of diagonals flipped is determined by the parameter m. If the Markov Chain defined by this process is rapidly mixing, then for m not too large, the 5000 combinatorial triangulations would form a sample picked uniformly at random from the set of all combinatorial triangulations on n points. The result was quite satisfying and lends support to the observation of Hiroshima et al. [9]: *of the 5000 combinatorial triangulations generated, only 1 is not Delaunay realizable.*

5   Understanding the Convergence

The angle adjustment process of the `FindAngles` algorithm starts with an initial angle assignment (a feasible solution to (C1)-(C5)) and modifies the angles iteratively in an effort to satisfy constraint (C6) as well. This process can can be viewed as moving from an inconsistent angle assignment in which the triangles cannot be "glued" together, toward a valid Delaunay triangulation. As a measure of how close to a valid Delaunay triangulation the current angle assignment is after an iteration, we define *error* as $\sum_{v \in I} |1 - F(v)|$, where I is the set of all inner vertices and $F(v)$ is as defined in Section 2.1. We are interested in the behavior of this error term as the iterations progress. The faster the error term converges to zero, the sooner the iterative process terminates. Figure 7 shows that the error falls rapidly initially and then converges more slowly as the algorithm gets closer to a valid Delaunay triangulation. While the decrease in error is largely monotonic, for a small percentage of iterations the error increases slightly. In Theorem 3 we prove that under certain initial conditions the error term strictly decreases. For any inner vertex v and real α, let $change(v, \alpha)$ be as described in Section 2.1.

Theorem 3. *Let v be an inner vertex of degree s and suppose that $F(v) \neq 1$. Further suppose that $F(v)$ is not smaller than the F-values of neighbors of v. Let α^* be such that performing $change(v, \alpha^*)$ makes $F(v) = 1$. Then performing $change(v, \alpha^*)$ decreases the error.*

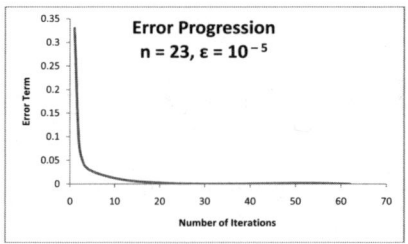 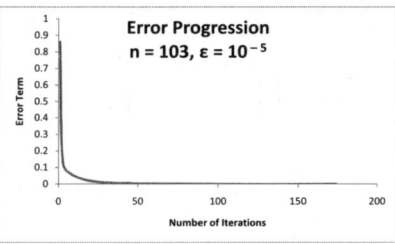

Fig. 7. This figure shows how the error term falls rapidly when the `FindAngles` algorithm is run with $\varepsilon = 10^{-5}$ on two separate input graphs, one of order $n = 23$ and one of order $n = 103$

Proof. There are two cases, depending on whether or not $F(v) < 1$. We first assume that $F(v) < 1$. Let ϕ_i (θ_i), $0 \leq i \leq s - 1$, be the cc-Facing (c-facing) angles about v (recall Figure 2). Group the terms in $F(v)$ as follows

$$F(v) = \left(\frac{\sin(\phi_0)}{\sin(\theta_1)}\right) \cdot \left(\frac{\sin(\phi_1)}{\sin(\theta_2)}\right) \cdots \left(\frac{\sin(\phi_{s-2})}{\sin(\theta_{s-1})}\right) \cdot \left(\frac{\sin(\phi_{s-1})}{\sin(\theta_0)}\right),$$

calling them $t_0, t_1, \ldots, t_{s-1}$ respectively. Note that the two angles involved in term t_j, namely ϕ_j and θ_{j+1}, are "facing" angles and therefore according to condition (C4) satisfy $\phi_j + \theta_{j+1} < 180$. Here (and in the remainder of the proof) we use modulo s arithmetic, so if $j = s - 1$, then $j + 1 \equiv 0$ and if $j = 0$, then $j - 1 \equiv s - 1$. Now let us view the quantity

$$\left(\frac{\sin(\phi_j + \alpha)}{\sin(\theta_{j+1} - \alpha)}\right)$$

as a function of α and denote it by $t_j(\alpha)$. So $t_j(0)$ is just t_j. Since $\phi_j + \theta_{j+1} < 180$, using Lemma 4.2 from Hiroshima et al. [9] we see that $t_j(\alpha)$ is monotonically increasing in the range $\alpha \in [0, \theta_{j+1})$. Therefore the product $\prod_{j=0}^{s-1} t_j(\alpha)$ is also monotonically increasing starting at

$$\prod_{j=0}^{s-1} t_j(0) = \prod_{j=1}^{s-1} t_j = F(v) < 1$$

and increasing to $+\infty$ as $\alpha \to \min\{\theta_0, \theta_1, \ldots, \theta_{s-1}\}$. Therefore there is some $0 < \alpha^* < \min\{\theta_0, \theta_1, \ldots, \theta_{s-1}\}$ such that $\prod_{j=0}^{s-1} t_j(\alpha^*) = 1$. In other words, performing the operation $change(v, \alpha^*)$ makes $F(v) = 1$.

Now fix an α, $0 < \alpha \leq \alpha^*$. Since $t_j(\cdot)$ is a monotonically increasing function in the range $[0, \alpha^*]$, each $t_j(\alpha)$ can be expressed as $t_j + \Delta_j$ for some $\Delta_j > 0$. Performing the operation $change(v, \alpha)$ results in $F(v)$ increasing from $\prod_{j=0}^{s-1} t_j$ to $\prod_{j=0}^{s-1}(t_j + \Delta_j)$. As $F(v)$ increases towards 1, we see a decrease in the contribution of v to the error by

$$\prod_{j=0}^{s-1}(t_j + \Delta_j) - \prod_{j=0}^{s-1} t_j$$

We can lower bound this quantity as follows:

$$\prod_{j=0}^{s-1}(t_j + \Delta_j) - \prod_{j=0}^{s-1} t_j > \Delta_0 \cdot \prod_{j\neq 0} t_j + \Delta_1 \prod_{j\neq 1} t_j + \cdots + \Delta_{s-1} \prod_{j\neq s-1} t_j = \sum_{j=0}^{s-1} \frac{\Delta_j}{t_j} F(v)$$

Thus the decrease in the error term due to v is greater than

$$\sum_{j=0}^{s-1} \frac{\Delta_j}{t_j} F(v). \tag{2}$$

The operation $change(v, \alpha)$ also affects $F(w_j)$ for each neighbor w_j, $0 \le j \le s-1$ of v. To understand the precise effect, note that each $F(w_j)$ contains the term $\frac{\sin(\theta_j)}{\sin(\phi_{j-1})}$ times other terms that do not involve any of the changed angles. Note that ϕ_{j-1}, which is a cc-facing angle about v, is a c-facing angle around w_j and similarly, θ_j, which is a c-facing angle about v, is a cc-facing angle round w_j. So $F(w_j)$ can be written as $\frac{1}{t_{j-1}} \times rest_j$, where $rest_j$ denotes the product of the other terms in $F(w_j)$. Performing the operation $change(v, \alpha)$ decreases $F(w_j)$ to

$$\frac{1}{(t_{j-1} + \Delta_{j-1})} \cdot rest_j$$

Therefore the change in $F(w_j)$ is

$$\frac{1}{t_{j-1}} \cdot rest_j - \frac{1}{(t_{j-1} + \Delta_{j-1})} \cdot rest_j = \frac{\Delta_{j-1}}{t_{j-1}(t_{j-1} + \Delta_{j-1})} \cdot rest_j = \frac{\Delta_{j-1}}{(t_{j-1} + \Delta_{j-1})} \cdot F(w_j)$$

The above change is an increase in the contribution of $F(w_j)$ to the error.

Now for each j, $0 \le j \le s-1$, we can "charge" the increase in the contribution to the error by $F(w_j)$ to the term $\frac{\Delta_{j-1}}{t_{j-1}} F(v)$ in Equation (2). This is because $F(v) \ge F(w_j)$, since v was chosen to maximize $F(v)$ and so we get:

$$\frac{\Delta_{j-1}}{t_{j-1}} \cdot F(v) \ge \frac{\Delta_{j-1}}{t_{j-1}} \cdot F(w_j) \ge \frac{\Delta_{j-1}}{(t_{j-1} + \Delta_{j-1})} \cdot F(w_j)$$

Thus the increase in the error due to w_j is less than the decrease in the error due to the term $\frac{\Delta_{j-1}}{t_{j-1}} F(v)$ in Equation (2). Summing this over all j, we obtain that the total increase in error due to the neighbors of v is less than the decrease in the error due to v. All of this is true for any $\alpha \in (0, \alpha^*]$ and in particular for $\alpha = \alpha^*$ as well.

The case in which $F(v) > 1$ is quite similar. Here we need to decrease $F(v)$ so we pick a negative α and perform a $change(v, \alpha)$. This results in an increase in $F(w_j)$-values, but as in the above proof the decrease in $F(v)$ decreases the error by more than total increase due to all of the $F(w_j)$'s. $\qquad\square$

6   Future Work

As an initial preprocessing step `FindAngles` looks for a feasible solution to (C1)-
(C5). We currently use the built in Mathematica function `LinearProgramming`
to solve this LP. We would like to replace this with a strictly local test for
feasibility. Currently vertices whose angles are adjusted are selected arbitrarily.
We may be able to improve convergence by selectively picking vertices based
on some threshold of their F-values. We would like to extend our algorithm to
run on all 3-connected planar graphs. Motivated by the desire to find a general
polynomial time algorithm for Delaunay realizability, we would like to study
more closely the behavior of the error term and identify an alternate definition
of error that decreases monotonically for all inputs.

References

1. Andreev, E.M.: On convex polyhedra in Lobacevskii space. Math. USSR
   Sbornik 10(3), 413–440 (1970)
2. Andreev, E.M.: On convex polyhedra of finite volume in Lobachevskii space. Math.
   USSR Sbornik 12(2), 255–259 (1970)
3. Ben-Chen, M., Gotsman, C., Gortler, S.: Routing with guaranteed delivery on
   virtual coordinates. In: Proceedings of the 18th Canadian Conference on Computational Geometry (CCCG 2006), pp. 117–120 (2006)
4. Ben-Chen, M., Gotsman, C., Wormser, C.: Distributed computation of virtual coordinates. In: Proceedings of the 23rd annual symposium on Computational geometry
   (SoCG 2007), pp. 210–219. ACM Press, New York, NY, USA (2007)
5. Bose, P., Morin, P.: Online routing in triangulations. In: Aggarwal, A.K., Pandu
   Rangan, C. (eds.) ISAAC 1999. LNCS, vol. 1741, pp. 113–122. Springer, Heidelberg
   (1999)
6. Collins, C.R., Stephenson, K.: A circle packing algorithm. Computational Geometry: Theory and Applications 25(3), 233–256 (2003)
7. Dhandapani, R.: Greedy drawings of triangulations. In: Proceedings of the 19th annual ACM-SIAM symposium on discrete algorithms (SODA 2008), SIAM, Philadelphia (2008)
8. Edelsbrunner, H., Shah, N.R.: Incremental topological flipping works for regular
   triangulations. In: Proceedings of the eighth annual symposium on Computational
   geometry (SoCG 1992), pp. 43–52. ACM, New York, NY, USA (1992)
9. Hiroshima, T., Miyamoto, Y., Sugihara, K.: Another proof of polynomial-time
   recognizability of Delaunay graphs. In: IEICE Transactions on Fundamentals of
   Electronics, Communications and Computer Sciences (IEICE 2000), April 2000,
   vol. 83(4), pp. 627–638 (2000)
10. Karp, B., Kung, H.T.: GPSR: Greedy perimeter stateless routing for wireless net-
    works. In: Proceedings of the 6th Annual ACM/IEEE International Conference on
    Mobile Computing and Networking (MobiCom 2000), pp. 243–254 (2000)
11. Knaster, B., Kuratowski, C., Mazurkiewicz, S.: Ein beweis des fixpunktsatzes für
    n-dimensionale simplexe. Fundamenta Mathematicae 14, 132–137 (1929)
12. Koebe, P.: Kontaktprobleme der konformen abbildung. Berichte über die Verhand-
    lungen d. Sächs. Akademie der Wissenschaften Leipzia 88, 141–164 (1936)
13. Mohar, B.: A polynomial time circle packing algorithm. Discrete Mathemat-
    ics 117(1–3), 257–263 (1993)

14. Pach, J., Agarwal, P.K.: Combinatorial Geometry. John Wiles & Sons, New York, NY, USA (1995)
15. Papadimitriou, C.H., Ratajczak, D.: On a conjecture related to geometric routing. Theoretical Computer Science 344(1), 3–14 (2005)
16. Santi, P.: Topology control in wireless ad hoc and sensor networks. ACM Computing Surveys 37(2), 164–194 (2005)
17. Smith, W.D.: Accurate circle configurations and numerical conformal mapping in polynomial time. NEC Research Institute, unpublished technical memorandum (December 1991)
18. Thurston, W.P.: The geometry and topology of 3-manifolds. Princeton University Notes, Princeton (1988)

Computing Branch Decomposition of Large Planar Graphs

Zhengbing Bian and Qian-Ping Gu

School of Computing Science, Simon Fraser University
Burnaby BC Canada, V5A 1S6
{zbian,qgu}@cs.sfu.ca

Abstract. A graph of small branchwidth admits efficient dynamic programming algorithms for many NP-hard problems on the graph. A key step in these algorithms is to find a branch decomposition of small width for the graph. Given a planar graph G of n vertices, an optimal branch decomposition of G can be computed in polynomial time, e.g., by the edge-contraction method in $O(n^3)$ time. All known algorithms for the planar branch decomposition use Seymour and Thomas procedure which, given an integer β, decides whether G has the branchwidth at least β or not in $O(n^2)$ time. Recent studies report efficient implementations of Seymour and Thomas procedure that compute the branchwidth of planar graphs of size up to one hundred thousand edges in a practical time and memory space. Using the efficient implementations as a subroutine, it is reported that the edge-contraction method computes an optimal branch decomposition for planar graphs of size up to several thousands edges in a practical time but it is still time consuming for graphs with larger size. In this paper, we propose divide-and-conquer based algorithms of using Seymour and Thomas procedure to compute optimal branch decompositions of planar graphs. Our algorithms have time complexity $O(n^3)$. Computational studies show that our algorithms are much faster than the edge-contraction algorithms and can compute an optimal branch decomposition of some planar graphs of size up to 50,000 edges in a practical time.

Keywords: Graph algorithms, branch-decomposition, planar graphs, algorithm engineering, computational study.

1   Introduction

The notions of branchwidth and branch decompositions are introduced by Robertson and Seymour [20] in relation to the more celebrated notions of treewidth and tree decompositions [18,19]. Branch/tree-decomposition based algorithms have been considered as efficient approaches for solving NP-hard problems on graphs of small branchwidth (or treewidth) [4,6,15]. To solve a problem, a branch/tree-decomposition based algorithm first computes a branch/tree decomposition with a small width and then applies a dynamic programming algorithm based on the decomposition. The second step usually runs exponentially

C.C. McGeoch (Ed.): WEA 2008, LNCS 5038, pp. 87–100, 2008.
© Springer-Verlag Berlin Heidelberg 2008

in the width of the decomposition computed in the first step. So it is extremely important to decide the branch/tree-width and compute the optimal decompositions. It is NP-complete to decide whether the width of a given general graph is at least an integer β if β is part of the input, both for branchwidth [22] and treewidth [3]. When the branchwidth (treewidth) is bounded by a constant, both the branchwidth and the optimal branch decomposition (treewidth and optimal tree decomposition) can be computed in linear time [7,9]. However, the linear time algorithms are not practical due to the huge constants behind the Big-Oh. The difficulty of computing a good branch/tree decomposition has been considered a hurdle for applying branch/tree-decomposition based algorithms in practice.

Recently, the branch-decomposition based algorithms have been receiving increasing attention for problems in planar graphs [10,11] because an optimal branch decomposition of a planar graph can be computed in polynomial time by Seymour and Thomas algorithm [22] and there is no huge hidden constant in the algorithm. Notice that it is open whether computing the treewidth of a planar graph is NP-hard or not. The result of the branchwidth implies a 1.5-approximation algorithm for the treewidth of planar graphs. Readers may refer to the recent papers by Bodlaender [8] and Hicks et al. [15] for extensive literature in the theory and application of branch/tree-decompositions.

Given a planar graph G of n vertices and an integer β, Seymour and Thomas give an algorithm (called ST Procedure for short in what follows) which decides if G has a branchwidth at least β in $O(n^2)$ time [22]. Using ST Procedure as a subroutine, they also give an edge-contraction algorithm which constructs an optimal branch decomposition of G. The edge-contraction algorithm calls ST Procedure $O(n^2)$ times and runs in $O(n^4)$ time. Gu and Tamaki [12] give an improved edge-contraction algorithm which calls ST Procedure $O(n)$ times and runs in $O(n^3)$ time to construct an optimal branch decomposition. Hicks proposes a divide-and-conquer algorithm called the cycle method to reduce the calls of ST Procedure for computing planar branch decompositions [13,14]. Since all known algorithms for branch decompositions use ST Procedure as a subroutine, the time and memory space required by ST Procedure limit the size of planar graphs for which the optimal branch decompositions can be computed in practice. Hicks reports that the edge-contraction algorithm of [22] can solve some instances of about 2,000 edges and the cycle method can solve some instances of about 7,000 edges in a practical time [13,14]. Very recently, there is a significant progress in the efficient implementation of ST Procedure [5]. Bian et al. show by computational study that the branchwidth of planar graphs of size up to one hundred thousand edges can be computed by ST Procedure in a practical time and memory space [5]. This result provides a powerful tool for computing branch decompositions for large planar graphs. It is reported that using the efficient implementations of ST Procedure, the edge-contraction algorithms of [12,22] can compute optimal branch decompositions for graphs of size up to 7,000 edges within 20 hours by a computer with a CPU speed of about 3GHz and a memory space of 2GByte [5]. However, it is still time consuming to use edge-contraction method to compute branch decompositions for larger instances.

Hicks reports that the divide-and-conquer approach is more practical to compute the branch decomposition of planar graphs [13,14]. In this approach, first the branchwidth β of a graph G is computed. Let S be a set of vertices that separates G into two subgraphs. Roughly speaking, a partition by S is valid if $|S| \leq \beta$, and each subgraph has branchwidth at most β (a formal definition on the valid partition will be given later). Next a valid partition of G is found. In this step, ST Procedure is used to test if each subgraph has branchwidth at most β. If a valid partition is found, then the branch decomposition of each subgraph is computed recursively. The branch decomposition of G is constructed from the decompositions of the subgraphs. How to find a valid partition efficiently is a key for this approach. Hicks proposes the cycle method for computing a valid partition [13,14]. Notice that there is no guarantee on the existence of a valid partition in a recursive step. The edge-contraction method is used to make a progress in the cycle method when a valid partition can not be found. In the worst case, the cycle method has time complexity $O(n^4)$. Computational results show that the cycle method is faster than the edge-contraction method by a factor of about $10 \sim 30$ in average for the Delaunay triangulation instances [14].

In this paper, we propose divide-and-conquer based algorithms for computing planar branch decompositions. Our algorithms are similar to the cycle method in finding a valid partition but make effort to balance the sizes of subgraphs. Our algorithms also use the edge-contraction method to make a progress when a valid partition can not be found. In the worst case, our algorithms run in $O(n^3)$ time. We tested our algorithms and the $O(n^3)$ time edge-contraction algorithm [12] on several classes of planar graphs. Computational results show that our algorithms are faster than the edge-contraction algorithm by a factor of $200 \sim 300$ for Delaunay triangulation instances of more than 5,000 edges. Using the more efficient implementations of ST Procedure of [5], our algorithms can compute optimal decompositions for some instances of size up to 50,000 edges in a practical time. Previous results of the cycle method [13,14] are obtained by a slower computer and a less efficient implementation of ST Procedure than those in this study. To compare our algorithms with the cycle method on a same platform, we implemented the unaltered cycle method [14] using the more efficient implementation of ST Procedure. Computational results show that our algorithms are faster than the unaltered cycle method by a factor of more than 10 for the Delaunay triangulation instances. Notice that our implementation of the unaltered cycle method is a straightforward one based on the information available in the published literature [13,14].

The results of this paper suggest that the optimal branch decompositions of large planar graphs can be computed in practice. Our divide-and-conquer algorithms are efficient tools for finding such branch decompositions. This may make the branch-decomposition based algorithms more attractive for many problems in planar graphs.

The rest of the paper is organized as follows. In Section 2, we give the preliminaries of the paper. Section 3 describes our algorithms. Computational results are presented in Section 4. The final section concludes the paper.

2 Preliminaries

Readers may refer to a textbook on graph theory (e.g., the one by West [23]) for basic definitions and terminology on graphs. In this paper, graphs are unweighted undirected graphs (i.e., each edge has a unit length) unless otherwise stated. Let G be a graph. We use $V(G)$ for the vertex set of G and $E(G)$ for the edge set of G. A *branch decomposition* of G is a tree T_B such that the set of leaves of T_B is $E(G)$ and each internal node of T_B has node degree 3. For each link e of T_B, removing e separates T_B into two sub-trees. Let E' and E'' be the sets of leaves of the subtrees. The width of e is the number of vertices of G incident to both an edge in E' and an edge in E''. The width of T_B is the maximum width of all links of T_B. The *branchwidth* $bw(G)$ of G is the minimum width of all branch-decompositions of G.

The algorithms of Seymour and Thomas [22] for branchwidth and branch decomposition are based on another type of decompositions called *carving decompositions*.

A carving decomposition of G is a tree T_C such that the set of leaves of T_C is $V(G)$ and each internal node of T_C has node degree 3. For each link e of T_C, removing e separates T_C into two sub-trees and the two sets of the leaves of the sub-trees are denoted by V' and V''. The width of e is the number of edges of G with both an end vertex in V' and an end vertex in V''. The width of T_C is the maximum width of all links of T_C. The *carvingwidth* $cw(G)$ of G is the minimum width of all carving decompositions of G. Notice that the carving decomposition is defined for more general graphs in [22]. The definition allows positive integer lengths on edges of the graphs. The width of e in T_C for the weighted graph is defined as the sum of lengths of edges with an end vertex in V' and an end vertex in V''.

Let G be a planar graph with a fixed embedding. Let $R(G)$ be the set of faces of G. The *medial graph* [22] $M(G)$ of G is a planar graph with an embedding such that $V(M(G)) = \{u_e | e \in E(G)\}$, $R(M(G)) = \{r_s | s \in R(G)\} \cup \{r_v | v \in V(G)\}$, and there is an edge $\{u_e, u_{e'}\}$ in $E(M(G))$ if the edges e and e' of G are incident to a same vertex v of G and they are consecutive in the clockwise (or counter clockwise) order around v. $M(G)$ in general is a multigraph but has $O(|V(G)|)$ edges.

Proposition 1. *(Seymour and Thomas [22]) Given a planar graph G of n vertices and an integer β, $bw(G) = cw(M(G))/2$, ST Procedure decides if $bw(G) \geq \beta$ by computing if $cw(M(G)) \geq 2\beta$ in $O(n^2)$ time, and an optimal carving decomposition of $M(G)$ can be translated into an optimal branch decomposition of G in $O(n)$ time.*

Seymour and Thomas give an edge-contraction method to compute an optimal carving decomposition of $M(G)$ [22]. The contraction of an edge e in $M(G)$ is to remove e from $M(G)$, identify the two end vertices of e by a new vertex, and make all edges incident to e incident to the new vertex. We denote by $M(G)/e$ the graph obtained by contracting e in $M(G)$. Given a 2-connected $M(G)$, the edge-contraction method computes an optimal carving decomposition of $M(G)$ by a sequence of edge contractions of $M(G)$ as follows: First the carvingwidth k of $M(G)$ is computed by ST Procedure. An edge e of $M(G)$ is *contractible* if the carvingwidth of $M(G)/e$ is at most k and $M(G)/e$ is 2-connected. Next, a contractible edge e of $M(G)$ is found by ST Procedure and $M(G)$ is contracted to graph $M(G)/e$. The contraction is repeated on $M(G)/e$ until the graph becomes one with three vertices. An optimal carving decomposition of $M(G)$ is constructed based on the sequence of edge contractions.

Proposition 2. *(Seymour and Thomas [22]) Let $e = \{x, y\}$ be a contractible edge of $M(G)$, x_e be the new vertex identifying $\{x, y\}$ in $M(G)/e$, and T'_C be an optimal carving decomposition of $M(G)/e$. Then the carving decomposition T_C obtained by adding links $\{x_e, x\}$ and $\{x_e, y\}$ to T'_C is an optimal carving decomposition of $M(G)$.*

A face $r \in R(G)$ and an edge $e \in E(G)$ are incident to each other if e is a boundary of r in the embedding. Notice that an edge e is incident to exactly two faces. For a face $r \in R(G)$, a vertex v is incident to r if v is an end vertex of an edge incident to r.

The *planar dual* G^* of G is defined as that for each vertex $v \in V(G)$, there is a unique face $r_v^* \in R(G^*)$; for each face $r \in R(G)$, there is a unique vertex $v_r^* \in V(G^*)$; and for each edge $e \in E(G)$ incident to r and r', there is a unique edge $e^* = \{v_r^*, v_{r'}^*\} \in E(G^*)$ which crosses e.

A *walk* in a graph G is a sequence of edges $e_1, e_2, ..., e_k$ of G, where $e_i = \{v_{i-1}, v_i\}$ for $1 \le i \le k$. A walk is *closed* if $v_0 = v_k$. The length of a walk is the number of edges in the walk. For two vertices u and v in a graph G, the distance $d(u, v)$ is the minimum length of all walks between u and v. The walk with distance $d(u, v)$ is a shortest path between u and v.

3   Divide-and-Conquer Based Algorithms

Following the divide-and-conquer approach used in the cycle method [13,14], we first describe a framework for our algorithms. Given a planar graph H with carvingwidth k, let C be a set of edges (cut set) that partitions H into subgraphs H_1 and H_2. For each H_i $(i = 1, 2)$, define H'_i to be the graph obtained by adding a new vertex v'_i and the edge set $\{\{u, v'_i\} | u \in V(H_i) \cap V(C)\}$ to H_i (see Fig. 1). Intuitively, H'_i is the graph of H_i and a vertex v'_i representing the part of H other than H_i. The partition by C is valid if $|C| \le k$, and each of H'_i has carvingwidth at most k. Below is the framework for our algorithms.

1. Given a planar graph G, compute the medial graph $M(G)$ and the carving-width k of $M(G)$ by ST Procedure and let $H = M(G)$.
2. If $|E(H)| > c$ (c is a constant)

- then try to find a valid partition of H:
  Partition H into subgraphs H_i ($i = 1, 2$) by a set C of edges with $|C| \leq k$.
  If every H'_i has carvingwidth at most k for $i = 1, 2$, then a valid partition
  is found.
- else compute the carving decomposition of H by enumeration.
3. If a valid partition is found
    - then goto Step 2 to compute the carving decomposition of every subgraph
      H'_i recursively; and construct the carving decomposition of H from the
      carving decompositions of the subgraphs.
    - else call an edge-contraction algorithm to contract an edge e of H such
      that the contracted graph H/e has carvingwidth at most k; goto Step 2
      to compute the carving decomposition of H/e; and construct the carving
      decomposition of H by Proposition 2.
4. Construct the branch decomposition of G from the carving decomposition
   of $M(G)$ (Proposition 1).

Lemma 1. *An optimal branch decomposition of G can be computed by the framework.*

Proof. By Proposition 2, if an optimal carving decomposition of H/e has been found then an optimal carving decomposition of H can be constructed. Assume that a valid partition of H is found and optimal carving decompositions T_1 and T_2 have been constructed for subgraphs H'_1 and H'_2 in the valid partition. We assume that T_1 has a leaf node u_1 corresponding to v'_1 and T_2 has a leaf node u_2 corresponding to v'_2, added in Step 2. Let $e_1 = \{u_1, w_1\}$ be the link of T_1 and $e_2 = \{u_2, w_2\}$ be the link of T_2. We get a carving decomposition T_C of H by first connecting T_1 and T_2 using a new link $\{w_1, w_2\}$ and then discarding links $\{u_1, w_1\}$ and $\{u_2, w_2\}$. Obviously, each internal node of T_C has degree three. Each link of $E(T_C) \setminus \{w_1, w_2\}$ has the same width as that of the corresponding link in T_1 or T_2. The width of link $\{w_1, w_2\}$ is $|C|$. Thus, T_C has width at most k and is an optimal carving decomposition of $H = M(G)$. By Proposition 1, T_C can be converted to an optimal branch decomposition of G. □

How to find a valid partition is a key on the efficiency of the divide-and-conquer algorithms. An obvious approach for finding such a partition is to compute a closed walk (cycle) W^* of length at most k in the planar dual $M(G)^*$ of $M(G)$. Let $E^*(W^*)$ be the set of edges in W^*. Let $R^*_{W^*}$ and $V^*_{W^*}$ be the sets of faces and vertices of $M(G)^*$ enclosed by W^*, respectively (see Fig. 1). Then the set of edges of $M(G)$ corresponding to the edges of $E^*(W^*)$ is a cut set between the subgraph of $M(G)$ with the vertex set and face set corresponding to $R^*_{W^*}$ and $V^*_{W^*}$, respectively, and the rest part of $M(G)$.

  In the cycle method [13,14], a closed walk is computed as follows. First, a face r^* of $M(G)^*$ is selected. Let $E^*_{r^*}$ be the set of edges incident to r^*. Next, a pair of vertices s^* and t^* incident to r^* is selected and a shortest path P^* that does not contain any edge of $E^*_{r^*}$ between s^* and t^* in $M(G)^*$ is computed. A path Q^* between s^* and t^* formed by edges of $E^*_{r^*}$ and path P^* give a closed walk W^* of $M(G)^*$. For a selected face r^*, the cycle method tries every pair of

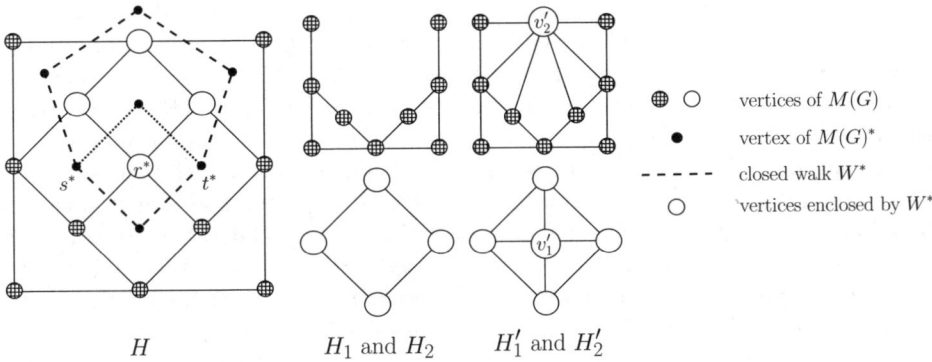

Fig. 1. Partition graph H into subgraphs by a cycle in H^*

vertices s^* and t^* incident to r^*. If a valid partition is found, then the method is applied recursively, otherwise the edge-contraction method is called.

Similar to the cycle method, our algorithms compute a closed walk W^* formed by paths Q^* and P^* between s^* and t^*. Our algorithms select the vertices s^* and t^* with the consideration on the sizes of subgraphs. Notice that the edges of $E_{r^*}^*$ is a closed walk. For vertices s^* and t^* incident to r^*, there are two paths Q_1^* and Q_2^* formed by the edges of $E_{r^*}^*$ between s^* and t^*. The partition may be balanced if there is a small difference between the lengths of Q_1^* and Q_2^*. Our first algorithm chooses the vertices s^* and t^* in an order that a smaller difference between the lengths of Q_1^* and Q_2^* is selected with a higher priority. We call this procedure the length-priority algorithm.

The cut set corresponding to a closed walk W^* partitions the input graph in a recursive step into two subgraphs. The size of a subgraph is the number of vertices in the subgraph. The partition is balanced if there is a small difference between the sizes of the two subgraphs. Our second algorithm chooses s^* and t^* in an order that a smaller difference between the sizes of the two subgraphs is selected with a higher priority. We call this procedure the size-priority algorithm.

In both algorithms, we try a constant number of pairs of vertices s^* and t^* incident to r^* in the order defined above. If a valid partition is found then the algorithms are applied recursively, otherwise an edge-contraction method is called. In both algorithms, the constant c in the framework is set to 3 and a subgraph in each partition has at least two vertices.

In the divide-and-conquer algorithms, we partition H into H_i $(i = 1, 2)$ and test if H_i' has carvingwidth at most k by ST Procedure. A test is called *positive* if H_i' has carvingwidth at most k, otherwise *negative*. Similarly, in the edge contraction method, we contract an edge e and test if H/e has carvingwidth at most k by ST Procedure. A test is called *positive* if H/e has carvingwidth at most k, otherwise *negative*.

Theorem 1. *Both the length-priority and size-priority algorithms compute an optimal branch decomposition of a planar graph G of n vertices in $O(n^3)$ time.*

Proof. By Lemma 1, the algorithms compute an optimal branch decomposition of G. The medial graph $H = M(G)$ has $|E(G)| = O(n)$ vertices. The carvingwidth of H can be computed in $O(n^2 \log n)$ time by ST Procedure (using a binary search). Because the branchwidth of G is $\beta = O(\sqrt{n})$ [11], the carvingwidth of H is $k = 2\beta = O(\sqrt{n})$. Since the carvingwidth of a graph is at least the maximum node degree of the graph, H and the subgraphs in each recursive step have node degree $O(\sqrt{n})$. Therefore, there are $O(n)$ pairs of s^* and t^* incident to a face r^* when we try to find a valid partition. It takes $O(n)$ time to compute a partition for each pair of s^* and t^*. Ordering $O(n)$ partitions takes $O(n \log n)$ time. Thus, both algorithms take $O(n^2)$ time to find and order the partitions for the $O(n)$ pairs of s^* and t^*. ST Procedure takes $O(n^2)$ time to test if a graph of n vertices has carvingwidth at least k. Since a constant number of partitions are tested by ST Procedure, the total time for deciding whether a valid partition can be found is $O(n^2)$. If a valid partition is not found, the edge contraction method is used to make a progress. This takes $O(n^2)$ time [12]. Let $T(n)$ be the time for computing an optimal carving decomposition of H with n vertices. Then

$$T(n) = \max\{T(n_1) + T(n_2) + O(n^2), T(n-1) + O(n^2)\},$$

where $T(n_i)$ $(i = 1, 2)$ and $T(n-1)$ are the time for computing optimal carving decompositions of H_i' and H/e, respectively. Since $n_1 \le n - 1$, $n_2 \le n - 1$, and $n_1 + n_2 = n + 2$, $T(n) = O(n^3)$. It takes $O(n)$ time to get a branch decomposition of G from the carving decomposition of H.   □

The bound of Theorem 1 is the worst case time complexity of the divide-and-conquer algorithms. If a valid partition is always found and sizes of the two subgraphs differ only in a constant factor in every recursive step, then the divide-and-conquer algorithms run in $O(n^2 \log n)$ time which is faster than the $O(n^3)$ time edge-contraction algorithm.

We call the length-priority and size-priority algorithms the 2-component method because, the input graph in each recursive step is partitioned into two subgraphs and ST Procedure is used to test the carvingwidth of each subgraph. The 2-component method can be generalized to the 2^i-component method $(i \ge 1)$. Given an input graph, we first choose one pair of s^* and t^* to partition the graph into two subgraphs. We call the subgraphs level-1 subgraphs. A subgraph is called a level-$(j+1)$ subgraph if it is obtained from a partition of a level-j $(j \ge 1)$ subgraph. In the 2^i-component method, we compute the level-i subgraphs (there are 2^i such graphs) by a sequence of partitions of the input graph. During the sequence of partitions, only one pair of s^* and t^* is used for each subgraph. We only check the sizes of the cut sets but do not check the carvingwidth for the level-j subgraphs for $j < i$. We use ST Procedure to check the carvingwidth for every level-i subgraph. If all level-i subgraphs have carvingwidth at most k, then the method is recursively applied to each level-i subgraph. If one level-j $(1 < j \le i)$ subgraph H' has carvingwidth greater than k then we test the level-$(j-1)$ subgraph from which H' is obtained. If all level-$(j-1)$ subgraphs have carvingwidth at most k then the method is applied recursively (notice that a level-$(j-1)$ subgraph H has carvingwidth at most k if all level-j

subgraphs obtained from H have carvingwidth at most k). If a level-1 subgraph has carvingwidth greater than k, then we give up the current pair of s^* and t^* and apply the method to the input graph on a different pair of s^* and t^*.

This generalization is motivated by the fact that testing the carvingwidth of large graphs by ST Procedure is the most time consuming part in finding the branch decompositions and some observations from the computational study: in most cases, a valid partition can be found in the first try and partitioning the input graph into smaller subgraphs can save the time used by ST Procedure. For constant i, the 2^i-component algorithms have time complexity $O(n^3)$.

The branch decomposition of a graph G which is not 2-connected can be easily constructed from the branch decompositions of its 2-connected components. So, the study of branch decomposition may be concentrated on 2-connected graphs.

4   Computational Results

We implemented our algorithms and the unaltered cycle method [13,14]. A number of efficient implementations of ST Procedure are reported in [5]. The implementations of ST Procedure with the best practical performances are used in our algorithms and the cycle method. The implementation of the cycle method is a straightforward one: The pair of vertices s^* and t^* is selected in an arbitrary order. If there are multiple shortest paths P^*'s between s^* and t^* in $M(G)^*$, an arbitrary one is used. Similarly, an arbitrary shortest path P^* is used for the length-priority and size-priority algorithms. We test our implementations on three classes of instances. Class (1) instances include Delaunay triangulations of point sets taken from TSPLIB [17]. The instances are provided by Hicks and are used as test instances in the previous studies [13,14]. The instances in Class (2) are generated by the LEDA library [2,16]. LEDA generates two types of planar graphs. One type of the graphs are the randomly generated maximal planar graphs and their subgraphs obtained from deleting some edges. Since the maximal planar graphs generated by LEDA always have branchwidth four, the subgraphs obtained by deleting edges from the maximal graphs have branchwidth at most four. The graphs of this type are not interesting for the study of branch decompositions. The other type of planar graphs are those generated based on some geometric properties, including Delaunay triangulations and triangulations of points uniformly distributed in a two-dimensional plane, and the intersection graphs of segments uniformly distributed in a two-dimensional plane. We report the results on the 2-connected intersection graphs. The instances in Class (3) are generated by the PIGALE library [1]. PIGALE randomly generates one of all possible planar graphs with a given number of edges based on the algorithms of [21]. We report the results on the 2-connected graphs generated by the PIGALE library.

We run the implementations on a computer with Intel(R) Xeon(TM) 3.06GHz CPU, 2GB physical memory and 4GB swap memory. The operating system is SUSE LINUX 10.0, and the programming language we used is C++.

Table 1. Computation time (in seconds) of several decomposition algorithms for Class (1) instances

Graphs G	$\|E(G)\|$	bw	EC_GT time	NT	Cycle time	NT	L_P time	NT	S_P time	NT	S4 time	NT
pr1002	2972	21	2667	102	369	37	155	34	150	63	271	129
rl1323	3950	22	6879	136	441	0	63	5	189	97	336	200
d1655	4890	29	13529	171	5958	806	295	34	218	28	402	59
rl1889	5631	22	29096	178	1896	527	130	0	115	1	90	2
u2152	6312	31	26092	192	2394	92	156	0	140	0	119	0
pr2392	7125	29	45728	271	5595	210	173	0	153	0	118	0
pcb3038	9101	40			6265	53	490	8	998	17	1899	36
fl3795	11326	25			8954	52	863	3	902	11	1190	22
fnl4461	13359	48			X	X	3795	31	2479	16	2441	16
rl5934	17770	41					2348	2	2585	6	3296	12
pla7397	21865	33					10291	88	3026	10	3376	21
usa13509	40503	63					25956	29	29539	79	50376	160
brd14051	42128	68					10536	19	31554	129	64802	263
d18512	55510	88					22378	44	X	X	X	X

4.1   Results for Instances in Class (1)

The computational results for Class (1) instances are reported in Table 1. In the table, $|E(G)|$ is the number of edges in the instance and thus the number of vertices in the medial graph $M(G)$ which is the input to the algorithms, bw is the branchwidth of the graph G, NT is the number of negative tests, $Cycle$ is the unaltered cycle method, L_P is the length-priority algorithm, S_P is the size-priority algorithm, and $S4$ is the 4-component algorithm with size-priority. For comparison, we include the running time of the $O(n^3)$ time edge-contraction method in column EC_GT (the data is taken from [5] which uses a computer of similar performance to the one we use in this paper, and the $O(n^3)$ algorithm itself is given in [12]). In the table, an "X" indicates that it requires more than 70,000 seconds to solve the instance and a blank indicates that we did not test the algorithms for that instance.

The data show that all divide-and-conquer algorithms (Cycle, L_P, S_P, and $S4$) are much faster than the edge-contraction algorithm. The length-priority and size-priority algorithms are faster than the edge-contraction method by a factor of $200 \sim 300$ for instances of more than 5,000 edges in this class. It is difficult to compare the data of our algorithms with those of the cycle method reported in previous studies [13,14], because computers of different speeds and different implementations of ST Procedure are used. To compare our algorithms with the cycle method on a same platform, we give a straightforward implementation of the unaltered cycle method using the same efficient ST Procedure used in our algorithms. Our algorithms are faster than the cycle method by a factor of at least 10 for instances of more than 5,000 edges. Notice that in average the cycle method is faster than the edge-contraction method by a factor of about 10 which is slightly smaller than that (10 ~ 30 in average) reported in previous studies [14]. Considering the fact that a more efficient edge-contraction algorithm is used in this study, our implementation of the cycle method has a similar performance as that used in the previous studies and our new algorithms are faster than the cycle method. For all instances which are solved within the 70,000 seconds time limit, the edge-contraction method is never used by any divide-and-conquer algorithm to make a progress, that is, a valid partition is always found in every recursive step.

There are two factors improving the running time of our algorithms. Both the length-priority and size-priority algorithms find more balanced partitions than the cycle method. This reduces the total running time in the divide-and-conquer approach. The other factor is that our algorithms have a smaller number of negative tests. In finding a valid partition, once a negative test happens, all divide-and-conquer algorithms try a different pair of s^* and t^* and the running time is increased. Also it takes more time for a negative test than a positive one. For Class (1) instances, the length-priority algorithm runs faster than the size-priority algorithm for large graphs while the size-priority algorithm does a better job for smaller graphs. Because the running time of the algorithms depends on both the size of the graphs and the number of negative tests, it may take a longer time to solve some instances than that for a larger graph. For example, Instance usa13509 requires a longer time than Instance brd14051 by the length-priority algorithm.

For Class (1) instances, the number of negative tests is non-trivial, especially for large graphs. This makes the 2^i-component ($i > 1$) algorithms less efficient, because using more than two components generally increases the number of negative tests and thus the total running time. As shown in Table 1, the 4-component algorithm is slower than the 2-component algorithms for most instances in this class.

4.2   Results for Instances in Classes (2) and (3)

Computational results for Classes (2) and (3) instances are given in Tables 2 and 3, respectively. In the tables, $S8$ is the 8-component algorithm with the size-priority. An "X" in the tables indicates that it takes more than 150,000 seconds to solve that instance. Similar to results for Class (1) instances, the edge-contraction method is never used by any divide-and-conquer algorithm to make a progress for Classes (2) and (3) instances.

It takes more time to find the branch-decomposition of a Class (2) instance than a Class (1) instance with a similar size by divide-and-conquer algorithms. This may be caused by the fact that Class (2) instances have smaller branchwidth than that of Class (1) instances. A larger branchwidth implies that a longer cycle is used in a valid partition and a longer cycle usually gives a more balanced partition. For Class (2) instances, the size-priority algorithm runs faster than the length-priority algorithm and is faster than the edge-contraction algorithm by a factor of about 50 ~ 150. Both the length-priority and size-priority algorithms are faster than the cycle method. Since the number of negative tests in the

Table 2. Computation time (in seconds) of several decomposition algorithms for Class (2) instances

Graphs G	$\|E(G)\|$	bw	EC_GT time	NT	Cycle time	NT	L_P time	NT	S_P time	NT	S4 time	NT	S8 time	NT
rand1160	2081	8	1749	34	53.2	0	46.3	0	29.9	0	23.1	0	18.4	0
rand1672	3047	10	4695	103	137	2	54.6	0	39.7	0	29.7	0	25.7	0
rand2780	5024	10	29073	147	2059	0	727	0	471	0	312	0	249	0
rand3857	7032	11	82409	281	1503	0	810	6	493	6	351	13	292	20
rand5446	10093	11			11474	17	3283	3	2361	3	1532	6	1205	9
rand8098	15031	13			12022	76	2783	1	1864	0	1465	0	1159	0
rand10701	20044	13			11782	9	4368	0	3699	0	2884	0	2475	0
rand15902	30010	14			68809	125	32409	0	19127	14	13240	29	11744	42
rand21178	40190	17			X	X	93897	0	54557	2	33429	4	26910	7
rand26304	50032	19					149570	0	85207	0	59907	0	47039	0

Table 3. Computation time (in seconds) of several decomposition algorithms for Class (3) instances

Graphs G	$\|E(G)\|$	bw	EC_GT time	NT	Cycle time	NT	L_P time	NT	S_P time	NT	S4 time	NT	S8 time	NT
PI855	1434	6	565	61	22.7	0	14.3	0	11.8	0	7.75	0	6.41	0
PI1277	2128	9	1563	101	107	1	47.5	1	25.9	0	17.5	0	14.6	0
PI1467	2511	6	3135	74	304	1	183	0	120	0	72	0	56.7	0
PI2009	3369	7	8127	90	253	0	142	0	115	0	64.8	0	55.7	0
PI2518	4266	8	17807	105	663	0	369	0	206	0	135	0	112	0
PI2968	5031	6	26230	162	2244	9	1235	6	773	3	488	7	423	11
PI3586	6080	8	49108	176	2340	1	1182	0	699	0	443	0	334	0
PI4112	6922	7	70220	132	10808	1	10817	0	5663	0	2973	0	2150	0
PI5940	10016	7	X	X	19770	0	18807	0	9517	0	5205	0	3782	0
PI8950	15097	10	X	X	33862	13	19216	1	11171	0	6871	0	4993	0
PI11974	20071	9	X	X	X	X	X	X	111747	0	61641	0	44479	0

divide-and-conquer algorithms for Class (2) instances is small, the 2^i-component ($i > 1$) algorithms are more efficient than the 2-component ones. Especially, the 8-component algorithm is faster than the edge-contraction, the cycle, and the 2-component size-priority algorithms by factors of about $100 \sim 200$, $5 \sim 8$, and 2, respectively.

It takes more time to find the branch-decomposition of a Class (3) instance than a Class (1) or Class (2) instance with a similar size by the divide-and-conquer algorithms, because Class (3) instances have a smaller branchwidth. As shown in the table, the branchwidth of the instances is small constants and does not increase in the size of the instances. In each valid partition of the divide-and-conquer algorithms, we get a small subgraph of a constant size and a large subgraph for most instances. This limits the speed-up by the 2-component divide-and-conquer algorithms to a constant factor. Similar to the results for Class (2) instances, the number of negative tests in the divide-and-conquer algorithms is small and the 2^i-component algorithms are faster than the 2-component ones. The 8-component algorithm is faster than the edge-contraction, the cycle, and the 2-component size-priority algorithms by factors of about $30 \sim 150$, $5 \sim 8$, and 2, respectively.

5   Concluding Remarks

Our divide-and-conquer algorithms can compute the optimal branch decompositions for Classes (1) and (2) instances of about 50,000 edges, and Class (3) instances of abut 20,000 edges in a practical time. This provides useful tools for applying the branch decomposition based algorithms to practical problems. It is still time consuming to compute optimal branch decompositions for very large planar graphs, especially for the graphs with small branchwidth by the current divide-and-conquer algorithms. On the other hand, the planar graphs with small branchwidth are more interesting for the branch-decomposition based algorithms because those algorithms usually run exponentially in the branchwidth of the graphs. An interesting future work is to design more efficient algorithms for very large planar graphs of small branchwidth. Using a better approach to make a valid partition balanced is one possible direction to get such algorithms.

All divide-and-conquer algorithms use the edge-contraction method to guarantee the branch decomposition can be found. However, the edge-contraction method has never been called in our computational study. It is interesting to prove that a valid partition can always be found in those algorithms efficiently.

Acknowledgment

The authors thank Dr. I.V. Hicks for providing the test instances of Class (1). The work was partially supported by the NSERC Research Grant of Canada.

References

1. Public Implementation of a Graph Algorithm Library and Editor (2008), http://pigale.sourceforge.net/
2. The LEDA User Manual, Algorithmic Solutions, Version 4.2.1 (2008), http://www.mpi-inf.mpg.de/LEDA/MANUAL/MANUAL.html
3. Arnborg, S., Corneil, D.G., Proskurowski, A.: Complexity of finding embedding in a k-tree. SIAM J. on Discrete Mathematics 8, 277–284 (1987)
4. Arnborg, S., Lagergren, J., Seese, D.: Easy problems for tree-decomposable graphs. Journal of Algorithms 12, 308–340 (1991)
5. Bian, Z., Gu, Q., Marzban, M., Tamaki, H., Yoshitake, Y.: Empirical study on branchwidth and branch decomposition of planar graphs. In: Proc. of the 9th SIAM Workshop on Algorithm Engineering and Experiments (ALENEX 2008), pp. 152–165 (2008)
6. Bodlaender, H.L.: A tourist guide through treewidth. Acta Cybernetica 11, 1–21 (1993)

7. Bodlaender, H.L.: A linear time algorithm for finding tree-decomposition of small treewidth. SIAM J. on Computing 25, 1305–1317 (1996)
8. Bodlaender, H.L.: Treewidth: Characterizations, applications, and computations. In: Fomin, F.V. (ed.) WG 2006. LNCS, vol. 4271, pp. 1–14. Springer, Heidelberg (2006)
9. Bodlaender, H.L., Thilikos, D.: Constructive linear time algorithm for branchwidth. In: Degano, P., Gorrieri, R., Marchetti-Spaccamela, A. (eds.) ICALP 1997. LNCS, vol. 1256, pp. 627–637. Springer, Heidelberg (1997)
10. Dorn, F., Penninkx, E., Bodlaender, H., Fomin, F.V.: Efficient exact algorithms for planar graphs: Exploiting sphere cut branch decompositions. In: Brodal, G.S., Leonardi, S. (eds.) ESA 2005. LNCS, vol. 3669, pp. 95–106. Springer, Heidelberg (2005)
11. Fomin, F.V., Thilikos, D.M.: Dominating sets in planar graphs: Branch-width and exponential speed-up. SIAM Journal on Computing 36(2), 281–309 (2006)
12. Gu, Q.P., Tamaki, H.: Optimal branch decomposition of planar graphs in $O(n^3)$ time. In: Caires, L., Italiano, G.F., Monteiro, L., Palamidessi, C., Yung, M. (eds.) ICALP 2005. LNCS, vol. 3580, pp. 373–384. Springer, Heidelberg (2005)
13. Hicks, I.V.: Branch decompositions and their applications. PhD Thesis, Rice University (2000)
14. Hicks, I.V.: Planar branch decompositions II: The cycle method. INFORMS Journal on Computing 17(4), 413–421 (2005)
15. Hicks, I.V., Koster, A.M.C.A., Kolotoğlu, E.: Branch and tree decomposition techniques for discrete optimization. In: TutORials in Operation Research: INFORMS–New Orleans 2005, pp. 1–29 (2005)
16. Mehlhorn, K., Näher, S.: LEDA: A Platform for Combinatorial and Geometric Computing. Cambridge University Press, New York (1999)
17. Reinelt, G.: TSPLIB-A traveling salesman library. ORSA J. on Computing 3, 376–384 (1991)
18. Robertson, N., Seymour, P.D.: Graph minors I. Excluding a forest. Journal of Combinatorial Theory, Series B 35, 39–61 (1983)
19. Robertson, N., Seymour, P.D.: Graph minors II. Algorithmic aspects of tree-width. Journal of Algorithms 7, 309–322 (1986)
20. Robertson, N., Seymour, P.D.: Graph minors X. Obstructions to tree decomposition. J. of Combinatorial Theory, Series B 52, 153–190 (1991)
21. Schaeffer, G.: Random sampling of large planar maps and convex polyhedra. In: Proc. of the 31st Annual ACM Symposium on the Theory of Computing (STOC 1999), pp. 760–769 (1999)
22. Seymour, P.D., Thomas, R.: Call routing and the ratcatcher. Combinatorica 14(2), 217–241 (1994)
23. West, D.B.: Introduction to Graph Theory. Prentice Hall Inc., Upper Saddle River, NJ (1996)

Experimental Evaluation of an Exact Algorithm for the Orthogonal Art Gallery Problem

Marcelo C. Couto, Cid C. de Souza*, and Pedro J. de Rezende**

Instituto de Computação
Universidade Estadual de Campinas — Campinas, Brazil
couto.marcelo@gmail.com, {cid,rezende}@ic.unicamp.br

Abstract. We consider the Orthogonal Art Gallery problem (OAGP) whose goal is to minimize the number of vertex guards required to watch an art gallery whose boundary is an n-vertex orthogonal polygon P. Here, we explore an exact algorithm for OAGP, which we proposed in [1], that iteratively computes optimal solutions to Set Cover problems (SCPs) corresponding to discretizations of P. While it is known [1] that this procedure converges to an exact solution of the original continuous problem, the number of iterations executed is highly dependent on the way we discretize P. Although the best theoretical bound for convergence is $\Theta(n^3)$ iterations, we show that, in practice, it is achieved after only a few of them, even for random polygons of hundreds of vertices. As each iteration involves the solution of an SCP, the strategy for discretizing P is of paramount importance. In this paper, we carry out an extensive empirical investigation with five alternative discretization strategies to implement the algorithm. A broad range of polygon classes is tested. As a result, we are able to significantly improve the performance of the algorithm, while maintaining low execution times, to the point that we achieve a fivefold increase in polygon size, compared to the literature.

1 Introduction

The classical *Art Gallery Problem* originally posed by Victor Klee in 1973 consists in determining the minimum number of guards sufficient to cover the interior of an n-wall art gallery [2]. Chvátal showed, in what became known as *Chvátal's Art Gallery Theorem*, that $\lfloor n/3 \rfloor$ guards are occasionally necessary and always sufficient to cover a simple polygon with n vertices [3].

Many variants of the art gallery problem have been studied in the literature. In this paper, we study the variation of the classical art gallery problem that deals specifically with orthogonal polygons (edges parallel to the x or y axis) where guards can only be placed on vertices that define the outer boundary of the gallery. This is called the *Orthogonal Art Gallery Problem* (OAGP) and

---

* Partially supported by CNPq – Conselho Nacional de Desenvolvimento Científico e Tecnológico – Grants # 307773/2004-3, 472504/2007-0 and FAPESP – Fundação de Amparo à Pesquisa do Estado de São Paulo – Grant # 107/97.
** Partially supported by CNPq – Grant # 201205/2005-0, 472504/2007-0.

C.C. McGeoch (Ed.): WEA 2008, LNCS 5038, pp. 101–113, 2008.

is an important subclass, due to most real life buildings and galleries being orthogonally shaped [4].

The earliest major result concerning this problem, due to Kahn *et al.* [5], states that $\lfloor \frac{n}{4} \rfloor$ guards are occasionally necessary and always sufficient to cover an orthogonal polygon with n vertices. Later, Schuchardt and Hecker proved that minimizing the number of guards in this variation is also NP-hard [6], settling a question that remained open for almost a decade [7].

Several placement algorithms have been proposed in the past, such as Edelsbrunner *et al.* [8] and Sack and Toussaint [7], which deal with the problem of efficiently placing exactly $\lfloor n/4 \rfloor$ guards covering a given orthogonal gallery.

On the other hand, in a recently revised manuscript, based on [9], Ghosh presents an $O(n^4)$ time approximation algorithm for simple polygons yielding solutions within a $\log n$ factor of the optimal. Further approximation results include Eidenbenz [10] who designed algorithms for several variations of terrain guarding problems and Amit *et al.* [11] who analyze heuristics with experimental evidence of good performance in covered area and in the number of guards.

Another approach tackled by Erdem and Sclaroff [12] and Tomás *et al.* in [13] consists of modeling the problem as a discrete combinatorial problem and then solving the corresponding optimization problem. The former discretize the interior of the polygon with a fixed grid, yielding an approximation algorithm and the latter gives empirical analysis of an exact method of successive approximations based on dominance of visibility regions.

Finally, in [1], we presented an exact algorithm to optimally solve the OAGP. In this algorithm, we iteratively discretize and model the problem as a classical *Set Cover problem* (SCP). Besides demonstrating the feasibility of this approach, we showed that, in practice, the number of iterations required to solve instances of up to 200 vertices was very small and that the resulting algorithm turned out to be quite efficient.

Our contribution. Though the number of iterations executed by the exact algorithm we proposed in [1] was shown to be polynomially bounded, its practical performance is much better depending on how the polygon is discretized. This becomes clearer when we notice that at each iteration an instance of SCP, a NP-hard problem, has to be solved at optimality, in our case, by an Integer Programming (IP) solver.

In this paper, we conduct a thorough experimental investigation concerning the trade-off between the number and nature of discretizing points and the number of iterations, analyzing the practical viability of each approach. Our test data, available in [14], includes multiple instances for each size of the vertex set, for various classes of orthogonal polygons with up to *a thousand* vertices.

The new experimental results significantly surpassed those we reported in [1]. This is due to the exploration of alternative discretization strategies, which allow us to address difficult instances as well as to handle a fivefold increase in the polygon size compared to the literature, while attaining low execution times.

Organization of the text. In the next section, we explain the basic ideas that support the algorithm. Section 3 is devoted to the description of the algorithm and the alternative strategies to discretize the polygon. Next, in Section 4 we give an account of the set up of the testing environment and present the different classes of instances used. Besides, following the recommendations of Johnson [15], McGeoch and Moret [16], Sanders [17] and Moret [18], we show an extensive experimental analysis of the algorithm implemented with multiple discretization strategies, including the evaluation of multiple measurements. Concluding remarks are drawn in the last section.

2 Basics

In an instance of the OAGP we are given an orthogonal simple polygon P that bounds an art gallery and we are asked to determine the minimum number and an optimal placement of vertex guards in order to keep the whole gallery under surveillance. Vertex guards are assumed to have a range of vision of $360°$.

The approach used by the algorithm described in Section 3 transforms the continuous OAGP into a discrete problem which, in turn, can be easily modeled as an instance of the SCP. In fact, for the last two decades, this has been the only known technique for transforming art gallery problems leading to efficient approximation algorithms. Below, we describe in detail the approach used by the algorithm, starting with some basic definitions.

An *n-wall orthogonal art gallery* can be viewed as a planar region whose boundary consists of an orthogonal simple polygon (without holes) P, i.e., one whose n edges are parallel to the x or y axis. The set or vertices of P are denoted by V and a vertex $v \in V$ is called a *reflex vertex* if the internal angle at v is greater then $180°$. Whenever no confusion arises, *a point in P* will mean a point either in the interior or on the boundary of P.

Any point y is said to be visible from any other point x if and only if the closed segment joining x and y does not intersect the exterior of P. The set $V(v)$ of all points visible from a vertex $v \in V$ is called the *visibility region of v*. In order to determine $V(v)$, we employ the linear time algorithm proposed by Lee [19] and extended by Joe and Simpson [20,21].

A set of points S is a *guard set* for P if for every point $p \in P$ there exists a point $s \in S$ such that p is visible from s. Hence, a *vertex guard set G* is any subset of vertices such that $\bigcup_{g \in G} V(g) = P$. In other words, a vertex guard set for P gives the positions of stationary guards who can oversee an entire art gallery of boundary P. Thus, the OAGP amounts to finding the smallest subset $G \subset V$ that is a vertex guard set for P.

From the above discussion one can see that the problem of finding the smallest vertex guard set for P can be seen as a *continuous* minimum set cover problem, where every visibility region $V(v), v \in V$ is a set and points $p \in P$ are elements of the set.

Notice that the term *continuous* is used here to denote the fact that there is an infinite number of elements to be covered in the SCP instance, as the points of P in the above definition comprise an infinite set. To cope with this, one can discretize the problem, generating a representative finite number of points in P so that the formulation becomes manageable. We now describe how the

solutions to successively refined discrete instances are guaranteed to converge to
an optimal solution to the original continuous problem. To this end, consider an
arbitrary discretization of P into a finite set of points $D(P)$. An IP formulation
of the corresponding SCP instance is shown below.

$$z = \min \sum_{j \in V} x_j$$

$$\text{s.t.} \sum_{j \in V} a_{ij} x_j \geq 1, \text{ for all } p_i \in D(P) \quad (1)$$

$$x_j \in \{0, 1\}, \text{ for all } j \in V$$

where the binary variable x_j is set to 1 if and only if vertex j from P is chosen
to be in the guard set. Moreover, given a point p_i in $D(P)$ and a vertex j of P,
a_{ij} is a binary value which is 1 if and only if $p_i \in V(j)$.

Given a feasible solution x for the IP above, let $Z(x) = \{j \in V \mid x_j = 1\}$.
Constraint (1) states that each point $p_i \in D(P)$ is visible from at least one
selected guard position in the solution and the objective function minimizes the
cardinality z of $Z(x)$. Clearly, as the set $D(P)$ is finite, it may happen that
$Z(x)$ does not form a vertex guard set for P. In this case, we must add a new
point inside each uncovered region and include these points in $D(P)$. A new SCP
instance is then created and the IP is solved again.

We are now able to describe the algorithm proposed in [1]. In the *preprocessing phase*, three procedures are executed. The first one computes the visibility
polygons for the points in V. The second one computes the initial discretization $D(P)$ and the third one builds the corresponding IP model. In the *solution
phase*, the algorithm iterates as described above, solving SCP instances for the
current discretization, until no regions remain uncovered.

We had shown in [1] that an upper bound on the number of iterations is
$O(n^4)$. This result was derived from the fact that the edges of the visibility
regions induce a subdivision of P which is comprised of $O(n^4)$ faces or *Atomic
Visibility Polygons* (AVPs). One point inside an AVP is enough to guarantee
that this entire AVP will be covered by the solution to the discretized problem.
Whence follows the upper bound on the number of iterations. However, it can
be derived from a result by Bose *et al.* [22] that $\Theta(n^3)$ is a tight bound on the
number of AVPs, improving the aforementioned worst case convergence result.

Moreover, the actual number of iterations that is required depends on how
many uncovered regions can be successively generated. As the cost of each iteration is related to the number of constraints in (1), an interesting trade-off
naturally sprouts and leads one to attempt multiple choices of discretization
schemes. On the other hand, any method of cleverly choosing the initial points
of the discretization will have a corresponding cost in preprocessing time, open-
ing another intriguing time exchange consideration. These questions are precisely
what we address next.

In Section 3 we consider several possible discretization schemes which lead to
the various performance analysis discussed in Section 4.

3   Discretization Strategies

The key point in the IP approach is to set up instances of the set cover problem that can rapidly be solved while minimizing the number of iterations required to attain an optimal solution to the original art gallery problem, within the least amount of time. However, one must take into account that sophisticated geometric properties used to build more efficient discretizations will generate a corresponding cost in preprocessing, possibly outweighing the benefits. Below, we discuss alternatives for the discretization of P.

Regular Grid. The first discretization strategy considered is to generate a dense grid inside the polygon in the assumption that few iterations might be required. This was the main venue for the experimentations described in [1]. Such grid is built with a step of size equal to the smallest gap in the x- and y-coordinates of the vertices of P. We also include all the vertices in this initial discretization.

As it turns out, for some polygons the number of grid points grows quadratically with the number of vertices, inflating the number of constraints in the formulation of the SCP which increases the time needed to solve each instance.

A summary of the outcome of the use of regular grids for two classes of polygons can be seen in Figure 3 and Table 1.

Induced Grid. Given the perception that reflex vertices are responsible for part of the difficulty of the problem, a natural discretization strategy to be considered is the grid induced by the edge extensions that intersect in the polygon. In this case, we generate fewer constraints than in the previous strategy while attempting to capture more of the intrinsic visibility information of the polygon. One might expect that this could lead to faster to solve instances of set cover while keeping the number of iterations low.

Just Vertices. In one extreme, given that all vertices of the polygon will have to be covered, we consider the rather sparse case where the starting discretization contains just the vertices of the polygon. Initially, this leads to quicker solutions to the set cover problem than the two previous approaches and has the benefit that each additional constraint comes really from "hard to see" regions. Since in this way we avoid any spurious grid points, one might envision that the potentially higher number of iterations could still be compensated by the smaller size of the SCP instances.

Complete Atomic Visibility Polygons. Recall that an AVP of a polygon P is any (convex) face of the subdivision of P induced by the visibility polygons of all its vertices. It then follows that if a guard set G covers the centroid of an AVP, then it must cover the entire AVP. Therefore, if G covers the centroids of every AVP of P, then G must be a guard set for P.

This suggests that we could solve the problem in a single iteration of the algorithm by building an instance of SCP from all these centroids. However, for sizeable instances, this approach would lead to an impractically large instance of up to $O(|V|^3)$ constraints, where V is the set of vertices of P (see [22]).

Nonetheless, as we will see, not all AVPs need to be represented in the set of constraints in order to guarantee a single iteration. Therefore, we do not need to consider this more costly discretization strategy.

Reduced Atomic Visibility Polygons. Following the previous discussion, we now show that we can significantly reduce the number of constraints required to guarantee that the algorithm will find the minimum number of guards necessary to cover P by solving a single instance of the set cover problem.

Firstly, given a vertex $v \in V$, an edge of the visibility polygon $V(v)$ is called a *visibility edge* of v. Furthermore, if it is not an edge of P, then it is called a *proper visibility edge* of v. It follows that an AVP is a face in the arrangement of visibility edges, interior to P. Hence, the edges of an AVP are either portions of edges of P or portions of proper visibility edges of vertices of P. An AVP $\mathcal{V}$ is called a *shadow AVP* if it is not visible from any of the vertices whose proper visibility edges spawn $\mathcal{V}$.

Let $G \subset V$ be a partial guard set for P and let U be a maximal connected region not covered by G. Note that U can be partitioned into a collection of AVPs. To see that at least one of these must be a shadow AVP, notice that if one side of a proper visibility edge of, say, vertex v_i that intersects U, is visible from v_i then the opposite side must not be. Hence, by successive partitioning U, at least one shadow AVP is bound to remain.

The Reduced AVP discretization strategy consists of taking all vertices of P plus the centroids of every shadow AVP. Since any guard set that covers all the points of this discretization cannot leave an uncovered region, it follows that no iterations will be required.

It remains to be experimentally analyzed which of these discretizing strategies will bring about the most benefit, timewise. This is done in the next section.

4   Computational Experiments

We now present an experimental evaluation of the several discretization strategies discussed in the previous section. We coded all variants of the algorithm described in earlier sections along with a visibility algorithm from [20]. The implementation was done in C++, compiled with GNU g++ 4.1, on top of CGAL 3.2.1, and used the IP solver Xpress v17.01.02. As for hardware, we used a desktop PC featuring a Pentium IV at 3.4 GHz and 1 GB of RAM running GNU/Linux 2.6.17.

4.1   Instances

We conducted the tests on a large number of instances downloadable from [14] and grouped into four different classes (see Figure 1). The first two of these classes are composed of n-vertex orthogonal polygons placed on an $n/2 \times n/2$ unit square grid and devoid of collinear edges, as suggested in [23] and the last two are based on a modified version of the von Koch curve (see [24]).

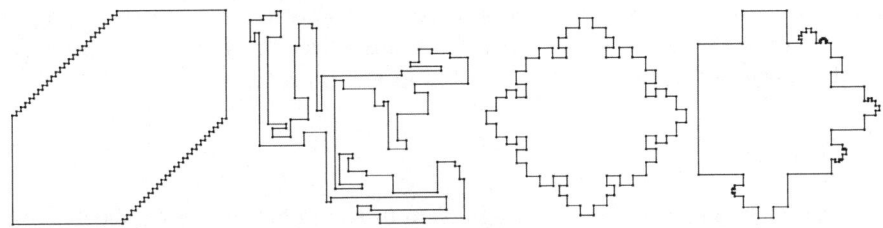

Fig. 1. Sample polygons with 100 vertices: FAT, Random, Complete von Koch and Random von Koch

(1) **FAT:** This class was introduced in [13] as an extreme scenario for the IP approach and also used in [1], where instances with up to 200 vertices were solved to optimality.

(2) **Random:** These are n-vertex randomly generated orthogonal polygons created using the algorithm proposed in [23].

(3) **Complete von Koch (CvK):** These polygons were generated based on a modified version of the von Koch curve. The fractal has a Hausdorff dimension of 1.34 and is generated, starting with a square, by recursively replacing each edge as shown in Figure 2, where $\overline{ar} = \overline{st} = \overline{ub}$ and $\overline{sr} = \overline{tu} = \frac{3}{4}\overline{ar}$.

Fig. 2. Levels of modified von Koch polygons

(4) **Random von Koch (RvK):** This class consists of randomized von Koch instances of up to level 4. Starting from a square, each of these instances is generated iteratively until the desired number of vertices is reached. In each iteration, we randomly choose an edge of the current polygon, with level smaller than 4, and decide in a random fashion whether we expand it or not.

The FAT and Random instances were generated for the number of vertices n in the ranges: [20, 200] with step 20, (200, 500] with step 50 and (500, 1000] with step 100. Similar sizes were chosen for the RvK class. The CvK class contains by construction only 3 instances with $n \in \{20, 100, 500\}$.

For our conclusions to be endowed with statistical significance, we had to decide on the sample size (number of instances generated), for each value of n, in the classes Random and RvK. To this end, we ran our algorithm on random instances, while varying the sample size s. We concluded that the variance of the results remains practically unchanged after $s \geq 30$ and, therefore, we decided to generate 30 instances for each value of n. It is worth noting that, up to scaling, only one instance is defined for a given n in the FAT class, hence no decision on sample size is needed in this case.

Thus, in total, our data set is composed of 1833 OAGP instances, having between 20 and 1000 vertices, i.e., our largest instances are five times the largest ones whose optimal solutions are reported in the literature.

4.2 Results

We now discuss the experimental evaluation of the different strategies described in Section 3. All values reported here are average results for 30 instances of each size, or 30 runs of the same instance, for FAT and CvK classes.

The FAT instances were introduced in [13] as an extremal scenario for the IP approach because of the larger number of constraints resulting from regular discretizations of P. Figure 3 displays the amount of time spent by the exact algorithm on the FAT class with each discretization strategy. It can be seen that there is a huge difference between the strategies, though all the discretizations lead to a solution in only one iteration. Notice that, in this case, the Regular and Induced Grids coincide, leading to the same running times. On the other hand, the *Reduced AVP* and *Just Vertices* discretizations are both composed of only the vertices of P, since FAT polygons have no shadow AVPs. Of course, the *Reduced AVP* strategy spent more time on the preprocessing phase, which causes the difference seen in the chart. However, the two strategies can deal with FAT polygons with up to 1000 vertices in reasonable time, going far beyond the results reported earlier for this class which are limited to 200-vertex polygons.

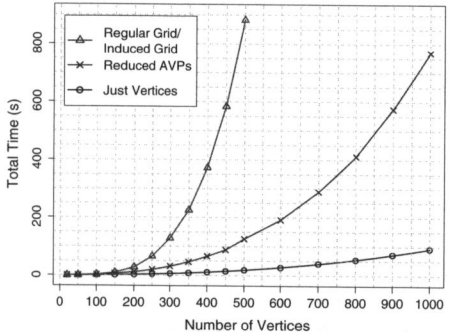

Table 1. Complete von Koch polygons

	Final $|D(P)|$			Total Time (s)		
# vertices	20	100	500	20	100	500
Reg. Grid	45	500	6905	0.05	1.57	92.37
Ind. Grid	24	205	1665	0.03	1.41	70.94
Red. AVPs	28	324	5437	0.07	3.14	143.93
Just Vert.	20	107	564	0.04	0.97	29.35

Fig. 3. Total time: FAT polygons

The usage of discretization strategies based on dense grids becomes more discouraging when we analyze the results in Table 1. This table displays the execution time and the size of the discretization of the strategies proposed in Section 3 for the CvK polygons. One can see that for these instances, the *Induced Grid* strategy has a better performance than the *Regular Grid* strategy. The size of the discretization produced by *Regular Grid* grows quadratically in the number of vertices, and thus inflates the number of constraints in the IP formulation increasing considerably the time necessary to optimally solve the SCP instances.

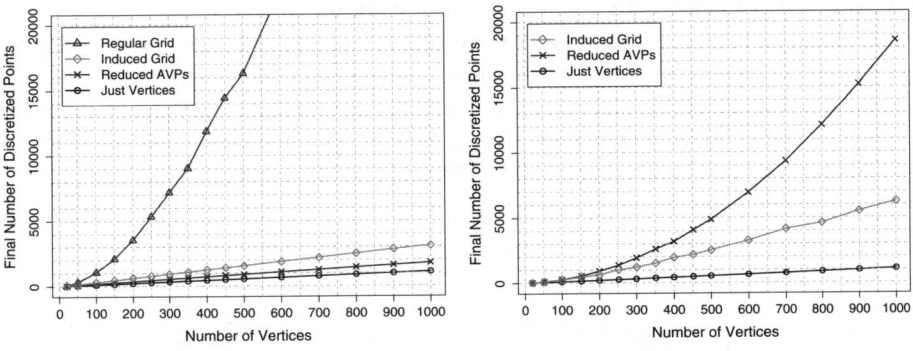

Fig. 4. Final discretization size: (a) Random polygons; (b) Random von Koch polygons

The *Reduced AVP* strategy has a poor behavior for CvK polygons since the number of shadow AVPs increases fast in this case. The *Just vertices* strategy is again the one that spends less time.

Figure 4 shows the amount of discretized points necessary for each strategy to achieve the optimal solution of OAGP for Random (in (a)) and RvK (in (b)) polygons. Especially from the Random case, one can see that the *Regular Grid* strategy rapidly becomes impractical due to the huge size of the discretization and, therefore, will no longer be analyzed for other classes of polygons. On the other hand, one can see that the *Reduced AVP* strategy still follows the same behavior of the CvK case for RvK instances, with the discretization size growing fast as the number of vertices of P increases. Nevertheless, this approach is very well-suited for random polygons. The curves corresponding to the *Just Vertices* strategy suggest that the set of vertices of the polygon is a good guess for the initial discretization since few new points are added to it to achieve the optimal solution of an OAGP instance for both classes of instances.

Figure 5 shows the number of iterations each strategy needs to achieve the optimal solution for both classes of random polygons. The chart in (a) displays

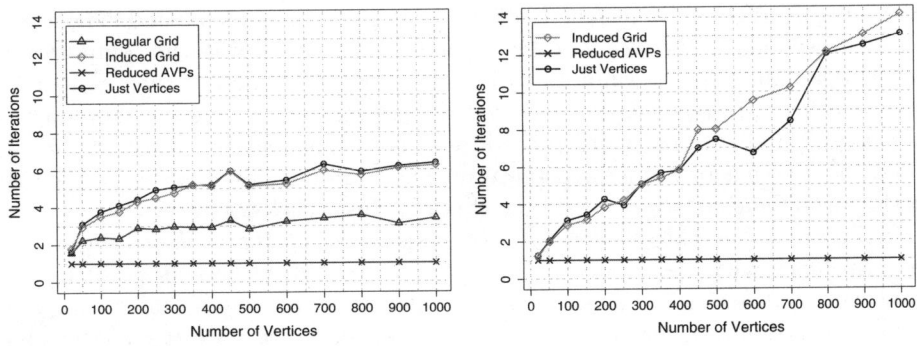

Fig. 5. Number of iterations: (a) Random polygons; (b) Random von Koch polygons

the expected behavior with the number of iterations increasing as the size of the discretizations decrease. Now, relative to the size of the input polygon, the number of iterations remains negligible when compared to the theoretical bound of $\Theta(n^3)$. In chart (b) relative to RvK polygons, the number of iterations increases a bit faster with the instance size but is still small. Somewhat surprisingly, in this case *Induced Grid* iterates slightly more than the *Just vertices* strategy.

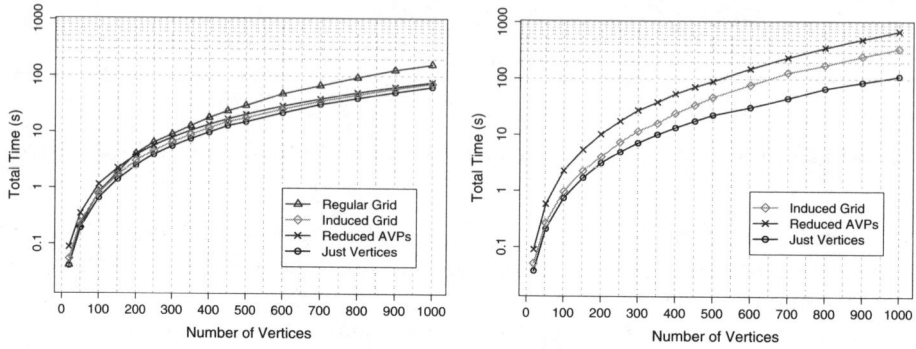

Fig. 6. Total time: (a) Random polygons; (b) Random von Koch polygons

Figure 6 shows the total amount of time, including the preprocessing and processing phases, to solve instances from the random classes. Notice that the curves are plotted in log × linear format and both charts are in the same scale. One can see that for Random polygons, all the strategies behave similarly except, as expected and explained before, the *Regular Grid*. The tendency of *Just vertices* strategy is very similar in both classes of polygons. This shows that, though we are solving harder instances in the RvK case, the strategy is robust.

We now turn our attention to the time spent by the algorithm in each phase for the discretization strategies. Recall that the preprocessing phase is composed of three procedures. The first one is common to all strategies and computes the visibility polygons. The second one computes the initial discretization and its cost is highly affected by the choice of the strategy to be implemented. The worst case corresponds to the *Reduced AVP* strategy since it requires the computation of all AVPs and the determination of the shadow AVPs and of their centroids. On the other extreme, we have the *Just vertices* strategy where no computation is needed. Finally, in the third procedure of the preprocessing phase one has to build the starting IP model and the time spent in doing so depends on the size of the discretization, which again benefits the *Just vertices* strategy.

In Figure 7 one can see that the time spent in the preprocessing phase is in accordance with the discussion above, the *Reduced AVP* strategy being the most time consuming for RvK. What is somehow surprising is that, though we are solving NP-hard problems in the solution phase, the majority of the time consumption refers to the preprocessing phase, which is entirely polynomial. The extraordinary developments in IP solvers together with the fact the SCP instances arising from OAGP are among the easy ones can explain this apparently counter intuitive behavior of the algorithm. Thus, for the *Reduced AVP* strategy to become competitive, a cleverer and faster procedure has to be developed to discard not only shadow AVPs but other ones. Comparing the size of the final discretizations of the different strategies shown earlier there seems to be room for such improvements.

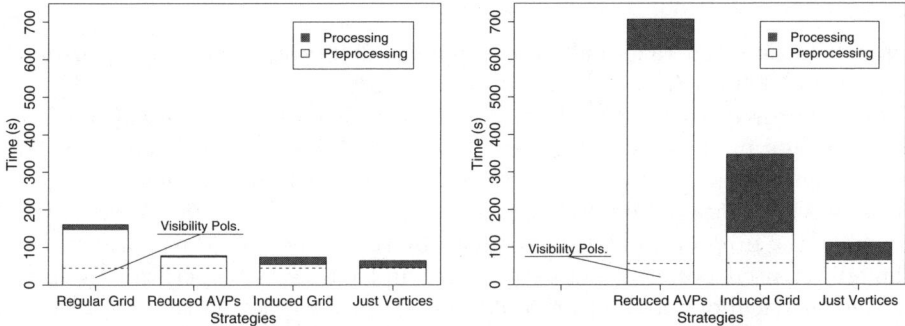

Fig. 7. Execution time for polygons of 1000 vertices: (a) Random polygons; (b) Random von Koch polygons. The lower part of the preprocessing time corresponds to the construction of the visibility polygons.

5   Conclusions and Remarks

In this paper, we conducted an experimental investigation of an exact algorithm for the Orthogonal Art Gallery problem (OAGP) proposed in [1] which relies on the discretization of the interior of the input polygon P and on the modeling of this simplified discrete problem as a Set Cover problem (SCP). The resulting SCP instance is solved to optimality by an IP solver and, if uncovered regions remain, additional constraints are included and the process is repeated. Clearly, the performance of the algorithm depends on the number of such iterations.

This work focused on different strategies to implement the discretization of P. Thorough experimentation was carried out to assess the trade-off between the number of iterations and time spent by the many variants of the algorithm that arise from the alternative discretization methods.

Our conclusion is that this exact algorithm is a viable choice to tackle instances of the OAGP, in light of the fact that the largest ones we solved were five times larger than those reported earlier in the literature.

The apparent advantage of a discretization which ensures an exact solution after a single iteration of the algorithm (like the *Reduced AVP* strategy) did not prove to be effective in practice. This became even clearer when we compared its results with those of the *Just Vertices* strategy which, represents the opposite extreme situation, as, in principle, it starts with the smallest "natural" SCP instance. However, as one can see from Table 2 this strategy leads to a very fast

Table 2. Total Time (seconds): *Just Vertices* strategy

	Polygons Classes			
n	Random	FAT	RvK	CvK
100	0.65	0.58	0.73	0.97
500	15.21	18.47	22.73	29.35
1000	64.13	92.41	111.55	♯

implementation that takes only few seconds of CPU time to solve OAGP instances with up to 1000 vertices.

The success of this exact algorithm clearly benefits from the extraordinary developments in IP solvers in recent years, which lead to the solution of large instances of SCP in a very small amount of time. Therefore, we believe that the *Reduced AVP* strategy can only become competitive with the *Just Vertices* strategy when the preprocessing time required by the former is significantly reduced. Though we used powerful data structures and packages to perform the necessary geometrical operations, we could not significantly lessen the preprocessing time which, for the largest instances tested here, correspond roughly to the time required by the IP solver to resolve ten instances of SCP.

A promising venue of further investigation lies in trying to identify inexpensive geometric properties that might lead to a set of constraints that capture the essence of the hardness of the problem, such as a significant reduction on the number of AVPs.

References

1. Couto, M.C., de Souza, C.C., de Rezende, P.J.: An exact and efficient algorithm for the orthogonal art gallery problem. In: Proc. of the XX Brazilian Symp. on Comp. Graphics and Image Processing, pp. 87–94. IEEE Computer Society, Los Alamitos (2007)
2. Honsberger, R.: Mathematical Gems II. Number 2 in The Dolciani Mathematical Expositions. Mathematical Association of America (1976)
3. Chvátal, V.: A combinatorial theorem in plane geometry. Journal of Combinatorial Theory Series B 18, 39–41 (1975)
4. Urrutia, J.: Art gallery and illumination problems. In: Sack, J.R., Urrutia, J. (eds.) Handbook of Computational Geometry, pp. 973–1027. North-Holland, Amsterdam (2000)
5. Kahn, J., Klawe, M.M., Kleitman, D.: Traditional galleries require fewer watchmen. SIAM J. Algebraic Discrete Methods 4, 194–206 (1983)
6. Schuchardt, D., Hecker, H.D.: Two NP-hard art-gallery problems for ortho-polygons. Mathematical Logic Quarterly 41, 261–267 (1995)
7. Sack, J.R., Toussaint, G.T.: Guard placement in rectilinear polygons. In: Toussaint, G.T. (ed.) Computational Morphology, pp. 153–175. North-Holland, Amsterdam (1988)
8. Edelsbrunner, H., O'Rourke, J., Welzl, E.: Stationing guards in rectilinear art galleries. Comput. Vision Graph. Image Process. 27, 167–176 (1984)

9. Ghosh, S.K.: Approximation algorithms for art gallery problems. In: Proc. Canadian Inform. Process. Soc. Congress (1987)

10. Eidenbenz, S.: Approximation algorithms for terrain guarding. Inf. Process. Lett. 82(2), 99–105 (2002)

11. Amit, Y., Mitchell, J.S.B., Packer, E.: Locating guards for visibility coverage of polygons. In: Proc. Workshop on Algorithm Eng. and Experiments, pp. 1–15 (2007)

12. Erdem, U.M., Sclaroff, S.: Automated camera layout to satisfy task-specific and floor plan-specific coverage requirements. Comput. Vis. Image Underst. 103(3), 156–169 (2006)

13. Tomás, A.P., Bajuelos, A.L., Marques, F.: On visibility problems in the plane - solving minimum vertex guard problems by successive approximations. In: Proc. of the 9th Int. Symp. on Artificial Intelligence and Mathematics (2006)

14. Couto, M.C., de Souza, C.C., de Rezende, P.J.: OAGPLIB - Orthogonal art gallery problem library, www.ic.unicamp.br/~cid/Problem-instances/Art-Gallery/

15. Johnson, D.S.: A theoretician's guide to the experimental analysis of algorithms. In: M.H.G., et al. (eds.) Data Structures, Near Neighbor Searches, and Methodology: Fifth and Sixth DIMACS Implem. Challenges, AMS, Providence, pp. 215–250 (2002)

16. McGeoch, C.C., Moret, B.M.E.: How to present a paper on experimental work with algorithms. SIGACT News 30 (1999)

17. Sanders, P.: Presenting data from experiments in algorithmics, pp. 181–196. Springer, New York (2002)

18. Moret, B.: Towards a discipline of experimental algorithmics. In: Proc. 5th DIMACS Challenge

19. Lee, D.T.: Visibility of a simple polygon. Comput. Vision, Graphics, and Image Process 22, 207–221 (1983)

20. Joe, B., Simpson, R.B.: Visibility of a simple polygon from a point. Report CS-85-38, Dept. Math. Comput. Sci., Drexel Univ., Philadelphia, PA (1985)

21. Joe, B., Simpson, R.B.: Correction to Lee's visibility polygon algorithm. BIT 27, 458–473 (1987)

22. Bose, P., Lubiw, A., Munro, J.I.: Efficient visibility queries in simple polygons. Computational Geometry 23(3), 313–335 (2002)

23. Tomás, A.P., Bajuelos, A.L.: Generating random orthogonal polygons. In: Conejo, R., Urretavizcaya, M., Pérez-de-la-Cruz, J.-L. (eds.) CAEPIA/TTIA 2003. LNCS (LNAI), vol. 3040, pp. 364–373. Springer, Heidelberg (2004)

24. Falconer, K.: Fractal Geometry, Mathematical Foundations and Applications, pp. 120–121. John Wiley & Sons, Chichester (1990)

Computing Multiple Watchman Routes*

Eli Packer

State University of New York at Stony Brook

Abstract. We present heuristics for computing multiple watchman routes. Given a polygon (with or without holes) and a parameter k, we compute a set of k routes inside the polygon such that any point inside the polygon is visible from at least one point along one route. We measure the quality of our solutions by either the length of the longest route or the sum of the route lengths, where the goal is to minimize each. We start by computing a set of static guards [2], construct k routes that visit all the static guards and try to shorten the routes while maintaining full coverage of the polygon. We implemented the algorithm and present extensive results to evaluate our methods, including a comparison with lower bound routes based on the idea of visiting large number of visibility-independent "witness points". Our experiments showed that for a large suite of input data our heuristics give efficient routes that are comparable with the optimal solutions.

Keywords: Watchman routes, Art gallery, Polygons, Arrangements.

1   Introduction

The *Art Gallery* problem is a famous computational geometry problem that has been extensively studied in the previous three decades. Presented in 1973 by Klee, the idea is to place a minimum number of point guards that collectively cover the interior of a given simple polygon[1]. Since then, numerous variants have been proposed and studied. Some of the most important results were the hardness proofs of the classic variant (point guards inside a simple polygon) [14] and others. These motivated the studying of approximations and heuristics.

As opposed to the classic variant which deals with static guards, the *watchman route* variant is concerned with guards that can translate along routes. The goal is similar: make any point inside the polygon visible by at least one point along one of the routes. This problem is motivated by many applications that involve security, surveillance, imaging, simulations and more. In this problem, as the number of guards is usually predetermined, the measure of the result is often the route lengths (the Euclidian metric is usually used here, but other metrics,

---

* This research has been supported by grants from the National Science Foundation (CCF-0528209, CCF-0729019).

[1] In the terminology of the art gallery study, point guards can see in any direction with no distance limit. A guard is said to cover a point inside the polygon if the line segment connecting them does not intersect any edge of the polygon.

C.C. McGeoch (Ed.): WEA 2008, LNCS 5038, pp. 114–128, 2008.
© Springer-Verlag Berlin Heidelberg 2008

such as the number of links in the routes, have been used too). Interestingly, the minimum watchman route (one guard) inside a simple polygon has an exact polynomial time solution [20]. Unfortunately, extending the problem to support holes inside the polygon or allowing more than one guard (when minimizing the longest route) make the corresponding decision problems hard (the hardness proofs use simple reductions from the TSP [8] and partition [15] problems respectively).

We study the k-watchman routes, $k \geq 1$, where multiple watchmen are allowed to translate inside a polygon (possibly with holes). Two natural goals in this case are to minimize the maximum length of any route and to minimize the sum of the route lengths. We denote the corresponding problems by $KWRP_m$ and $KWRP_s$ respectively ($KWRP$ stands for k-watchman routes inside a polygon). We note that these problems are well motivated: one motivation behind $KWRP_m$ is to minimize the time it takes to cover the polygon while the motivation of $KWRP_s$ could be to save the total energy or frames taken by the entire system. We show later that it is even hard to give any meaningful proven approximation to both measures (Theorem 1).

Our Contribution. We propose heuristics for computing watchman routes that cover polygons (with or without holes), for both $KWRP_m$ and $KWRP_s$. While it is impossible to develop exact or even approximate polynomial time algorithms for both problems unless $P = NP$, we conduct an extensive experimental analysis of their performance. We show that our heuristics work well in practice for many kinds of polygons, and in some cases compare to lower bounds obtained by the idea of independent witness point set. As far as we know, our work is the first attempt to conduct a systematic experimentation with watchman route heuristics.

As we mentioned above, the 1-watchman route problem has been optimally solved with a polynomial time algorithm. However, to the best of our knowledge, neither implementation nor experiments have ever been reported. Perhaps the reason is that the algorithm is not easy to implement. Further, it is not clear if this algorithm will suffer from robustness problems. Hence, our work is also targeted for this specific and important variant, as when we set $k = 1$ we compute the 1-watchman route.

Related Work. Good references for the various art gallery problems are [17,19,22]. Two detailed reports that provide valuable information about the watchman route problem and present algorithms for restricted versions are [16,17]. The 1-watchman route problem has been extensively studied. After a few publications that were found to have flaws, Tan et al. [21] gave an $O(n^4)$ time algorithm where the starting point is given and finally Tan [20] presented an $O(n^5)$ algorithm for the general case, which followed [5]. Other interesting variants that have been studied are the minimum-link watchman route [1,3] and Pursuit-Evasion [18] and others [6,11,13].

The rest of this paper is organized as follows. In the next section we provide background information. In Section 3, we present our algorithm and discuss its implementation and performance. In Section 4 we present our experiments with the software we have implemented. We conclude and propose ideas for future research in Section 5.

2   Preliminaries

Given a polygon P (possibly with holes)[2] and a point $p \in P$, we denote by $\mathcal{V}(p)$ (P is omitted for simplicity) the set of points inside P that are visible from p.[3] It is easy to observe that $\mathcal{V}(p)$ is a star-shaped polygon and it is termed the *visibility polygon* of p. A set of points $S \subset P$ is said to cover P if $\bigcup_{p \in S} \mathcal{V}(p) = P$. The classic art gallery problem is to find a smallest such set.

Over the years numerous variations of this problem have been proposed and studied. One of these variations allows guards to translate inside the polygon along predefined routes. In this case, the guards are often termed *watchmen* or *mobile guards* and their routes are termed *watchman routes*. Let w be a watchman with route R_w inside a polygon P (P is omitted for simplicity). Let $\mathcal{V}(w) = \bigcup_{p \in R_w} \mathcal{V}(p)$ be defined similarly to the visibility polygons of the static guards. The goal here is to cover P as well, namely to find a set of watchmen S such that $\bigcup_{w \in S} \mathcal{V}(w) = P$. In this context, the size of S is usually given (we denote it by k) and the measure (or quality) of the solution involves the length of the routes. Two popular measures are the length of the longest route (corresponds to $KWRP_m$) and the sum of route lengths (corresponds to $KWRP_s$). More formally, the measure of $KWRP_m$ and $KWRP_s$ are $max_{w \in S} L(w)$ and $\Sigma_{w \in S} L(w)$ where $L(w)$ is the length of route R_w. Table 1 summarizes the complexity of computing watchman routes for simple polygons. Table 2 summarizes results for constrained polygons that were established in [16][4]. In these tables MinSum and MinMax refer to the problems of minimizing the sum and minimizing the maximum, respectively. Both tables are borrowed from [16], while we updated the the complexity of the MinSum problem in Table 1 for two or more guards. We note that when the polygon may have holes, both MinSum and MinMax problems become NP-hard for any number of watchmen.

Next we show that the related decision problems of $KWRP_m$ and $KWRP_s$ cannot have any k-approximation for any $k < n$ (unless $P = NP$).

Theorem 1. *$KWRP_m$ and $KWRP_s$ can have no polynomial approximation algorithms unless $P = NP$.*

Proof. Suppose there is such a polynomial algorithm $\mathcal{A}$ (for either $KWRP_m$ or $KWRP_s$) with running time $O(\Gamma)$. We show how to solve the classic art gallery

---

[2] From now on, by P we refer to any polygon.

[3] Two points are visible to each other if the line segment that connects them does not intersect any edge of the polygon.

[4] In this table two kinds of polygons are defined as follows: (1) A polygon is an *alp* if it is monotone and one of the chains in the partition is a line segment parallel to the x-axis. (2) A polygon is a *street* if its boundary can be partitioned into two chains, each of which are guard sets for the polygon.

Table 1. Complexity of computing sets of watchman routes of various sizes inside a simple polygon

Optimization criterion	Number of watchman routes			
	1	2	...	arbitrary
MinSum	P	NP-hard	NP-hard	NP-hard
MinMax	P	unknown	unknown	NP-hard

Table 2. Complexity of computing sets of watchman routes of any size, for some classes of polygons

Optimization criterion	Polygon classes				
	Spiral	Histogram	Alp	Street	Simple
MinSum	P	P	P	NP-hard	NP-hard
MinMax	P	P	unknown	NP-hard	NP-hard

problem (denoted by $\mathcal{AG}$) optimally in polynomial time using $\mathcal{A}$, thus contradicting the above assumption, unless $P = NP$. It can be easily verified that any polygon P can be guarded by k static guards if and only if $\mathcal{A}$ returns routes of zero length, given k as a parameter. Given a polygon P with n vertices, the art gallery theorem states that $\lfloor \frac{n}{3} \rfloor$ guards are always sufficient to guard P. It follows that the solution to $\mathcal{AG}$ can be found by searching the minimum k for which the solution of $\mathcal{A}$ contains only routes of zero length. It would require $O(\log(n)\Gamma)$ time. It follows that unless $P = NP$, such a polynomial approximation algorithm cannot exist.

Given two points $p_1, p_2 \in P$, we say that p_1 and p_2 are *independent* if there is no point $g \in P$ such that both $p_1 \in \mathcal{V}(g)$ and $p_2 \in \mathcal{V}(g)$. Let S be a set of pairwise independent points in P of size m. It follows that m static guards are necessary (but may not be sufficient) to guard P. Hence, computing independent sets is a convenient tool to find lower bounds for the art gallery problem and used in [2] for that purpose. We use an analogous idea in our work to compute lower bounds. Recall that k is the number of watchmen. By considering all partitions of S into k groups, and then computing the routes that cover all of the points in each group while optimizing the given problem (either $KWRP_m$ or $KWRP_s$), we find lower bounds. Although we did not design or implement any polynomial heuristic to carry out this task in this work, we use this idea to find lower bounds, and use them in our experimental evaluation.

3   Algorithm

The following is a high-level description of our heuristics.

We continue with a detailed description of the steps. We also describe the data structures that we use and analyze the complexity of the heuristics.

COMPUTE WATCHMAN ROUTES

Input: A polygon P (possibly with holes), k (number of watchmen) and an indication whether to perform $KWRP_m$ or $KWRP_s$

Output: A set of k watchman routes inside P that cover its interior

Measure: The length of the longest route (for $KWRP_m$) or the sum of lengths of the routes (for $KWRP_s$)

(a) Compute a static guard set S with heuristic A_1 of [2]. (A_1 is one of the three proposed heuristics. It was found efficient in time and produced good results.)

(b) Construct the visibility graph $\mathcal{U}$ of $S \cup V$, where V is the set of vertices of P.

(c) Using $\mathcal{U}$, compute the pairwise shortest paths between any $s_1, s_2 \in S$ inside P (denoted by $\mathcal{Z}$).

(d) Construct the minimum spanning tree of S inside P (denoted by $\mathcal{T}$) where the distance between any pair of points is computed from $\mathcal{Z}$.

(e) Split $\mathcal{T}$ into k subtrees, $\mathcal{T}_1$-$\mathcal{T}_k$.

From step (f) and on, we work on each subtree independently.

(f) Construct a Hamiltonian route (let $\mathcal{R}$ denote an arbitrary one).

(g) Substitute vertices along $\mathcal{R}$ with others that shorten the length of $\mathcal{R}$ and maintain full coverage.

(h) Remove redundant vertices of $\mathcal{R}$ by connecting their adjacent vertices (we say that a vertex is redundant if when we connect its two adjacent vertices with their shortest path, the polygon remains fully covered).

3.1   Computing a Static Guard Set (Step a)

We start by computing a static guard set S [2]. The idea is that routes cover the polygon if they visit the static guards. The time to find the static guard set using heuristic A_1 is $O(n^3)$ where n denotes the size of the input [2].

3.2   Constructing the Visibility Graph (Step b)

We construct the visibility graph in $O(n^2)$ time, or in $O(n \log n + b)$ time where b is the number of arcs of the visibility graph [12].

3.3   Computing the Pairwise Shortest Paths (Step c)

We use the Floyd-Warshall algorithm to compute all pairs of the shortest paths in $O(n^3)$ time [10].

3.4   Constructing the Minimum Spanning Tree (Step d)

Using Prim's algorithm, we compute the minimum spanning tree in $O(n^2)$ time [10].

3.5   Splitting the Minimum Spanning Tree into k Subtrees (Step e)

We split $\mathcal{T}$ into k subtrees, $\mathcal{T}_1$-$\mathcal{T}_k$. If the problem is $KWRP_s$ we do it by simply removing the longest $k-1$ edges of $\mathcal{T}$. The heuristic is more involved if the problem is $KWRP_m$. By a reduction from the partition problem [15], the corresponding optimization problem is hard. We partition $\mathcal{T}$ by removing edges. The goal is to minimize the weight of the heaviest subtree. We remove edges by using ideas from parametric search (see below).

Implementation Details. If the problem is $KWRP_s$, we need to remove the longest $k-1$ edges. Finding them takes $O(n \log n)$ time by sorting. If the problem is $KWRP_m$, we use parametric search for finding the subtrees. We perform a binary search on the values 0 to the weight of the tree (denoted by W; Note that $W = O(2^n)$) and stop when the interval on which we search becomes very small (smaller than a predefined constant). Thus, we perform $O(\log W)$ iterations. For each iteration, we do the following. Suppose the current weight we test is W'. We visit the tree in a bottom-up fashion and remove edges once the tree below them has weight larger than W'. It is optimal for the current iteration because if the edge that we remove is below, then its removal creates a smaller subtree and is clearly wasteful. Each iteration traverses the tree in $O(n)$ time. Together, this step takes $O(n \log W) = O(n^2)$ time.

3.6   Constructing Hamiltonian Routes (Step f)

We use the ideas from the algorithm of Christofides in order to approximate the optimal route in $O(n^{2.5} \log^4 n)$ time [9].

3.7   Substituting Vertices (Step g)

We try to substitute some of the vertices along each route $\mathcal{R}$ with others that shorten its length. The vertices of $\mathcal{R}$ can be partitioned into two groups: the first (denoted by $\mathcal{R}_g$) contains vertices that belong to the static guard set and the second (denoted $\mathcal{R}_a$) contains the rest, namely the vertices of P along the paths that connect two vertices of $\mathcal{R}_g$. The idea here is to replace vertices of $\mathcal{R}_g$ by others in order to shorten the length of $\mathcal{R}$. For any $p_1, p_2 \in P$, let $\alpha(p_1, p_2)$ be the shortest path from p_1 to p_2 inside P. Let $v \in \mathcal{R}_g$ and let $u, w \in \mathcal{R}_g$ be the two vertices before and after v along $\mathcal{R}$ (excluding the vertices of R_a; Note that if $|R_g| = 2$ then $u = w$). Let $z \in P$ be some point. The *replacement* of v by z (denoted by $R(v, z)$) is defined as modifying $\mathcal{R}$ by removing $\alpha(u, v)$ and $\alpha(v, w)$ from $\mathcal{R}$ and inserting $\alpha(u, z)$ and $\alpha(z, v)$ instead. Note that $\mathcal{R}$ remains closed after performing this operation.

Let H be the set of reflex vertices of P. We extend the edges that are adjacent to the vertices of H into the interior of P, until they hit the boundary of P (denoted by @P). See an example in Figure 1(g). We denote these extensions by Q. Let $\mathcal{G}$ be the arrangement of $(@P) \cap Q$. We call $\mathcal{G}$ the *extension arrangement* of P. For each vertex $v \in \mathcal{R}$, let $f(v)$ be the face of $\mathcal{G}$ that contains it (if v is a vertex of $\mathcal{G}$, $f(v)$ will be the set of adjacent faces). Based on the properties of the cells in extension arrangements, any vertex $w \in @f(v)$ (@$f(v)$ is the boundary of $f(v)$) has a good chance to maintain full coverage while replacing v (by performing $R(v, w)$). In case we find vertices on @$f(v)$ that both maintain full coverage when they replace v and shorten $\mathcal{R}$, we modify $\mathcal{R}$ by replacing v with the one that minimizes the length of $\mathcal{R}$. We then replace v by the new vertex in $\mathcal{R}_g$. We iterate this process and work on more cells of $\mathcal{G}$, as long as any improvement is achieved.

Implementation Details. We maintain an arrangement $\mathcal{B}$ for a union of visibility polygons and P. Note that in such an arrangement we can easily mark faces as covered or not by the visibility polygons. We initialize $\mathcal{B}$ with the set of visibility polygons of the vertices of $\mathcal{R}_g$. Then for each $v \in \mathcal{R}_g$ we find the face $F(v)$ that contains it in the extension arrangement $\mathcal{G}$ (using point location). We then test whether we can replace v by any vertex of $v' \in F(v)$ in the following way. We remove $\mathcal{V}(v)$ from $\mathcal{B}$ and insert $\mathcal{V}(v')$, while maintaining the information whether a face is covered by the visibility polygons of v' or any other vertex of $\mathcal{R}_g - \{v\}$. We then check whether P is fully covered. If so and the corresponding route is shorter, we perform a replacement and update $\mathcal{B}$ accordingly. We iterate this process until no improvement is detected. Since there are $O(n)$ vertices on the routes, together they are tested for replacement with $O(n^2)$ vertices of $\mathcal{B}$. Since each test takes $O(n^2)$ time, the total time is $O(n^4)$.

3.8   Removing Vertices (Step h)

Let $u, v, w \in \mathcal{R}_g$ be defined as in step **g** above. The *removal* of v is defined as removing $\alpha(u, v)$ and $\alpha(v, w)$ from $\mathcal{R}$, and inserting $\alpha(u, w)$ instead. The idea of this step is to perform removal of vertices if this operation maintains full coverage (note that it necessarily shortens $\mathcal{R}$). Let T be the vertices obtained when intersecting $\alpha(u, w)$ and $\mathcal{G}$. We check whether $\bigcup_{p \in T} \mathcal{V}(p)$ contains $\mathcal{V}(v)$. If so, we perform the removal of v and maintain full coverage. We iterate this process until no vertices can be removed.

Implementation Details. The routes contain $O(n)$ vertices. Each vertex can be removed at most once and each removal check requires a test with $O(n)$ visibility polygons for coverage. To carry out this test, we use an arrangement data structure similarly to step **g**, and use similar ideas. The induced arrangement for each vertex is thus of complexity $O(n^2)$. Hence, the total time is $O(n^3)$.

Figure 1 illustrates the execution of all the steps of heuristic $KWRP_m$.[5]

3.9   Total Complexity

Combining all steps, we get that the time of our heuristic is $O(n^4)$. The space requirement is dominated by the resources required to maintain the various arrangements, which results in an $O(n^2)$ space.

We note that by analyzing the performance of our experiments, we observed that the asymptotic time was always much smaller (bounded by $O(n^3)$ or less).

---

[5] Static guards from step (a) are marked as red discs. Note that some of the routes are not simple close chains. Some routes or parts of them sometime degenerate to polygonal chain or contain chains that connect two parts of the routes. The idea is that the watchman walks on them in both directions. It follows that some of the vertices along the routes may be of degree $d > 2$. In order to remove any possible ambiguity, we use arrows to clarify. In some cases some of the watchman routes degenerate to points (the watchmen essentially become static). We represent them with a disc around the points. Note that the above elements appear in most of the figures of this report.

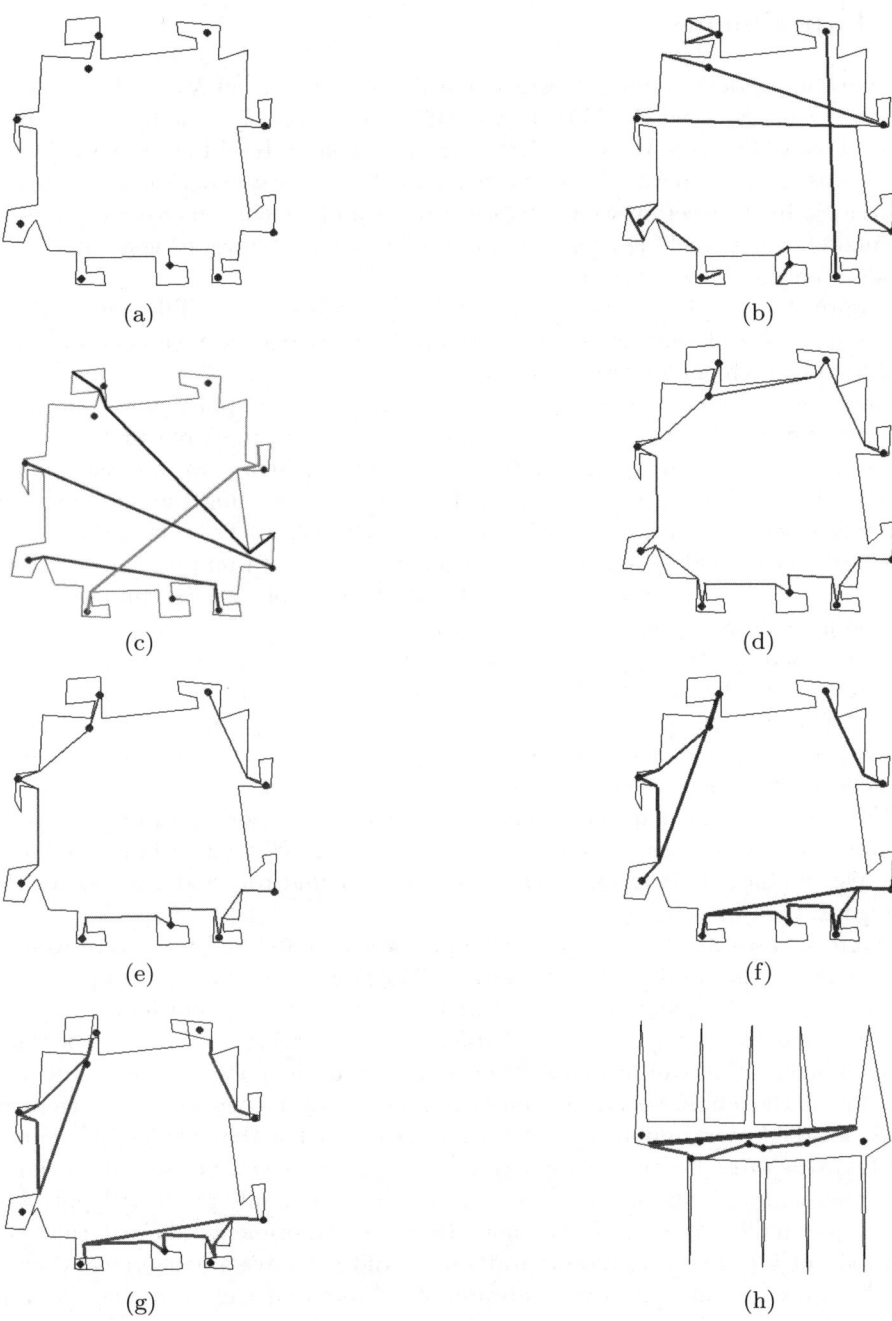

Fig. 1. Illustration of all the steps of $KWRP_m$. For clarity, the data in subfigures (b) and (c) are partially presented. The blue segments in Figure (g) are some of the extension edges that affected the output (see Section 3). In Figure (h), the green and the brown routes represent the situation before and after performing step (h), respectively (so the final route consists of the top segment).

4   Experiments

We have implemented our heuristics on a PC with Microsoft Visual C++ .NET (version 7.1). We used the libraries openGL and CGAL [7]. The tests were performed on a Microsoft Windows XP workstation on an Intel Pentium 4 3.2 GHz CPU with 2GB of RAM. We have performed extensive experiments with our heuristics. In this section we report our results and conclusions from our experiments. Our tests include user-generated polygons and polygons generated by a random polygon generator [4].

Figure 2 shows the routes obtained by our software for different kinds of polygons with different values of k. Figure 3 compares the results of different number of watchmen on two polygons.

Since the k-watchman route problem is hard even to approximate, it is frequently difficult to evaluate the results by comparing it to optimal solutions or even solutions that approximate the optimum. Instead, we use the idea of independent witnesses (see Section 2). However, both finding the maximum independent set and finding the shortest routes that visit a set of points (for $k > 1$) are hard (the second by a simple reduction from the partition problem). Given a set of independent points, we need to find an optimal solution for routes that see all the independent points (by either minimizing the longest route or minimizing the sum or route length). We made this task easier by constructing polygons manually such that the partition of independent points to watchmen was evident. Of course, this restricts the shape of the available polygons. We note that in many cases the lower bound measure was smaller than the optimum. Figure 4 depicts two instances, each with lower and upper bounds.

We note that our heuristics were always within a factor of 5 from the lower bound. Most were even within a factor of 2 and many were within a factor of 1.5. We emphasis that in many cases the lower bound routes did not cover the polygons, thus were not tight.

Even in cases where we could not find lower bounds easily, we were usually satisfied with the results obtained with our software on many types of polygons. In many cases our results looked optimal or very close to be optimal.

There are many parameters that affect the time taken to run our software. The main ones that can be quantified are the size of the polygons, the number of watchmen, the number of static guards and the size of the extension arrangement $\mathcal{G}$. As for the latter, in polygons with wide areas (such as the polygon in Figure 1 (a)-(g)) $\mathcal{G}$ is large because many extension edges intersect. On the other hand, polygons with narrow passages (as the common randomly-generated polygon) result in smaller sizes of $\mathcal{G}$. We note that our experiments showed that the selection of k (number of mobile watchmen) did not have any significant effect on the time: running times were always very close, and moreover, they had no correlation with k. Thus, we ignore this parameter here. In our results the time refers to an average of four runs, with $k = \{1, 2, 3, 4\}$. In Figures 5, 6 and 7 we plot the time as a function of several parameters. In each figure there are two graphs, for $KWRP_m$ and $KWRP_s$. Figure 5 shows the results of many kinds of polygons. We devote separate graphs for random polygons in Figure 6

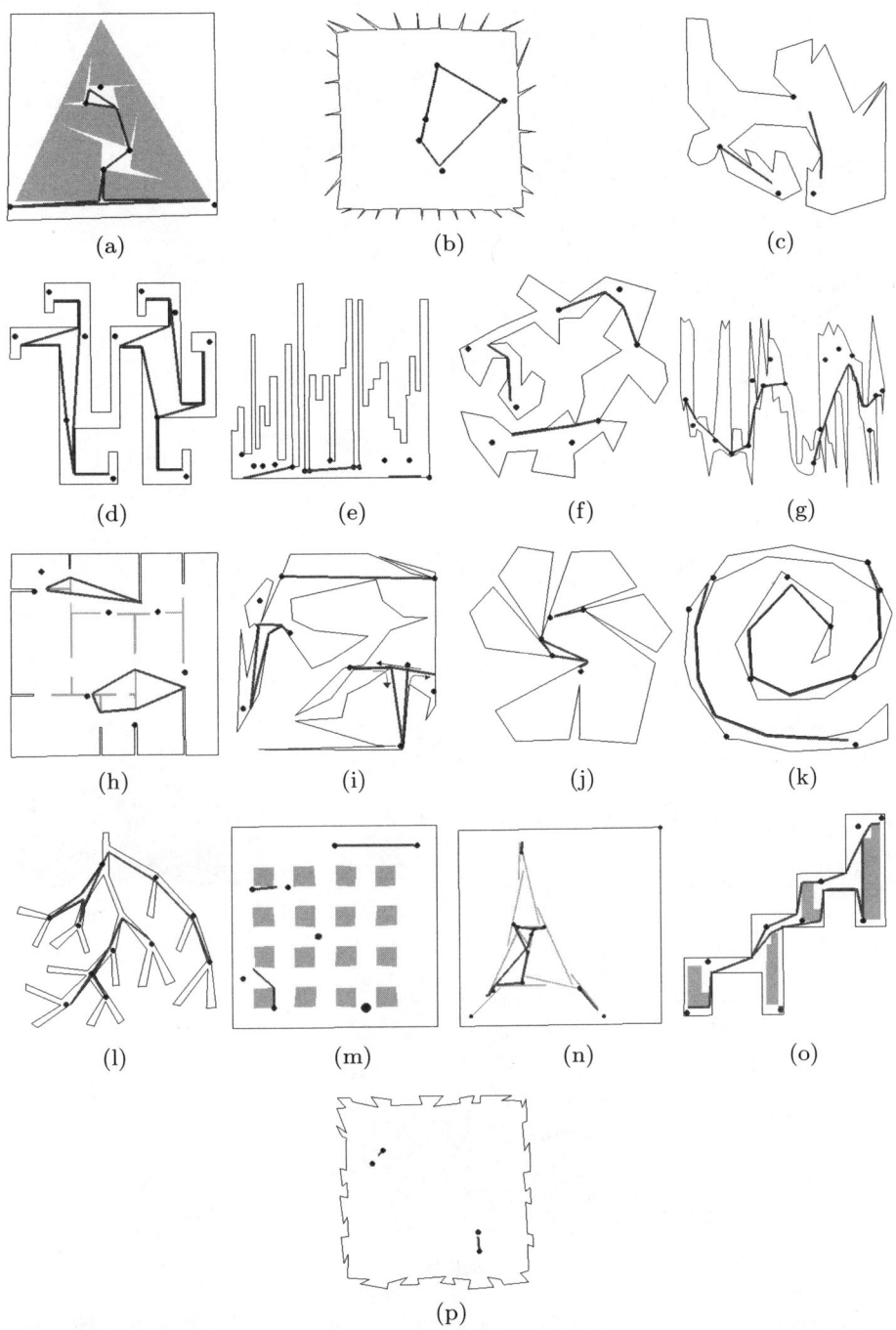

Fig. 2. Experiment snapshots obtained with our software on different kinds of polygons. Subfigures (a) and (b) show results for $k = 1$. Subfigures (c)-(l) shows results of $KWRP_m$ while subfigures (m)-(p) shows results of $KWRP_s$.

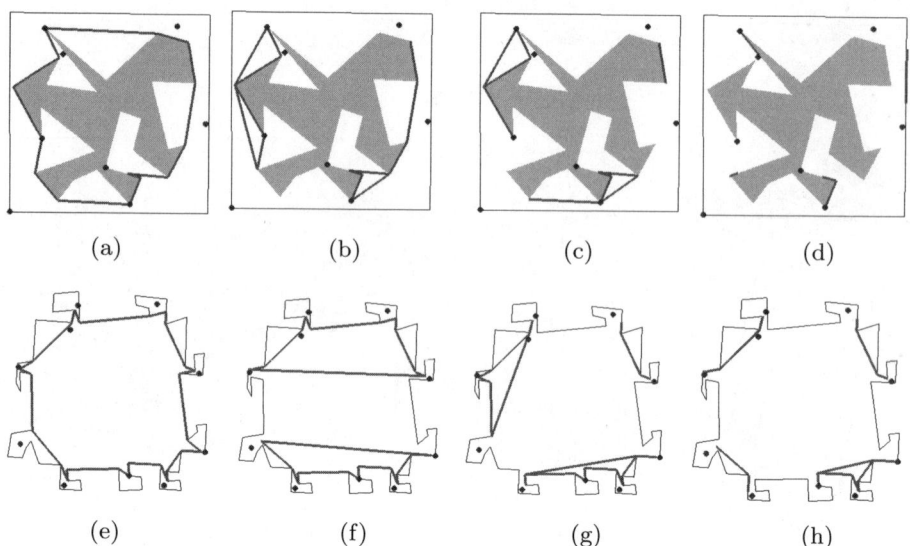

Fig. 3. Results for different number of watchmen

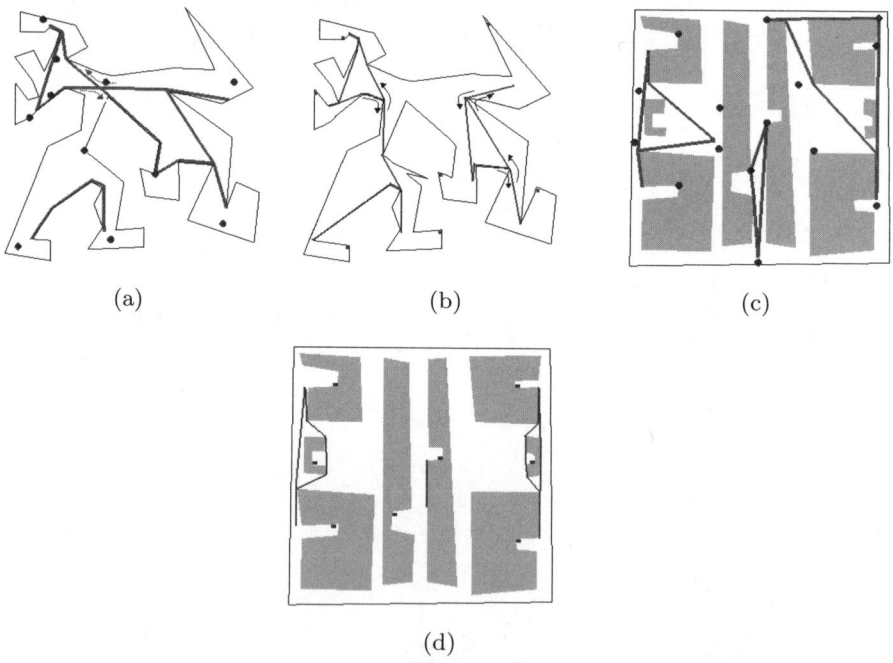

Fig. 4. Comparing the results with lower bounds. Subfigures (a) and (c) depict upper bounds obtained with our software, while Subfigures (b) and (d) depicts corresponding lower bounds where the small squares represent independent witness points.

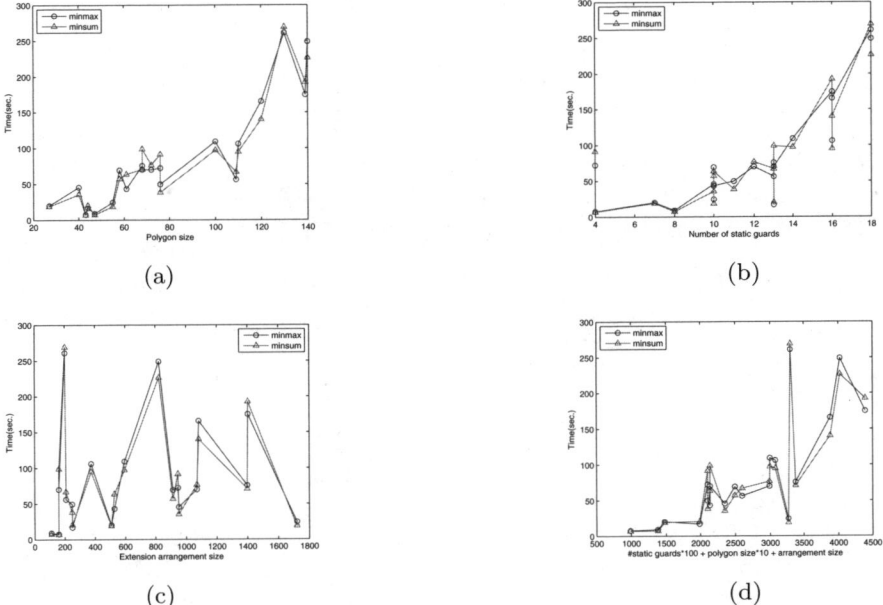

Fig. 5. Time as a function of different parameters for different polygons

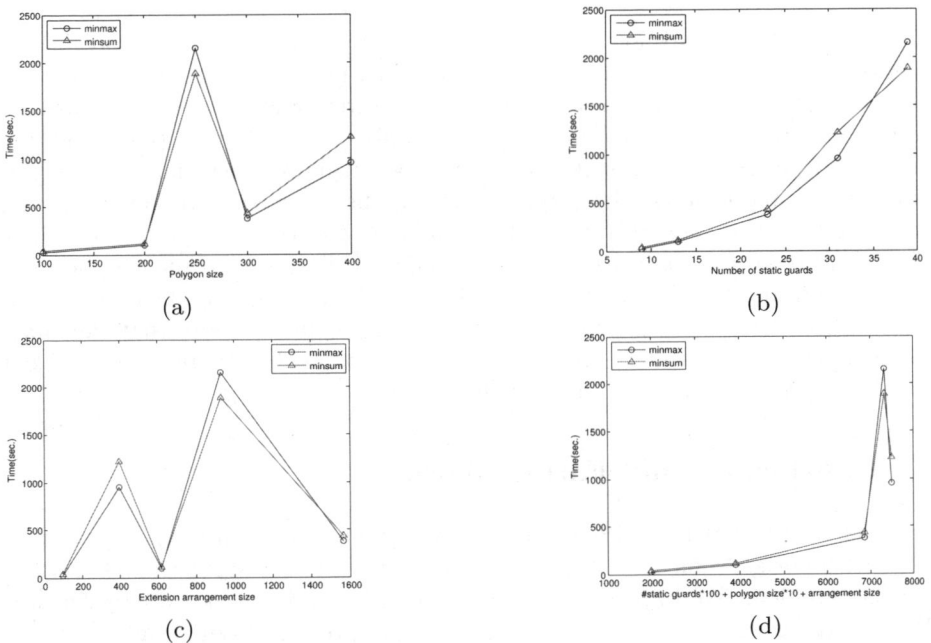

Fig. 6. Time as a function of different parameters for randomly generated polygons

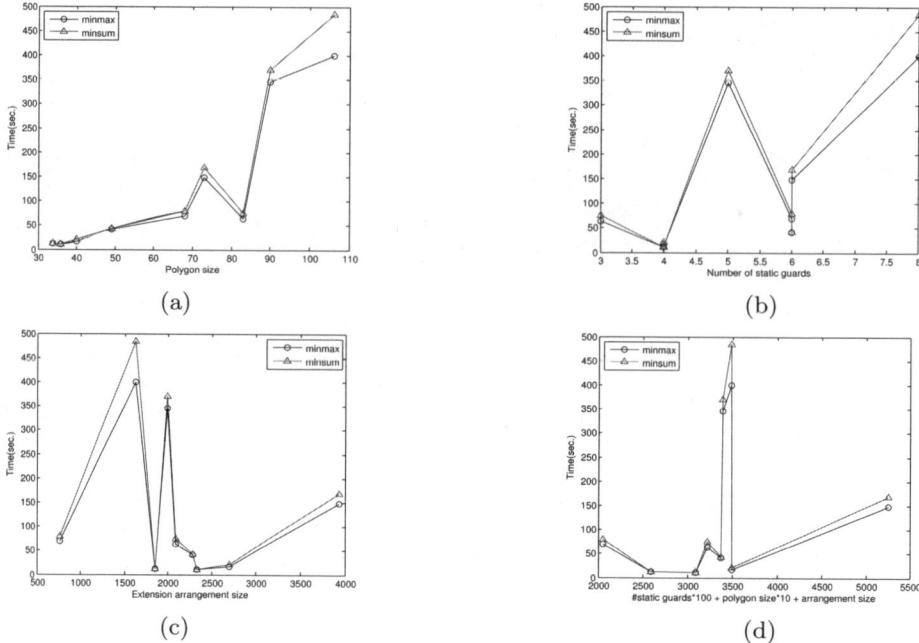

Fig. 7. Time as a function of different parameters for spike box polygons

(Figure 2(i) is a snapshot of a random polygon) and spike box polygons in Figure 7 (Figure 2(b) is a snapshot of a spike box polygon). These two classes of polygons are different in the following aspects. First, spike box polygons usually results in large $\mathcal{G}$ while random polygons usually results in small $\mathcal{G}$. Second, the size of independent sets in spike box polygons is usually much smaller than this size in random polygons. The graphs show that there is a correlation between the time and both the polygon's size and the number of static guards (although there are exceptions). On the other hand, we could not find any correlation between the size of $\mathcal{G}$ and the time. Note that spikes in one one graph are usually a result of stronger dependency on another parameter. Finally, we note that the times for $KWRP_m$ and $KWRP_s$ were usually very similar and there is no evidence for either one to be faster than the other in any set of polygons.

5 Conclusions and Future Work

We presented heuristics for constructing k-watchman routes inside polygons, possibly with holes. As far as we know, this is the first attempt to develop heuristics for this problem. We implemented these heuristics and conducted experiments. We tested our software with many polygons and presented our results in detail. In limited cases we were even able to evaluate our results by comparing with lower bounds, and obtained a bound of factor-5 approximation. Moreover, many other results in many kinds of polygons look efficient and not far from optimal.

We are currently investigating possible directions to improve our heuristics. Next we briefly summarize some. We note that these ideas seem to require solutions that are NP-hard, very challenging, do not have clear heuristics, and seem to take much processing time.

- Possibly start with a different set of static guards that cover the polygon (even if it is not optimal for the static problem). The idea is that if the radius of this set is small, it may lead to shorter routes.
- Locally change the minimum spanning tree such that the Hamiltonian routes become shorter.
- Try also different kinds of splits to the minimum spanning tree. We observed that multiple subtrees that share a vertex can improve the results in specific cases. Sometimes combining this idea with adding Steiner vertices can improve the results further.

Finally, we propose the following directions for future research.

- Find ways to improve the time bounds of the different steps of our heuristics.
- Explore ideas for practical implementations of lower bounds.
- Develop heuristics for other kinds of watchman route problems.
- Find a way to prove efficient lower bounds.
- Develop approximation algorithms for restricted versions.

Acknowledgments. The author thanks Esther Arkin, Alon Efrat, Joseph Mitchell, Girishkumar Sabhani and Valentin Polishchuk for interesting and helpful discussions on related problems.

References

1. Alsuwaiyel, M.H., Lee, D.T.: Finding an approximate minimum-link visibility path inside a simple polygon. Inf. Proc. Lett. 55(2), 59–79 (1995)
2. Amit, Y., Mitchell, J., Packer, E.: Locating guards for visibility coverage of polygons. In: Workshop on Algorithm Engineering and Experiments, ALENEX (2007)
3. Arkin, E., Mitchell, J., Piatko, C.: Minimum-link watchman tours. Technical Report, University at Stony Brook (1994)
4. Auer, T., Held, M.: Heuristics for the generation of random polygons. In: Proc. 8th Canad. Conf. Computat. Geometry (1996)
5. Carlsson, S., Jonsson, H., Nilsson, B.J.: Finding the shortest watchman route in a simple polygon. In: ISAAC: 4th International Symposium on Algorithms and Computation (formerly SIGAL International Symposium on Algorithms), Organized by Special Interest Group on Algorithms (SIGAL) of the Information Processing Society of Japan (IPSJ) and the Technical Group on Theoretical Foundation of Computing of the Institute of Electronics, Information and Communication Engineers (IEICE)) (1993)
6. Carlsson, S., Nilsson, B.J., Ntafos, S.C.: Optimum guard covers and m-watchmen routes for restricted polygons. In: Workshop on Algorithms and Data Structures, pp. 367–378 (1991)

7. The CGAL User Manual, Version 3.1 (2004), www.cgal.org
8. Chin, W., Ntafos, S.: Optimum watchman routes. Inform. Process. Lett. (1988)
9. Christofides, N.: Worst-case analysis of a new heuristic for the travelling salesman problem. Report 388, Graduate School of Industrial Administration, CMU (1976)
10. Cormen, T.H., Leiserson, C.E., Rivest, R.L., Stein, C.: Introduction to Algorithms, 2nd edn. The MIT Press, Cambridge (2001)
11. Gewali, L., Ntafos, S.C.: Watchman routes in the presence of a pair of convex polygons. Information Sciences 105(1-4), 123–149 (1998)
12. Ghosh, S.K., Mount, D.M.: An output-sensitive algorithm for computing visibility graphs. SIAM J. computing, 888–910 (1991)
13. Icking, C., Klein, R.: The two guards problem. International Journal of Computational Geometry and Applications 2(3), 257–285 (1992)
14. Lee, D.T., Lin, A.K.: Computational complexity of art gallery problems. IEEE Trans. Info. Th IT-32, 276–282 (1986)
15. Mitchell, J., Wynters, E.: Watchman routes for multiple guards. In: Proc. 3th Canad. Conf. Computat. Geometry, pp. 126–129 (1991)
16. Nilsson, B.: Guarding art galleries - methods for mobile guards. PhD thesis, Lund University (1995)
17. O'Rourke, J.: Art gallery theorems and algorithms. Oxford University Press, Oxford (1987)
18. Park, S.-M., Lee, J.-H., Chwa, K.-Y.: Visibility-based pursuit-evasion in a polygonal region by a searcher. In: Orejas, F., Spirakis, P.G., van Leeuwen, J. (eds.) ICALP 2001. LNCS, vol. 2076, Springer, Heidelberg (2001)
19. Shermer, T.: Recent results in art galleries. In: Proc. of the IEEE, pp. 1384–1399 (1992)
20. Tan, X.: Fast computation of shortest watchman routes in simple polygons. Inf. Proc. Lett. 77, 27–33 (2001)
21. Tan, X., Hirata, T., Inagaki, Y.: An incremental algorithm for constructing shortest watchman routes. Int. J. Comput. Geometry Appl (IJCGA) 3(4), 351–365 (1993)
22. Urrutia, J.: Art gallery and illumination problems. In: Sac, J., Urrutia, J. (eds.) Handbook of Computational Geometry, pp. 973–1027. Elsevier Science Publishers, Amsterdam (2000)

Engineering Parallel In-Place Random Generation of Integer Permutations

Jens Gustedt

INRIA Nancy – Grand Est, France
Jens.Gustedt@loria.fr

Abstract. We tackle the feasibility and efficiency of two new parallel algorithms that sample random permutations of the integers $[M] = \{1, \ldots, M\}$. The first reduces the communication for p processors from $O(M)$ words ($O(M \log M)$ bits, the coding size of the permutation) to $O(M \log p / \log M)$ words ($O(M \log p)$ bits, the coding size of a partition of $[M]$ into M/p sized subsets). The second exploits the common case of using pseudo-random numbers instead of real randomness. It reduces the communication even further to a use of bandwidth that is proportional to the used real randomness. Careful engineering of the required subroutines is necessary to obtain a competitive implementation. Especially the second approach shows very good results which are demonstrated by large scale experiments. It shows high scalability and outperforms the previously known approaches by far. First, we compare our algorithm to the classical sequential data shuffle algorithm, where we get a speedup of about 1.5. Then, we show how the algorithm parallelizes well on a multicore system and scales to a cluster of 440 cores.

1 Introduction and Overview

Generating random permutations is costly. One issue that causes this high cost is the use of (pseudo-)random number generators (*PRG*), but it is not the only one: the random (!) memory read pattern of the classical shuffling algorithm, see Moses and Oakford [15], Durstenfeld [6] and also Knuth [13, Sec. 3.4.2], implies cache misses for almost all memory accesses. Thus the performance is in general dominated by the CPU to memory latency. Neither augmenting the speed of the CPU nor the memory bandwidth would improve the performance, only augmenting the frequency of the interconnection bus would do.

On a modern architecture our implementation of that shuffling algorithm achieves an amortized run time of roughly 250 cycles per shuffled 64 bit integer. We think that this is not improvable by much, because of the difficulties mentioned above. In addition, for commodity architectures the performance growth is nowadays only assured by augmenting the parallelism of the CPUs, via multiple processors, cores or pipelines or via hyperthreading. In such parallel settings the random shuffle algorithm doesn't scale, it is inherently sequential. Anderson [2] gave a parallelization of the classical shuffling but which has to schedule conflicting `swap` operations, a strategy that turns out to be only efficient for few processors, see Cong and Bader [4]. Several subsequent attempts have been made to tackle and experiment this problem differently in parallel

C.C. McGeoch (Ed.): WEA 2008, LNCS 5038, pp. 129–141, 2008.
© Springer-Verlag Berlin Heidelberg 2008

and/or distributed settings, see *e.g* the work of Cong and Bader [4] Czumaj et al. [5], Goodrich [8], Guérin Lassous and Thierry [9], Gustedt [12], Sanders [17].

The most important use of random permutations of data are probably simulations and statistical measurements where they make the simulation (respectively measurement) independent from a particular ordering data may have been generated (or appeared). In many cases it is in fact not necessary to effectively permute the data by itself but it is sufficient to compute a random integer permutation. These usually have more compact encoding than the data and may directly serve as ranks of the individual data items. Thereby an implementation may avoid the costly repetitive shuffling of application data.

Such a strategy has already been used or described in various contexts, as *e.g* in VLSI design by Alpert et al. [1], for combinatorial algorithms by Cohen [3], or for non-parametric Monte Carlo testing of time series, see Dwass [7] and Nichols and Holmes [16]. In view of the results of the present paper such a strategy might in particular bring important gains when applied to a distributed setting. Here it can entirely avoid the re-distribution of data between different processors.

In Gustedt [10] and [12] we have shown that random shuffling of data can be realized with *linear* resource usage, CPU time as well as bandwidth, and this for a large variety of paradigms, in particular parallel, distributed and out-of-core computation. For permutations of just the integers $\{1, \dots, M\}$, we give two new algorithms that take advantage of the restriction to that specific problem in [11]. Here in this paper we tackle the engineering aspects of the two later algorithms and report on benchmarks. These show that the second of these algorithms is able to simultaneously take advantage of two types of parallelism, SMP (provided by multicore processors) and cluster parallelism.

The first approach tackles bandwidth requirements (as opposed to latency). An information theoretic lower bound which is sublinear in the number of bits of the input size for the communication can given by a counting argument. In fact entropy compression can be used that asymptotically realises this bound. This can be done when assuming *full randomness*, i.e when all random decisions that we make are given by an abundant sequence of random bits.

For the second algorithm, we place ourselves in the common case that we use pseudo-random numbers (instead of real randomness). The lower bound doesn't hold any more: the amount of solutions is limited by the state space of the PRG. Therefore the minimum information that has to be transferred is just that, the state space of the PRG.

Sec. 2 introduces the general framework, wherein 2.1 and 2.2 briefly explain the two new algorithms that are the subject of this paper. Sec. 3 introduces the engineering part of the present work, namely the basis of the implementation, parXXL, the explicit use of integer types of different widths, and special floating point capacities. Then it focuses on the implementation of a range coder, Sec. 3.1, and universal hash functions, Sec. 3.2, that are needed as subroutines to implement the algorithms effectively. In Sec. 4, we then report on large scale experiments that prove the efficiency and practicability of our approach in different settings: sequential execution, for parallel execution with multi-processor multi-core machines and for the distributed setting of clusters.

Procedure. `ParIntPerm(m, p, ν)`: Parallel Random Integer Permutation

> **Input**: Non-negative integers m (local size), p (amount of processors) and $ν$ the id of the processor
>
> **Output**: Table $V = V[1], \ldots, V[m]$ such that the sets of all $V[i]$ on all processors represent a permutation of the integers $1, \ldots, p \cdot m$.
>
> matrix All processors collectively choose $A = (a_{i,j})$, the communication matrix
>
> partition **begin**
>
> > Create a table V' with $V'[1] = (ν - 1) \cdot m + 1, \ldots, V'[m] = ν \cdot m$
> >
> > Randomly partition the elements of V' into blocks $B'_{ν,1}, \ldots, B'_{ν,p}$ of size
> >
> > $a_{ν,1}, \ldots, a_{ν,p}$
> >
> > starting at $V[1]$, **for** $j = 1, \ldots, p$ **do** set $B_{j,ν}$ to the next block of size $a_{j,ν}$ in V
>
> **end**
>
> exchange **for** $j = 1, \ldots, p$ **do** copy the block $B'_{ν,j}$ to the block $B_{j,ν}$ of processor j
>
> local mix Permute V

---

2 Randomized Distributed Shuffling and the Generation of Integer Permutations

Procedure `ParIntPerm` gives an algorithm that on each processor first partitions a source table V' according to a communication matrix A. Our contribution in [10] was to show that it is possible to sample such a matrix separately and still obtain a uniformly distributed random permutation of the items with a resource usage that is linear and equally shared between the processors. Procedure `ParIntPerm` then exchanges these blocks between all processors, and locally permutes the parts that were received at the end.

2.1 Reducing Communication under Full Randomness Assumptions

One main bottleneck for implementing this algorithm are the bandwidth requirements for **exchange**. In [11] we show that the information theoretic lower bound for this is $O(M \log p)$ bits and this bound can be achieved asymptotically by using range encoding. This is done by separating out the 'bits' that are to be send to each individual other processor Q. We do that by taking all elements that go to Q in ascending order (using table T) and by encoding this sequence by the difference between successive elements. Procedure `CompressPartition` summarizes such a procedure that does this encoding 'on the fly' for all target processors.

Each individual difference d that we compute in **differ** can be large. But if we look at the total sequence of such differences that a target processor will receive from all others we see that their average is $M/m = p$.

In **cram** we encode a segment between two occurring elements with an alphabet of two symbols ('0' and '1'), namely by inserting d '0's followed by a '1'. Because we know that '1's only occur with a probability of $1/p$ we can use range encoding, see e.g Martin [14], to encode the overall sequence for any source or target processor Q with $O(m \cdot \log p)$ bits. For the linearity of the algorithm we have to adapt the range coding in **cram** such that it encodes several '0's at once, resulting in amortized constant time per execution of **cram**. The details of this will be presented in Sec. 3.1.

Procedure. CompressPartition(o, m, p, a, P)

Input: Non-negative integers o (start offset), m (local size) and p (amount of processor)
$a = (a[1], \ldots, a[p])$ with $m = \sum_i a[i]$, the row of the communication matrix
$P = (P[1], \ldots, P[m])$ with $\{x \mid x = P[i] \text{ for some } i\} = \{y \mid o < y \le o + m\}$;
Output: compressed streams $(C[1], \ldots, C[p])$, $C[i]$ representing a part of P of size $a[i]$.
Use o to compute $T = (T[1], \ldots, T[m])$ such that $T[i]$ is the target processor for $P[i]$
Initialize $C = C[1], \ldots, C[p]$ to all empty
Initialize $V = V[1], \ldots, V[p]$ to all 0
foreach $i = 1, \ldots, m$ **do**

differ
cram

| Set $t = T[i]$ the target processor of element i
| Set $d = i - V[t]$, the difference of i to the previous element for processor t
| Append d '0's and a '1' to $C[t]$
| Set $V[t] = i$

2.2   Generating Permutations in Place

Procedure GenPermBlock presents a new algorithm that replaces large parts of the communication in ParIntPerm. The main idea is that instead of communicating an already partitioned (or permuted) integer table, this first phase is "*emulated*" directly on each of the target processors. Supposing that block i of the source data would have been permuted by permutation π_i, we communicate the inverse $\mu_i = \pi_i^{-1}$ such that each target processor is able to compute the elements that it would have received from block i. Then a locally computed permutation γ_j is used to write the generated elements for block j of the target data in random order.

GenPermBlock uses universal hash functions as a tool for the local permutations. Besides that, the other source of randomness is the communication matrix. As a consequence the amount of randomness that is used by the algorithm is related to the number of blocks into which the problem is subdivided. If we want this to be tunable, the dependency from an architectural parameter such as p alone is not desirable. GenPermBlock avoids this by dividing the problem into more blocks, b per processor.

3   Engineering

The implementation of the algorithms was undertaken with parXXL[1], a C++-library that allows experimenting and benchmarking of unmodified code on different types of architectures, parallel machines or clusters.

To give an idea of what we are heading for, let us look at the performance of the sequential shuffling algorithm that is implemented in parXXL. This implementation is already quite efficient, since it uses some prefetching techniques to circumvent the latency problems that were mentioned in the introduction.

That implementation needs about 140 *ns* per item for a 64bit integer permutation on a 1.8 *GHz* PC (i86_64 architecture). This corresponds to roughly 250 clock cycles.

---

[1] http://parxxl.gforge.inria.fr/

Procedure. $\texttt{GenPermBlock}\,(m, p, b, \nu, U, \bar{a}, \bar{O})$ Generate a random integer permutation in place

---

Input: Non-negative integers m (block size), p (amount of processors), b (blocks per
   processor) and ν the id of the processor.
Input: $U = U[1], \ldots, U[p \cdot b]$ states of universal hash functions μ_i on $1, \ldots, m$
foreach $j = 1, \ldots, b$
   | **Input**: $a_j = a_j[1], \ldots, a_j[p \cdot b]$ a column of the $p \cdot b \times p \cdot b$-communication matrix.
   | **Input**: $O_j = O_j[1], \ldots, O_j[p \cdot b]$ the offset of the part of block i going to block j.
   | **Output**: $V_j = V_j[1], \ldots, V_j[m]$, the local part of the target permutation.

foreach $i = 1, \ldots, p \cdot b$ **do** Initialize μ_i from $U[i]$
foreach $j = 1, \ldots, b$ **do**
   | Set $t = 0$, this will step through elements in block j
   | Sample a new universal hash function γ_j
   | **foreach** $i = 1, \ldots, p \cdot b$ **do**
   | | Set $o = (m(i - 1)) + 1$ the overall offset of block i
   | | **foreach** $k = O_j[i], \ldots, O_j[i] + a_j[i] - 1$ **do**

preimage | Set $k^{-1} = \mu_i(k)$, the pre-image of k under $\pi_i = \mu_i^{-1}$
generate | Set $K = o + \mu_i(k^{-1})$, the element that would have been sent
permute | Set $t' = \gamma_j(t)$, the final position of K
store | $V_j[t'] = K$
            | $t = t + 1$

These numbers are basically against what we have to compete with an alternative implementation and which should also enable us to judge the parallel efficiency: the time processor product per item should not exceed these 140 ns by much.

We also will have to take the time for the sampling of the communication matrix A into account. The computing time for that is dominated by draws of a hypergeometric distribution which takes about 1 μs in the same setting, based on the standardized PRG $\texttt{jrand48}$[2]. Since the size of that matrix grows quadratic in the number of buckets in which we split the problem, we will have to be careful not to subdivide the problem too much. The implementations that are described here are based on the matrix generation that is already found in parXXL. Unfortunately it is not yet completely parallelized, which we will see to be an issue for the benchmarks, see Sec. 4.

Since we will implement algorithms that go down to the bit level of the represented data another issue that has to be handled carefully is the wordsize of the target architecture. Even talking about "*the*" wordsize is generally not possible. Modern hybrid architectures may use different constants for different types of addressing, e.g 36 bits for physical addressing, 48 bits for virtual addressing, and 64 bits to represent pointers. Arithmetic can be performed with varying efficiency if the data are 32 bit integers ($\texttt{uint32\_t}$), 64 bit integers ($\texttt{uint64\_t}$), floating point numbers ($\texttt{double}$), or of some platform specific register vectors, such as the $\texttt{i386}$'s SSE registers.

To be able to realistically represent large integer permutations we will assume that the *final* output will be a table of $\texttt{uint64\_t}$. But arithmetic on this type may be slow (in particular division and modulo) and storage (and bandwidth) might be wasted if we

---

[2] http://opengroup.org/onlinepubs/007908799/xsh/drand48.html

represent small numbers with it. For the implementation we therefore distinguish the target data type from intermediate ones that are used during the computation, in particular `double` for range encoding and `uint32_t` for universal hash functions. We provide a generic C++ `templates` implementation that depends on two *type* parameters, one for the target type and one for the intermediate type. This enabled us to chose them easily in function of the target architecture.

3.1   Range Encoding

Range encoding (Martin [14]) is a particular case of entropy encoding that is asymptotically optimal. That is, it encodes a string over an alphabet Σ according to the probability $P(\sigma)$ of the occurrence of the individual symbols $\sigma \in \Sigma$. Under the assumption of independence of the occurrence of the symbols, the length of the encoding tends towards the information theoretic optimum.

It views the encoded string (the code) as a big binary number C. Its name comes from the fact that during the encoding phase it works with a lower and upper bound C^- and C^+ that define a *range* within the final code will be found. Each occurrence of a new symbol $\sigma \in \Sigma$ restricts the actual range to a new range with a size that is proportional to $P(\sigma)$.

The particularity in our context for the range encoding needed for `CompressPartition` is that we need to encode long runs of '0's efficiently. A commonly used trick to cope with that is to add artificial symbols to Σ that represent long runs. Whereas such an approach is fast on the coding side, it requires a binary search for the encoded artificial symbol on the decoding side. Thus it has some overhead that is proportional to the logarithm of the length of the run.

To avoid such a logarithmic factor, we use `doubles` to represent the 'interesting' part of the bounds, i.e that part of the bounds that are yet subject to change during encoding or decoding. IEEE `doubles`[3] are normalized to have 52 bits in the mantissa, from which we use 48 for our implementation. They have the advantage that their order of magnitude is automatically maintained in the exponent and that is accessible through cheap bit operations. By that an estimation of the length of the next run can easily be obtained by an integer logarithm operation, on the decoding site.

3.2   Families of Universal Hash Functions

Since the goal of our implementation is first of all to show the potential of the approach we chose some relatively simply universal hash functions:

- The universal hash functions must be fast.
- They must be independent for all processors.
- They must allow for a controlled trade-off between their state-size and their efficiency.

A simple well-known such family is given by an arithmetic progression:

$$\Pi^{\rho}_{\alpha,\beta}(x) := \alpha \cdot x + \beta \quad (\text{mod } \rho) \tag{1}$$

---

[3] http://grouper.ieee.org/groups/754/

Procedure. $\mathrm{uhash}_{\alpha,\beta}^{\rho,m}(x)$ universal hash function with twist 1.

Input: Non-negative integers x (input), ρ (prime), m (domain), such that $x < m \leq \rho$
α (factor) and β (additive shift), with $0 < \alpha < m$ and $\beta < m$
Output: Non-negative integer $y < m$, such that for all $x_1 \neq x_2 < m$,
$$\mathrm{uhash}_{\alpha,\beta}^{\rho,m}(x_1) \neq \mathrm{uhash}_{\alpha,\beta}^{\rho,m}(x_2)$$
repeat $x = \Pi_{\alpha,\beta}^{\rho}(x)$ **until** $x < m$
return x

Where ρ is a prime number and $0 < \alpha < \rho$, $0 \leq \beta < \rho$ are some fixed parameters. Since by definition α and ρ are mutually prime, it is easy to see that for any such choices $\Pi_{\alpha,\beta}^{\rho}$ is a permutation on $\{0, \ldots, \rho - 1\}$. In addition, if we fix ρ, the choice of α and β gives us $\rho \cdot (\rho - 1)$ distinct functions $\Pi_{\alpha,\beta}^{\rho}$: for two distinct choices of β the images of $x = 0$ are distinct, and for two distinct choices of α_1 and α_2 the image $y = \alpha_1\alpha_2 + \beta$ (mod ρ) has different pre-images, namely α_2 and α_1.

We need universal hash functions that operate on any interval $[0, \ldots, m - 1]$, not only for prime numbers. Procedure $\mathrm{uhash}_{\alpha,\beta}^{\rho,m}$ generalizes the family $\Pi_{\alpha,\beta}^{\rho}$ to general m by simply following a cycle of the permutation $\Pi_{\alpha,\beta}^{\rho}$ that might lead outside of the range $[0, \ldots, m - 1]$ until it leads back into it. Again, it is easy to see that $\mathrm{uhash}_{\alpha,\beta}^{\rho,m}$ defines a permutation and that for fixed m and ρ these permutations are all pairwise distinct.

In our implementation we chose the prime number ρ deterministically based on m and on the ID of the processor. This ensures that all these prime numbers are different for all processors, and that all processors may compute them without the need to exchange them. Only the constants α and β are chosen randomly for each processor and are then exchanged.

Choosing p prime numbers can be done efficiently if we restrict ourselves to the case where $p \ll m/\ln m$. We just test the values of $m, m + 1, \ldots$ for primality. Because of the known density of prime numbers of about $1/\ln x$ we are sure to find enough prime numbers in the range $[m, 2m)$ with an amortized computational overhead that does not exceed $O(m)$ on all processors.

The computational cost of $\mathrm{uhash}_{\alpha,\beta}^{\rho,m}$ in our context is dominated by the number of evaluations of $\Pi_{\alpha,\beta}^{\rho}$. Since as a whole we could have to run through all cycles of the permutation this number may be ρ. So, $\mathrm{uhash}_{\alpha,\beta}^{\rho,m}$ could be very expensive in our setting, if the range m and the prime number ρ were of different orders of magnitudes. But fortunately, as seen above, we may restrict ourselves to the case that $\rho < 2m$ and thus the total number of calls to $\mathrm{uhash}_{\alpha,\beta}^{\rho,m}$ *per source processor*[4] is $O(m)$.

Another important issue is to obtain a competitive implementation of $\Pi_{\alpha,\beta}^{\rho}$. Here the non-trivial operation is taking the modulus. Platforms differ greatly on the efficiency of that operation not only between different CPUs but also on the same CPU for data

---

[4] This amortization only holds per source processor, an individual target processor could be overcharged when he would have to run through a lot of cycles. It is possible to avoid such a potential imbalance by computing and communicating these cycles of the permutations in advance. We will see below this was not relevant for the experiments, so such a strategy was not implemented.

Table 1. Platform summary

platform	type	compiler version	nodes	per node			
				processor cores	speed	cache	RAM
damogran	laptop	gcc 4.1.3	1	Intel x86_64 2	1.80 GHz	2 MiB	3.86 GiB
grelon	cluster	gcc 4.1	120	Intel x86_64 4	1.60 GHz	4 MiB	1.97 GiB

types of different width. In particular on the target platforms, all equipped with i86 processors, modulus for 32bit integers is quite efficiently done by a single instruction in some clock cycles. For 64bit integers this might be synthesized in software and take much longer. Therefore it was crucial for the success of the implementation to eventually split the problem in more than p sub-ranges, as was presented in GenPermBlock. Hereby we ensure that the local indices for each block do not exceed 32 bit, i.e that the blocks have less than 2^{32} elements.

4   Experiments

We present experiments on two different platforms, one a laptop computer "*damogran*" and the other a compute cluster "*grelon*", part of Grid5000[5]. The algorithms that were implemented are a variant of the classical shuffling algorithm, our parallel data permutation algorithm of [10], and the in place generation algorithm of this paper with different strategies for the block sizes but with fixed hash strategy $\text{uhash}_{\alpha,\beta}^{\rho,m}$.

The programs were benched in a "reasonable" range of problem sizes: the maximal value M^+ is generally the size that still fits into the platform's RAM. From there other smaller values corresponding to $M^+/2^{i/2}$ for some values of i were also tested. Each data point in the graphs corresponds to the average over 20 runs. In addition, some figures show error bars for the computed variance of the results, but in most cases the variance is so small that this is not noticeable.

To emphasize on the scaling properties and proportions, the results are represented in doubly-logarithmic scale. Data points are chosen such that every second point roughly corresponds to a doubling in size (or processors), *i.e* each step is about $\sqrt{2}$ from the previous.

4.1   Compression by Range Encoding

Before we come to parallel run times, Fig. 1 shows the computing time of an entropy encoding. Here the measurement is quite involved since we first have to benchmark the encoding algorithm together with the random process that generates the data, then we have to benchmark the process without the encoding and the difference is then taken as the time for encoding. As we see the sum of encoding and decoding is between 180 and 200 *ns*, much slower than the random shuffling itself. If used as a compression technique for communication, this corresponds to a throughput of 35 to 40 *MB/s* on the network link, too restrictive in most of today's computing environments to pay off. Therefore

---

[5] http::/www.grid5000.fr

we did not push the implementation of that setting further and did not integrate this encoding scheme into the parallel setting.

4.2 In Place Generation

Sequential and SMP performance. Fig. 2 shows a comparison of the different permutation programs on *damogran*. We see that shuffling takes about 110 to 140 *ns* per item. The parallel data permutation algorithm for two processors slows down to about 170 *ns*. In fact the break even point for this parallel algorithm lays be-

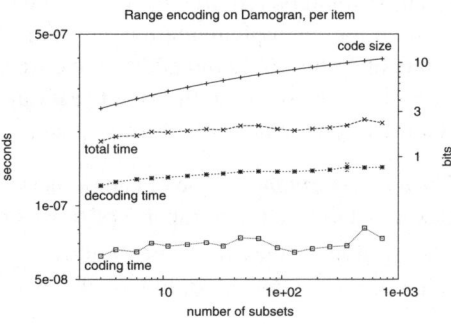

Fig. 1. Compression by range encoding, sequential

tween 3 and 4 processors, so the parallelization for this restricted parallelism of only two cores is not yet worth it, see also Gustedt [10].

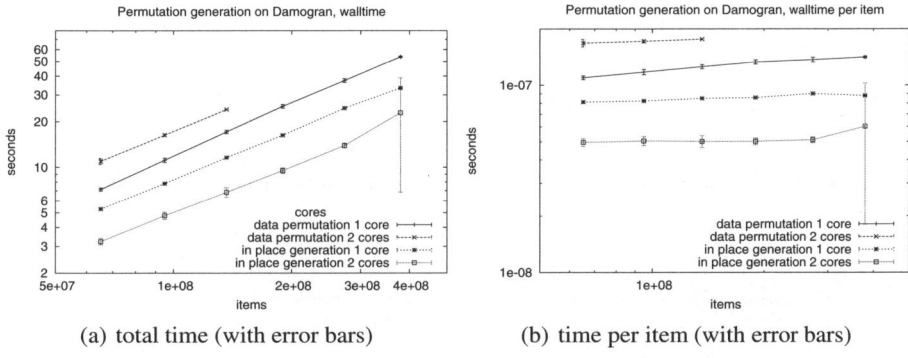

(a) total time (with error bars)   (b) time per item (with error bars)

Fig. 2. Run time comparison on bi-core

Compared to that, the new generation algorithm with 310 to 590 blocks already shows a speedup when only executed on 1 core (80 to 90 ns) and improves to 50 ns when run on 2 cores. Fig. 3 plots the speedup and slowdown values for the possible comparisons. Observe also that already for the smallest value of 2^{26} items the total amount of permutations in the sample space is about $e^{26 \cdot 3.258} \approx e^{84.7} \approx 2^{122.2}$. So a pseudo-random generator with a state of at least

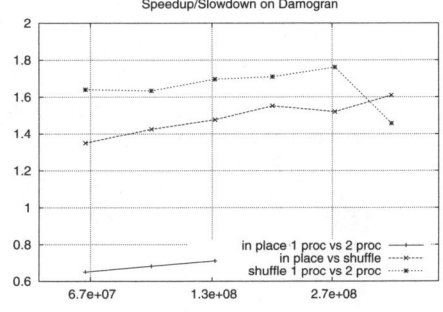

Fig. 3. Speedup or slowdown on bi-core

123 bits would be required to cover the whole sample space. The `rand48` routines that are used for the implementation have the advantage that they are quite fast but only hold a state of 48 bits. Without additional cost, our in place generation here would be able to take advantage of some thousand real random bits (a hash function state per block) as obtained from devices like Linux' `/dev/random`.

Cluster performance. The cluster experiments on *"grelon"* follow two different strategies to determine the number b of blocks per processor of `GenPermBlock`. The first strategy uses a heuristic value of about $\frac{\sqrt{M}}{\log_2 M}$ that is meant to warrant that the computation of matrix doesn't dominate the problem. The other strategy is to fix b to 1024.

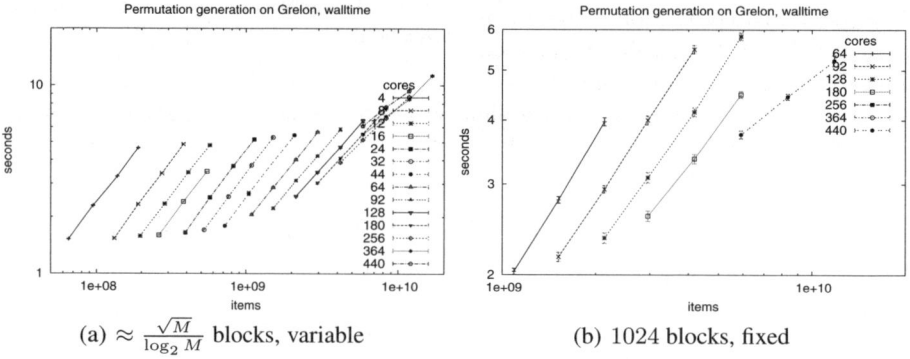

(a) $\approx \frac{\sqrt{M}}{\log_2 M}$ blocks, variable   (b) 1024 blocks, fixed

Fig. 4. Cluster experiments with two different strategies for the block sizes, total times with error bars

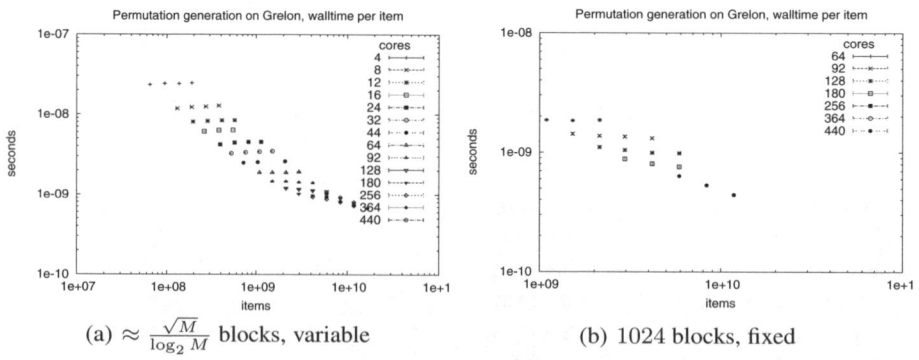

(a) $\approx \frac{\sqrt{M}}{\log_2 M}$ blocks, variable   (b) 1024 blocks, fixed

Fig. 5. Cluster experiments with two different strategies for the block sizes, amortized times per item

Fig. 4 gives the average running times for experiments within two orders of magnitude for the problem size and for the number of processors. The plots show very good

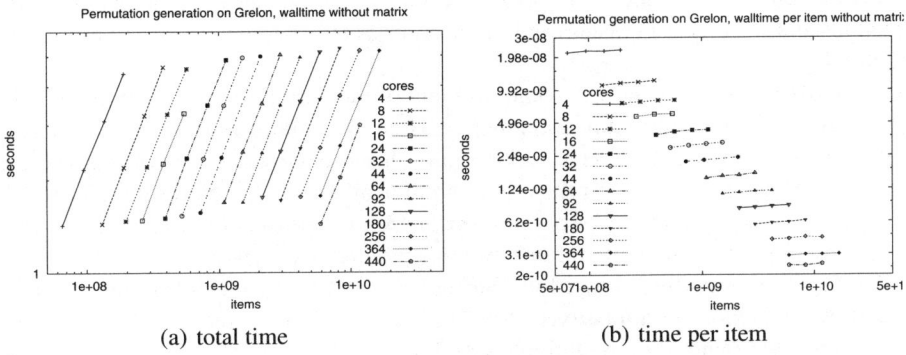

(a) total time   (b) time per item

Fig. 6. Cluster experiments with variable block size, times without matrix generation

scaling of the programs, the progression of the time with the data size are straight lines and the error ranges are invisible.

As the problem sizes concern different orders of magnitude a direct comparison by means of "speedup" plots as given above is not possible. Instead, Fig. 5 shows the same data as before but now the computing times are given as *seconds per data item*. The plots are mainly horizontal lines, meaning that in fact the total running times are linear in the number of items. This holds up to 440 processor cores, where the limit of the scalability seems to be attained. In fact for such large problems the running time for the matrix generation starts to dominate the execution time of the current implementation, in [11] we give some plots for the matrix generation. The in place generation itself (Fig. 6) without the matrix generation behaves very regular.

5   Conclusion and Outlook

With the present work we show that for the generation of integer permutations there are alternatives to the classical data shuffling algorithm and to other more sophisticated redistribution algorithms. The shuffling algorithm doesn't scale in the first place; it is inherently sequential and is not suited to a computing world that consists of distributed multicore machines. Redistribution of (generated) data on the other hand doesn't use the information-theoretical potential. Both share the problem that they generally use PRGs to draw random positions of items. The state space of these PRGs is easily underspecified and does not allow to cover the whole sample space.

Our first approach of using range encoding to limit communication to the information-theoretic bound shows to be as compute consuming as the shuffling algorithm itself. So it can only be competitive in a restrictive setting where bandwidth is very limited compared to computing power, probably less then 10 *MB/s* for today's platforms.

The second approach avoids the initial (and artificial) generation of the identity permutation and generates the target permutation in place. The only data that is communicated between the processors are parts of a communication matrix and state vectors of universal hash functions. This approach improves over the previously know ones, in sequential and in parallel. The parallel implementation shows real and effective speedups and sizeups, for clusters and multi-cores as they become more and more dominant today.

The experiments also showed some problem of the current implementation, namely the generation of the communication matrix. Here, in a future work the parallelization will be driven further to be able to tackle multi-cluster environments of perhaps several thousands of nodes or cores. At the other end of available architectures it could also be possible to use instruction level parallelism to speed up the evaluations of $\text{uhash}_{\alpha,\beta}^{\rho,m}$ by regrouping several such operations into one parallel instruction.

Another limitation that showed up during this work is the lack of accepted quality measures for random permutations. What would be a good statistical test that a family of generated random permutations would have to pass? The lack of such a quality measure also made it pointless for the time being to try other hash functions, such as an iterated version of $\text{uhash}_{\alpha,\beta}^{\rho,m}$. An interesting future study could be to compare the gains of randomness that are obtained by more complicated hash functions and/or by augmenting the amount of blocks into which the problem is subdivided.

Acknowledgement

The experimental part of this research has been undertaken on the Nancy site of the Grid5000 project which has partially been financed by the Lorraine Region.

References

[1] Alpert, C.J., Huang, J.H., Kahng, A.B.: Multilevel circuit partitioning. IEEE Transactions on Computer-Aided Design of Integrated Circuits and Systems 17(8), 655–667 (1998), http://vlsicad.ucsd.edu/Publications/Journals/j34.pdf
[2] Anderson, R.J.: Parallel algorithms for generating random permutations on a shared memory machine. In: Proceedings of the second annual ACM symposium on Parallel algorithms and architectures (SPAA 1990), pp. 95–102. ACM, New York (1990)
[3] Cohen, E.: Estimating the size of the transitive closure in linear time. In: Goldwasser, S. (ed.) 35th Annual Symposium on Foundations of Computer Science, Santa Fe, New Mexico, USA, 20-22 November 1994, pp. 190–200. IEEE, Los Alamitos (1994), http://csdl2.computer.org/comp/proceedings/sfcs/1994/6580/00/0365694.pdf
[4] Cong, G., Bader, D.A.: An empirical analysis of parallel random permutation algorithms on SMPs. In: Proc. 18th ISCA International Conference on Parallel and Distributed Computing Systems (PDCS 2005) (2005)
[5] Czumaj, A., Kanarek, P., Kutylowski, M., Lorys, K.: Fast generation of random permutations via networks simulation. Algorithmica 21(1), 2–20 (1998)
[6] Durstenfeld, R.: Algorithm 235: Random permutation. Commun. ACM, 420 (1964)
[7] Dwass, M.: Modified randomization tests for nonparametric hypotheses. Annals of Mathematical Statistics 28, 181–187 (1957)
[8] Goodrich, M.T.: Randomized fully-scalable BSP techniques for multi-searching and convex hull construction. In: Saks, M., et al. (eds.) Proceedings of the eighth annual ACM-SIAM Symposium on Discrete Algorithms, pp. 767–776. SIAM, Society of Industrial and Applied Mathematics, Philadelphia (1997)

[9] Guérin Lassous, I., Thierry, É.: Generating random permutations in the framework of parallel coarse grained models. In: Proceedings of OPODIS 2000. Studia Informatica Universalis, vol. 2, pp. 1–16 (2000)

[10] Gustedt, J.: Efficient Sampling of Random Permutations. Journal of Discrete Algorithms 6(1), 125–139 (2008), http://hal.inria.fr/inria-00000900/en/

[11] Gustedt, J.: Sublinear Communication for Integer Permutations. Technical Report RR-6403, INRIA (December 2007), http://hal.inria.fr/inria-00201503/en/

[12] Gustedt, J.: Randomized Permutations in a Coarse Grained Parallel Environment [extended abstract]. In: auf der Heide, F.M. (ed.) Fifteenth Annual ACM Symposium on Parallelism in Algorithms and Architectures (SPAA 2003), San Diego, CA, June 2003, pp. 248–249. ACM Press, New York (2003)

[13] Knuth, D.E.: The Art of Computer Programming. Seminumerical Algorithms, 1st edn., vol. 2. Addison-Wesley, Reading (1981)

[14] Martin, G.N.N.: Range encoding: an algorithm for removing redundancy from a digitised message. In: The Video & Data Recording Conference, Southampton, UK (March 1979), http://www.compressconsult.com/rangecoder/rngcod.pdf.gz

[15] Moses, L.E., Oakford, R.V.: Tables of Random Permutations. Stanford University Press (1963)

[16] Nichols, T.E., Holmes, A.P.: Nonparametric permutation tests for functional neuroimaging: A primer with examples. Human Brain Mapping 15, 1–25 (2001), http://www.fil.ion.ucl.ac.uk/spm/doc/papers/NicholsHolmes.pdf

[17] Sanders, P.: Random permutations on distributed, external and hierarchical memory. Inf. Process. Lett. 67(6), 305–309 (1998)

Parallel Partition Revisited

Leonor Frias[*,**] and Jordi Petit[***]

Departament de Llenguatges i Sistemes Informàtics
Universitat Politècnica de Catalunya
lfrias@lsi.upc.edu

Abstract. In this paper we consider parallel algorithms to partition an array with respect to a pivot. We focus on implementations for current widely available multi-core architectures. After reviewing existing algorithms, we propose a modification to obtain the minimal number of comparisons. We have implemented these algorithms and drawn an experimental comparison.

1   Introduction

The partitioning of an array is a basic building block of many key algorithms, as quicksort and quickselect. Partitioning an array with respect to a pivot x consists of rearranging its elements such that, for some splitting position s, all elements at the left of s are smaller than x, and all other elements are greater or equal than x. It is well known that an array of n elements can be partitioned sequentially and in-place using exactly n comparisons and m swaps, where m is the number of greater elements than x whose original position is smaller than s.

In this paper we consider the problem of partitioning an array in parallel, focusing on current widely available multi-core architectures.

Several algorithms have been proposed to partitioning in parallel [1,2,3,4,5]. In this paper, we consider a simple algorithm by Francis and Pannan [2], a fetch-and-add based algorithm by Tsigas and Zhang [3] and a variation of the former in the MCSTL library [5]. These algorithms, which we survey in Sect. 2, seem suitable for a practical multi-core implementation. However, in order to avoid too much synchronization, they perform more than n comparisons and m swaps. Though very different in nature, they can be divided into three main phases: a) A sequential setup of each processor's work, b) a parallel main phase in which most of the partitioning is done, and c) a cleanup phase, which is usually sequential.

In this paper we show that these algorithms disregard part of the work done in the main parallel phase when cleaning up. In order to overcome this drawback, we propose an alternative parallel cleanup phase that uses the whole comparison information of the parallel phase. A small static order-statistics tree is used to

---

[*] Supported by grant number 2005FI 00856 of the *Agència de Gestió d'Ajuts Universitaris i de Recerca* with funds of the European Social Fund.

[**] Partially supported by Spanish project ALINEX (ref. TIN2005-05446).

[***] Partially supported by FET proactive integrated project 15964 (AEOLUS) and by Spanish project FORMALISM (ref. TIN2007-66523).

C.C. McGeoch (Ed.): WEA 2008, LNCS 5038, pp. 142–153, 2008.

efficiently locate the elements to be swapped and to swap them in parallel. With this new method, we obtain scalable parallel partitioning algorithms that achieve an optimal number of comparisons. We provide a detailed analysis.

We have implemented and evaluated all these algorithms, both with their original cleanup and with our cleanup. Besides, the implementation is provided according to the specification of the `partition` function of the Standard Template Library (STL) of the C++ programming language [6]. Previously, only F&A implementation was available in the MCSTL library [5]. Our goal is to get a comparison of their behavior when executed on a currently inexpensive widely available parallel machine, namely a machine with two quad-core processors.

The paper is organized as follows. In Sect. 2, we present the considered algorithms. Then, in Sect. 3, we present our cleanup algorithm. Then, we present our implementation of the previous algorithms and the experimental results in Sect. 4 and 5 respectively. We sum up the conclusions of this work in Sect. 6.

2   Previous Work and a Variant

In this section, we present an overview of the partitioning algorithms we consider in this paper. In the following, the input consist of an array of n elements and a pivot. p processors are available and we assume $p \ll n$. Besides, we disregard some details as rounding issues for the sake of simplicity.

Strided Algorithm. The STRIDED algorithm by Francis and Pannan [2] works as follows:

1. **Setup:** The input is (conceptually) divided into p pieces of size n/p. The pieces are not made of consecutive elements, but one of every p elements instead. That is, the i-th piece is made up of elements $i, i + p, i + 2p, \dots$.
2. **Main phase:** Each processor i, in parallel, gets a piece, applies sequential partitioning on it, and returns its splitting position v_i.
3. **Cleanup:** Let $v_{min} = \min\{v_i : 1 \le i \le p\}$ and $v_{max} = \max\{v_i : 1 \le i \le p\}$. It holds that all the elements at the left of v_{min} and at the right of v_{max} are already well placed with respect to the pivot. In order to complete the partition, sequential partition is applied to the range (v_{min}, v_{max}).

The main phase takes $\Theta(n/p)$ parallel time. For random inputs, the cleanup phase is expected to take constant time. However, [2] did not state that in the worst-case it takes $\Theta(n)$ time and thus, there is no speedup. E.g. If the pieces are made exclusively of either smaller or greater elements than the pivot and these are alternated, then, $v_{min} \le p$ and $v_{max} \ge n - p$, and $|(v_{min}, v_{max})| = \Theta(n)$.

Blocked Algorithm. Accessing elements with stride p as in STRIDED, can provoke a high cache miss ratio. We propose BLOCKED to overcome this problem. It uses blocks of b elements instead of individual elements. Each block in the piece is separated by stride p blocks. If $b = 1$, BLOCKED is equal to STRIDED.

F&A Algorithms. Heidelberger *et al.* [1] proposed a parallel partitioning algorithm in the PRAM model in which elements from both ends of the array are taken using fetch-and-add instructions. Fetch-and-add instructions (atomically increment a variable and return its original value) were introduced in [7] and are useful, for instance, to implement synchronization and mutual exclusion.

In a first approach, exactly one element is taken at a time and so, at the end of the parallel phase, the array is already partitioned. In this case, n fetch-and-add operations are used. In a second approach, the algorithm is generalized to blocks: a block of b elements is acquired at each fetch-and-add instruction. So, the number of fetch-and-add instructions is n/b. However, in this case, some sequential cleanup remains to be applied after the parallel phase.

Later, Tsigas and Zhang [3] presented a variant of the second approach for multiprocessors. More recently, a further variant has been included in the MC-STL library [5]. In the latter, the cleanup phase is partially done in parallel.

Let us now briefly describe these F&A algorithms:

1. **Setup:** Each processor takes two blocks, one from each end of the array. Namely, one left block and one right block.
2. **Main phase:** While there are blocks, each processor applies the so-called *neutralize* method to its two blocks. The neutralize method consists on applying the sequential partitioning algorithm to the array made by (conceptually) concatenating the right block to the left. However, the left and right pointers to the current elements cannot cross the borders of a block. When a left (right) block is completely processed (i.e. neutralized), a fresh left (right) block is acquired and processed.
3. **Cleanup:** At this point, at most p blocks remain unneutralized. Each author presents a different cleanup algorithm:
   — In [3], while unneutralized blocks remain, one block is taken from each end and neutralization is applied to them. Then, the unneutralized blocks are placed between the neutralized blocks. At most p blocks need to be swapped and this is done sequentially. Finally, sequential partition is applied to the range of blocks with unprocessed elements.
   — In [5], all unneutralized blocks are placed between the neutralized blocks. Then, the parallel partitioning algorithm is applied recursively to this range. The number of processors is divided by two in each call until there is only one processor or block. Finally, the remaining range is partitioned sequentially.

The main parallel phase takes $\Theta(n/p)$ parallel time. The cleanup phase takes $\Theta(bp)$ sequential time in [3]. Rather, in [5], it takes $\Theta(b \log p)$ parallel time.

3   The New Parallel Cleanup Phase

In this section, we present our cleanup algorithm. It avoids extra comparisons and swaps the elements fully in parallel. We have applied it on the top of STRIDED, BLOCKED and F&A. First, we introduce the terminology. Then, we present the data structure on which the algorithm relies. It follows the cleanup algorithm itself. Finally, we analyze the resulting STRIDED, BLOCKED and F&A algorithms.

Terminology. In the following, we shall use the following terms to describe our algorithm. A *subarray* is the basic unit of our algorithm and data structure. The *splitting position* v of an array is the position that would occupy the pivot after partitioning. A *frontier* separates a subarray in two consecutive parts that have different properties. A *misplaced element* is an element that must be moved by our algorithm. We denote by m the total number of misplaced elements and by M the total number of subarrays that may have misplaced elements.

The Case of BLOCKED. In this case, subarrays correspond with exactly one of the p pieces. Moreover, v can be easily known after the parallel phase. The frontier of a subarray corresponds with the position that would occupy the pivot after partitioning this subarray. Thus, a frontier defines a left and a right part. A misplaced element corresponds either to an element smaller than the pivot that is on the right of v (misplaced on the right) or to an element greater than the pivot that is on the left of v (misplaced on the left). The total number of subarrays that may have misplaced elements (M) is at most p.

The Case of F&A. In this case, a subarray corresponds to one block. The frontier separates a processed part from an unprocessed part. The processed part of left blocks is the left part and the processed part of right blocks is the right part. Though v is unknown after the parallel phase, it holds that v is in some range $V = [v_{\mathrm{beg}}, v_{\mathrm{end}}]$. A misplaced element corresponds either with a processed element that is in V or with an unprocessed element that is not in V.

We consider now the array of elements made by left blocks and the array of elements made by right blocks. We denote by β the global border that separates the left and right blocks (i.e. the point where no more blocks could be obtained). In left blocks, a misplaced element on the left is an unprocessed element in a position smaller than v_{beg} and a misplaced element on the right is a smaller element than the pivot in a position greater or equal than v_{beg}. Let μ_l be the total number of misplaced elements in left blocks and rank those misplaced elements starting to count from the leftmost towards the right. In right blocks, a misplaced element on the left is an element greater than the pivot in a position smaller or equal than v_{end} and a misplaced element on the right is an unprocessed element in a position greater than v_{beg}. Let μ_r be the total number of misplaced elements in right blocks and rank those misplaced elements starting to count from the rightmost towards the left. Note that $m = \mu_l + \mu_r$ and that the total number of subarrays that may have misplaced elements M is at most $2p$, because there at most p blocks that may contain unprocessed elements and there at most p blocks that may contain misplaced processed elements.

3.1   The Data Structure

We use a complete binary tree with M leaves (or the next power of two if M is not a power of two) to know which pairs of elements must be swapped. This tree

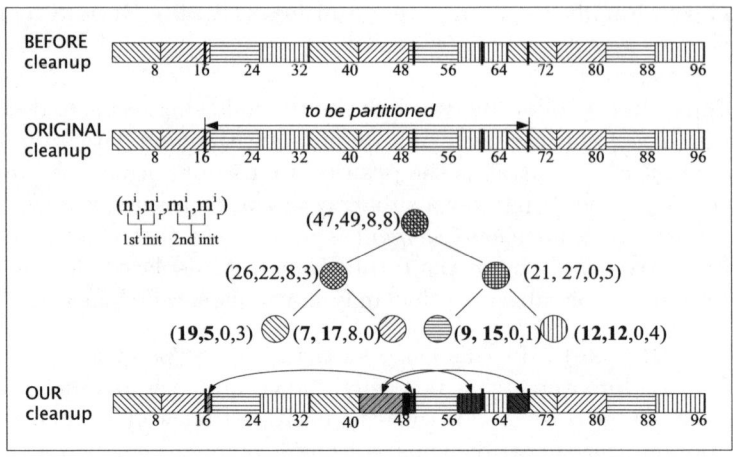

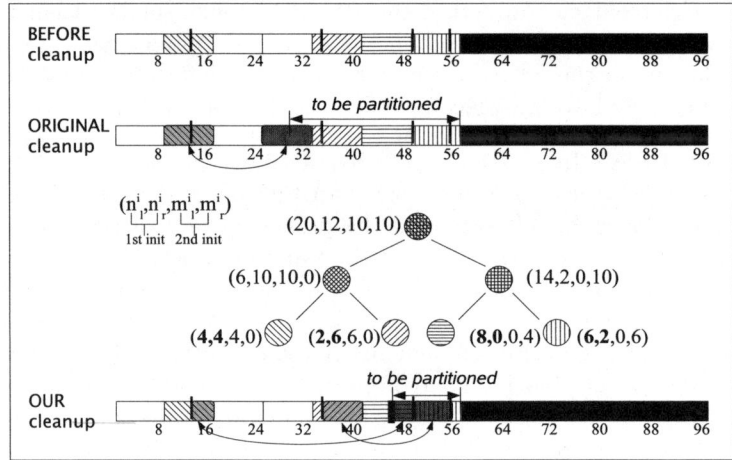

Fig. 1. Example of our data structure

is shared by all processors and is stored in an array (like a heap, which provides easy and efficient access to the nodes).

Each leaf stores information of the i-th subarray. Specifically, how many elements are misplaced to the left and to the right of its frontier (m_l^i and m_r^i) and how many elements are in the left and in the right to its frontier (n_l^i and n_r^i). The internal nodes accumulate the information of their children but do not add any new information. In particular, the root stores the information of the array made of all the subarrays in the leaves.

So, our tree data structure can be considered as a special kind of order-statistics tree in which the internal nodes have no information by themselves. An order-statistics tree (see e.g. [8, Sect. 14]) perform rank operations efficiently using the information of the size of the subtrees.

Figure 1 shows two instances of our tree data structure.

3.2   The Algorithm

Tree Initialization Phase. In this phase the tree is initialized. Specifically, two bottom-up traversals of the tree are needed. Only the first initialization of the leaves depends on the partition algorithm used in the main parallel phase.

1. **First initialization of the leaves.** In the case of BLOCKED, the leaves values n_l^i and n_r^i for each subarray i can be trivially computed during the parallel phase. In the case of F&A, the left (right) blocks that contain unprocessed elements can be easily known after the parallel phase. The left (right) blocks that contain misplaced elements but have already been processed can only be located between the left (right) unprocessed blocks. In order to locate the latter efficiently, we sort the unneutralized blocks with respect to the block position in the array. Then, we iterate on left (right) blocks (sequentially) to the left (right) of the border β until p neutralized blocks have been found or the leftmost (rightmost) unneutralized block has been reached.
2. **First initialization of the non-leaves.** Using a parallel reduce operation, each internal node computes its n_l^j and n_r^j values from its children. As a result, the root stores the number of left and right elements in the whole array. Thus, the splitting point v can be directly deduced.
3. **Second initialization of the leaves.** The leaves get the values m_l^i and m_r^i using n_l^i, n_r^i and v. At this point, it may turn out that some subarrays have no misplaced elements. This does not disturb the correctness of our algorithm.
4. **Second initialization of the non-leaves.** The number of misplaced elements for the internal nodes are computed using a second parallel reduction operation on m_l^j and m_r^j fields.

Parallel Swapping Phase. In this phase, our tree data structure is queried so that the misplaced elements can be swapped in parallel and no comparisons are needed. This phase is independent of the specific partitioning algorithm.

The total number of misplaced elements is used to distribute the work equally among the processors. A range of ranks $[r_i, s_i)$ of misplaced elements to swap is assigned to each processor. The elements are swapped in ascending rank. Specifically, the j-th misplaced left element is swapped with the j-th misplaced right element. To locate the first pair of elements to swap, respective rank queries are made to the tree. That is, a query is made for the r_i left misplaced element and another for the r_i right misplaced element. Misplaced elements are swapped as long as the rank s_i is not reached. If the rank s_i has not yet been reached but the current subarray has no more misplaced elements, the next subarray is fetched. Let c_i be the position in the tree corresponding to the current subarray. Then, the next subarray of left misplaced elements is in $c_i + 1$ and the next subarray of right misplaced elements is in $c_i - 1$.

Completion Phase. This phase depends on the specific partitioning algorithm. In the case of BLOCKED, the whole array has already been partitioned and we are done. In the case of F&A, some unprocessed elements may remain. When this happens, V is not empty and includes exclusively all the unprocessed elements.

In order to obtain a valid partition, we apply $\log p$ times our parallel partitioning algorithm recursively in V using blocks of half their original size until b elements or less remain. Sequential partition is applied to those remaining elements. Note that in each recursive call, the size of the problem is at most half the previous because at least p blocks have been fully processed.

3.3   Cost Analysis

Theorem 1. BLOCKED *and* F&A *perform exactly n comparisons when using our cleanup algorithm.*

Proof. The tree initialization and the parallel swapping phases perform no comparisons for both BLOCKED and F&A.

In the case of BLOCKED, the completion phase is empty. Therefore, no comparisons are performed during cleanup for BLOCKED and thus, comparisons are only performed during the main parallel phase, which are exactly n.

In the case of F&A, after the first main parallel phase $n - |V|$ elements have been compared and $|V|$ have remained unprocessed. In the next recursive step, V is the input. Besides, elements can only be compared during a certain parallel phase and at most once. All the elements must be eventually compared because our algorithm produces a valid partition. Thus, our cleanup algorithm makes exactly $|V|$ comparisons, and n comparisons are needed as a whole.

Lemma 1. *The tree initialization phase takes $\Theta(\log p)$ parallel time for* BLOCKED *and* F&A.

Proof. The algorithm-independent part takes $\Theta(\log p)$ parallel time because all the work is done in parallel, and is dominated by the two parallel reductions, which can be performed in logarithmic parallel time [9].

In the case of BLOCKED, the algorithm-dependent part takes constant parallel time because each leaf can be initialized trivially and in parallel.

In the case of F&A, the algorithm-dependent part takes $\Theta(\log p)$ parallel time, because $2p$ elements are sorted and this takes $\Theta(\log p)$ parallel time using p processors [10].

Thus, in both cases, the total cost is $\Theta(\log p)$ parallel time.

Lemma 2. *The parallel swapping phase performs exactly $m/2$ swaps and requires $\Theta(m/p)$ parallel time. In the case of* F&A*, this parallel time is $O(b)$.*

Proof. There are m misplaced elements after the main parallel phase. The parallel swapping phase swaps pairs of misplaced elements so that their final position is not misplaced. Therefore, $m/2$ swap operations are needed. Besides, the pairs are evenly divided among the p processors. Thus swapping all of them takes $\Theta(m/p)$ parallel time. In the case of F&A, $m \le 2bp$, thus parallel swapping takes $O(b)$ parallel time.

Theorem 2. *The cleanup phase takes $\Theta(m/p + \log p)$ parallel time for* BLOCKED *and the whole partition takes $\Theta(n/p + \log p)$ parallel time.*

Table 1. Summary of costs for BLOCKED and F&A algorithms

	BLOCKED					
	total comparisons		total swaps		parallel time	
	original	tree	original	tree	original	tree
main	n		$\leq n/2$		$\Theta(n/p)$	
cleanup	$v_{max} - v_{min}$	0	$m/2$	$m/2$	$\Theta(v_{max} - v_{min})$	$\Theta(m/p + \log p)$
total	$n + v_{max} - v_{min}$	n	$\leq \frac{n+m}{2}$	$\leq \frac{n+m}{2}$	$O(n)$	$\Theta(n/p + \log p)$

	F&A					
	comparisons		swaps		parallel time	
	original	tree	original	tree	original	tree
main	$n - |V|$		$\leq \frac{n-|V|}{2}$		$\Theta(n/p)$	
cleanup	$\leq 2bp$	$|V|$	$\leq 2bp$	$\leq m/2 + |V|$	$\Theta(b \log p)$	$\Theta(\log^2 p + b)$
total	$\leq n + 2bp$	n	$\leq \frac{n-|V|}{2} + 2bp$	$\leq \frac{n+m}{2} + |V|$	$\Theta(n/p + b \log p)$	$\Theta(n/p + \log^2 p)$

Proof. From Lemmas 1 and 2 follows that the cleanup phase takes $\Theta(m/p + \log p)$ parallel time for BLOCKED. Given that $m = O(n)$, the whole BLOCKED algorithm takes $\Theta(n/p + \log p)$ parallel time in the average and in the worst-case.

Theorem 3. *Consider $p \leq b$. The cleanup phase takes $\Theta(\log^2 p + b)$ parallel time for F&A and the whole partition takes $\Theta(n/p + \log^2 p + b)$ parallel time.*

Proof. F&A takes $T(n, p) = \Theta(n/p) + C(b, p)$, where $C(b, p)$ is the cost of our cleanup algorithm. $C(b, p) = b + \log p + T'(b/2, \log p)$ parallel time, and T' is defined by the following recurrence:

$$T'(b, i) = \begin{cases} O(3b + \log p) + T'(b/2, i - 1) & \text{if } i > 1, \\ O(2b/p) & \text{otherwise.} \end{cases}$$

There are $2\beta p$ blocks at the beginning of each recursive step and $\log p - 1$ recursive steps are needed. Thus, $C(b, p) = O(\log^2 p + b)$ parallel time.

Theorem 3 improves previous bounds for F&A (provided $\log p \leq b$, which is of practical relevance).

Table 1 summarizes worst-case results for BLOCKED and F&A algorithms.

4   Implementation

Since implementations of STRIDED were not available, we have resorted to implement it ourselves. We have also implemented our BLOCKED variant, which improves STRIDED cache performance. As for F&A, we have taken its implementation from MCSTL 0.7.3-beta and we have implemented it ourselves.

Our implementation of F&A and the one in MCSTL differ in the following: a) ours statically assigns the initial work and, so, avoids mutual exclusion here; b) ours does not use `volatile` variables and critical regions are slightly simpler; and c) ours avoids redundant comparisons using a better book-keeping.

On the other hand, we have implemented our cleanup algorithm on the top of the previous four algorithms.

The implementation is available at http://www.lsi.upc.edu/~lfrias. It uses C++ and OpenMP. Besides, it follows the specification of the `partition` function of the STL, so that it can be used instead of the sequential implementation.

5   Experimental Analysis

We have analyzed STRIDED, BLOCKED, and F&A with and without our cleanup algorithm.

The tests have been run on a machine with 4 GB of main memory and two sockets, each one with an Intel Xeon quad-core processor at 1.66 GHz with a shared L2 cache of 4 MB shared among two cores. Thus, there are 8 cores in total. We have used the GCC 4.2.0 compiler with the -O3 optimization flag.

All tests have been repeated 100 times; figures show averages.

Basic Evaluation. We have first analyzed the speedup of partitioning in parallel a large number of random integers. The speedup is always measured with respect to the sequential partition algorithm of the STL. The block size b has been set to 10^4 (see the reason below).

Figure 2 shows the results. In this figure, Strided refers to our implementation of STRIDED, BlockedStrided refers to our implementation of BLOCKED, F&A_MCSTL refers to the MCSTL implementation of F&A, F&A refers to our own implementation of F&A. We add the suffix _tree to the previous labels to refer to the algorithm with our modified parallel cleanup phase.

These results show that F&A is better than BLOCKED, which is better than STRIDED. Whereas the speedup of STRIDED is nonexistent for more than two threads, BLOCKED performs reasonably well and our F&A implementation achieves some better results than the MCSTL F&A. Besides, using our cleanup phase maintains the same speedups for STRIDED and BLOCKED and improves slightly the speedup of F&A, making it almost perfect for up to 4 threads.

The awful performance of STRIDED is due to its high cache miss ratio; its behavior clearly contrasts with BLOCKED (which uses blocks of elements rather than individual elements).

In order to understand the loss of performance when using more than 5 threads we have devised two new experiments. The first one reproduces the previous experiment but uses a slower comparison function. Its results are shown in Fig. 3. In this case, all algorithms show similar behavior and excellent speedups with up to 8 threads. Specifically, there is not much of a difference whether our cleanup phase is used or not. Our second experiment has consisted in measuring the speedup of a trivial parallel program to compute the sum of two arrays. The resulting speedups (not shown) are also not optimal for the biggest number of threads. So, we can conclude that memory bandwidth is limiting the efficiency of the partitioning algorithms, which are demanding with regard to I/O.

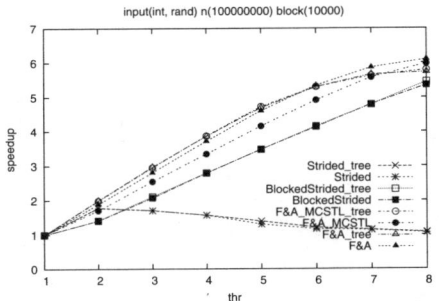

Fig. 2. Parallel partition speedup, $n = 10^8$ and $b = 10^4$

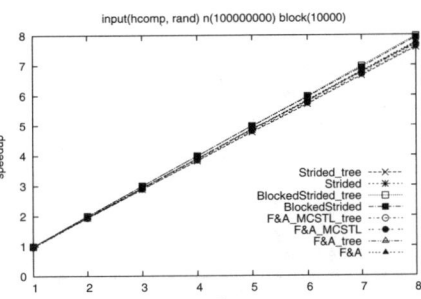

Fig. 3. Parallel partition speedup for costly $<$, $n = 10^8$ and $b = 10^4$

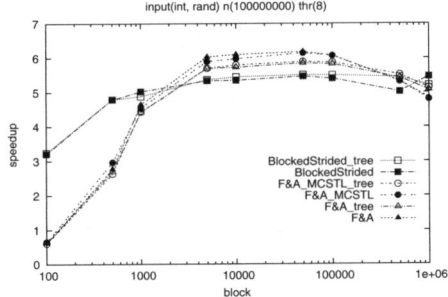

Fig. 4. Parallel partition with varying block size, $n = 10^8$ and num_threads $= 8$

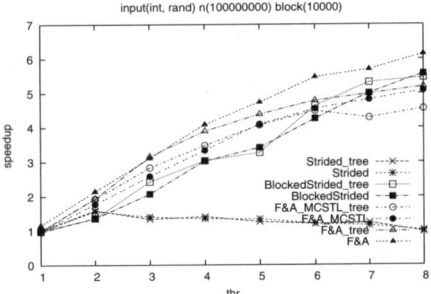

Fig. 5. Parallel quickselect speedup, $n = 10^8$ and $b = 10^4$

Influence of Block Size. Several algorithms rely on a block size parameter b. In order to determine its optimal value, we have run various tests, with 8 threads and different values of b. The results in Fig. 4 show that, except for very small block sizes, the performance is not much affected. Besides, given that for smaller input sizes, big block sizes are not convenient, our selection has been $b = 10^4$.

Operations Count. In order to analyze the behavior of the cleanup phase in more detail, we have counted swap and comparison operations. Figures 6 and 7 show respectively the number of extra comparisons and swaps with respect to the sequential implementation. They are depicted divided by the block size.

Figure 6 gives an experimental proof of Theorem 1. Combining our cleanup algorithm with the original MCSTL algorithm does not achieve the optimality in the number of comparisons, because this implementation makes extra comparisons whenever a new block is fetched in the main parallel phase. Specifically, our experiments show that two comparisons are repeated per block in the average.

Figure 7 shows that our cleanup algorithm does not need more swaps than the original cleanup algorithms. Essentially, the same number of extra swaps are needed. In the case of F&A, we could not give such an equality analytically.

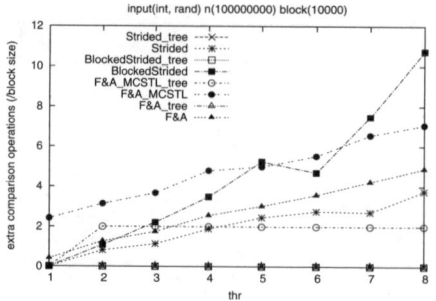
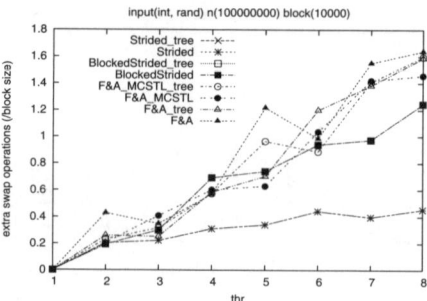

Fig. 6. Number of extra comparisons, $n = 10^8$ and $b = 10^4$

Fig. 7. Number of extra swaps, $n = 10^8$ and $b = 10^4$

As a by product, these results show that the number of misplaced elements resulting from the parallel phase is really small, no matter the partitioning algorithm. In particular, STRIDED is the algorithm that performs less extra operations (for a random input of integers). However, its performance is the worst because of its bad cache usage.

Application: Quickselect. Quicksort and quickselect are typical applications of partitioning. As quicksort offers two (not exclusive) ways to be parallelized — parallelizing the partitioning and parallelizing the independent work by divide and conquer—, we found more interesting analysing quickselect. For this test, we have made that the STL `nth_element` function calls the parallel partitioning algorithms in this paper instead of the sequential (unless the array is small).

The results are shown in Fig. 5. These are coherent with those of partition but different given that the relative behavior between the algorithms changes slightly with the size of the input. First, our F&A implementation advantage increases. Second, our cleanup algorithm harms a little F&A based quickselect. Indeed, in our experiments we have observed that for a big number of threads and as input gets smaller, using our cleanup algorithm with F&A is counterproductive. Finally, Fig. 5 shows that the simple BLOCKED algorithm performs quite well.

6   Conclusions

In this paper we have presented, implemented and evaluated several parallel partitioning algorithms suitable for multi-core architectures.

From an algorithmic point of view, we have described a novel cleanup parallel algorithm that does not disregard comparisons made during the parallel phase. This cleanup has successfully been applied to three partitioning algorithms: STRIDED, BLOCKED (a cache-aware implementation of the former) and F&A. In the case of STRIDED and BLOCKED, a benefit of our cleanup is reducing its parallel time in the worst case from $\Theta(n)$ to $\Theta(n/p + \log p)$. In the case of F&A, we have shown how to modify it to reduce its parallel time from

$\Theta(n/p + b\log p)$ to $\Theta(n/p + \log^2 p)$. Unlike their original versions, these algorithms perform the minimal number of comparisons when using our cleanup phase.

As automatic parallelization is still limited, and as parallel programming is hard and expensive, the use of parallel libraries is a simple way to benefit from multi-core processors. From this engineering perspective, we have contributed carefully designed implementations of the afore mentioned algorithms that are compliant with the specification of STL `partition`.

Finally, and from an experimental point of view, we have conducted an evaluation to compare those algorithms and implementations. According to our experiments, the partitioning algorithm of choice is F&A, because it scales nicely. Moreover, our implementation performs slightly better than the one in MCSTL. However, the results also show that, in practice, the benefits of our cleanup algorithm are limited. This happens because the number of misplaced elements after the parallel phase is very small.

Our experiments also show that I/O between the memory and the processor limits the performance achieved by parallel implementations as the number of threads increases. It remains to be further investigated how these results change for a bigger number of available cores or/and memory bandwidth.

References

1. Heidelberger, P., Norton, A., Robinson, J.T.: Parallel quicksort using fetch-and-add. IEEE Trans. Comput. 39(1), 133–138 (1990)
2. Francis, R.S., Pannan, L.J.H.: A parallel partition for enhanced parallel quicksort. Parallel Computing 18(5), 543–550 (1992)
3. Tsigas, P., Zhang, Y.: A simple, fast parallel implementation of quicksort and its performance evaluation on SUN enterprise 10000. In: 11th Euromicro Workshop on Parallel, Distributed and Network-Based Processing (PDP 2003), pp. 372–381 (2003)
4. Liu, J., Knowles, C., Davis, A.: A cost optimal parallel quicksorting and its implementation on a shared memory parallel computer. In: Pan, Y., Chen, D.-x., Guo, M., Cao, J., Dongarra, J. (eds.) ISPA 2005. LNCS, vol. 3758, pp. 491–502. Springer, Heidelberg (2005)
5. Singler, J., Sanders, P., Putze, F.: The Multi-Core Standard Template Library. In: Kermarrec, A.-M., Bougé, L., Priol, T. (eds.) Euro-Par 2007. LNCS, vol. 4641, pp. 682–694. Springer, Heidelberg (2007)
6. International Standard ISO/IEC 14882: Programming languages — C++. 1st edn. American National Standard Institute (1998)
7. Gottlieb, A., Grishman, R., Kruskal, C.P., McAuliffe, K.P., Rudolph, L., Snir, M.: The NYU ultracomputer - designing a MIMD, shared-memory parallel machine. In: ISCA 1998: 25 years of the international symposia on Computer architecture (selected papers), pp. 239–254. ACM Press, New York, NY, USA (1998)
8. Cormen, T.H., Leiserson, C., Rivest, R., Stein, C.: Introduction to algorithms, 2nd edn. The MIT Press, Cambridge (2001)
9. Kumar, V.: Introduction to Parallel Computing. Addison-Wesley, Boston, MA, USA (2002)
10. JáJá, J.: An introduction to parallel algorithms. Addison-Wesley, Redwood City, CA, USA (1992)

Broadword Implementation of Rank/Select Queries

Sebastiano Vigna

Dipartimento di Scienze dell'Informazione, Università degli Studi di Milano, Italy

Abstract. Research on succinct data structures (data structures occupying space close to the information-theoretical lower bound, but achieving speed similar to their standard counterparts) has steadily increased in the last few years. However, many theoretical constructions providing asymptotically optimal bounds are unusable in practise because of the very large constants involved. The study of practical implementations of the basic building blocks of such data structures is thus fundamental to obtain practical applications. In this paper we argue that 64-bit and wider architectures are particularly suited to very efficient implementations of rank (counting the number of ones up to a given position) and select (finding the position of the i-th bit set), two essential building blocks of all succinct data structures. Contrarily to typical 32-bit approaches, involving precomputed tables, we use pervasively *broadword* (a.k.a. SWAR—"SIMD in A Register") programming, which compensates the constant burden associated to succinct structures by solving problems in parallel in a register. We provide an implementation named `rank9` that addresses 2^{64} bits, consumes less space and is significantly faster then current state-of-the-art 32-bit implementations, and a companion `select9` structure that selects in nearly constant time using only access to aligned data. For sparsely populated arrays, we provide a simple broadword implementation of the Elias–Fano representation of monotone sequences. In doing so, we develop broadword algorithms for performing selection in a word or in a sequence of words that are of independent interest.

1 Introduction

A *succinct* data structure (e.g., a succinct tree) provides the same (or a subset of the) operations of its standard counterpart, but occupies space that is asymptotically near to the information-theoretical lower bound. A classical example is the $(2n + 1)$-bit representation of a binary tree with n internal nodes proposed by Jacobson [1]. Recent years have witnessed a growing interest in succinct data structures, mainly because of the explosive growth of information in various types of text indexes (e.g., large XML trees).

In this paper we discuss practical implementations of two basic building blocks— *rank* and *select*. Given an array B of n bits, we are interesting in *ranking* the i-th position (computing the number of ones up to that position) and *selecting* the i-th bit set to one.

It is known that with an auxiliary data structure occupying $o(n)$ bits it is possible to answer both rank and select queries in constant time (see, e.g., [2] and references therein for an up-to-date overview). A complementary approach discards the bit vector altogether, and stores explicitly the positions of all ones in a *fully indexable dictionary*, which represents a set of integers making it possible to access the k-th element of the

C.C. McGeoch (Ed.): WEA 2008, LNCS 5038, pp. 154–168, 2008.
© Springer-Verlag Berlin Heidelberg 2008

set in increasing order, and to compute the number of elements of the set smaller than a given integer. These two operations correspond to selection and ranking over the original bit vector: by using succinct dictionaries, it is possible to reduce significantly the space occupancy with respect to an explicit bit vector in the sparse case.

We start from concerns similar to those of González, Grabowski, Mäkinen and Navarro [3]: it is unclear whether these solutions are usable in practise. The asymptotic notation is often hiding constants so large that before the asymptotic advantage actually kicks in, the data structure is too large. In this case, it is rather fair to say that the result is interesting mathematically, but has little value as a data structure.

This problem is made even worse by the fact that succinct data structure are exactly designed for very large data sets, which are useless if the access to the data is slow. For instance, the authors of [3] argue that word-aligned, $O(n)$ solutions are extremely more efficient than the optimal counterparts, and that for perfectly reasonable data sizes they actually occupy less space. To solve locally (wordwise) rank and select the author use *population counting* techniques—precomputed tables containing, say, the number of bits set to one in each possible byte.

In this paper we depart from this approach, arguing that on modern 64-bit architecture a much more efficient approach uses *broadword programming*. The term "broadword" has been introduced by Don Knuth in the fascicle on bitwise manipulation techniques of the fourth volume of *The Art of Computer Programming* [4]. Broadword programming uses large (say, more than 64-bit wide) registers as small parallel computers, processing several pieces of information at a time. An alternative, more traditional name for similar techniques is SWAR ("SIMD Within A Register"), a term coined by Fisher and Dietz [5]. One of the first techniques for manipulating several bytes in parallel were actually proposed by Lamport [6].

For instance, a broadword algorithm for *sideways addition* (counting the number of ones in a register—of course, part of computing ranks) was presented in the second edition of the textbook "Preparation of Programs for an Electronic Digital Computer", by Wilkes, Wheeler, and Gill, in 1957. One of the contributions of this paper is a broadword counterpart to select bits in a word.

The main advantage of broadword programming is that we gain more speed as word width increases, with almost no effort, because we can process more data in parallel. Note that, in fact, broadword programming can even be used to obtain better asymptotic results: it was a basic ingredient for the success of *fusion trees* in breaking the information-theoretical lower bound for integer sorting [7].

Using broadword programming, we are able to fulfil at the same time the following apparently contradictory goals:

- address 2^{64} bits[1]
- use less space;
- obtain faster implementations.

A second concern we share with the authors of [3] is that of minimising cache misses, as memory access and addressing is the major real bottleneck in the implementation of rank/select queries on large-size arrays. To that purpose, we *interleave* data from

---

[1] All published practical implementations we are aware of address 2^{32} bits; this is a serious limitation, in particular for compressed structures.

different tables so that usually a single cache miss is sufficient to find all information related to a portion of the bit array (we wish to thank one of the anonymous referees for pointing out that the idea already appeared in [8]).

We are also very careful of avoiding tests whenever possible. Branching is a very expensive operation that disrupts speculative execution, and should be avoided when possible. All the broadword algorithms we discuss contain no test and no branching.

We concentrate on 64-bit and wider architecture, but we cast all our algorithms in a 64-bit framework to avoid excessive notation: the modification for wider registers are trivial. We have in mind modern processors (in particular, the very common Opteron processor) in which multiplications are extremely fast (actually, because the clock is slowed down in favour of multicores), so we try to use them sparingly, but we allow them as constant-time operations. While this assumption is debatable on a theoretical ground, it is certainly justified in practise, as experiments show that on the Opteron replacing multiplications by shifts and additions, *even in very small number*, is not competitive.

The C++/Java code implementing all data structures in this paper is available under the terms of the GNU Lesser General Public License at http://sux.dsi.umimi.it/.

2 Notation

Consider an array b of n bits numbered from 0. We write b_i for the bit of index i, and define

$$\mathrm{rank}_b(p) = \sum_{0 \le i < p} b_i \qquad 0 \le p \le n,$$

that is, as the number of ones up to position p, excluded, and

$$\mathrm{select}_b(r) = \max\{\, p < n \mid \mathrm{rank}_b(p) \le r \,\}, \qquad 0 \le r < \mathrm{rank}_b(n),$$

that is, as the position of the one of index r, where ones are numbered starting from 0. When b is clear from the context, we shall omit it.

Note that in the literature there is some variation in the choice of indexing (starting from one or zero) and in the exact definition of these two primitives (including or not the one at position p in $\mathrm{rank}(p)$).

To be true, we couldn't find the 0-based definitions given above in the literature, but they are extremely natural for several reasons:

- As it always happen with modular arithmetic, starting with 0 avoids falling into "off-by-one hells". This consideration is of course irrelevant for a theoretical paper, but we are in a different mindset.
- In this way, $\mathrm{rank}(p)$ can be interpreted as $\mathrm{rank}[0 \ldots p)$—counting the ones in the semiopen interval $[0 \ldots p)$. Counting from zero and semiopen intervals are extremely natural in programming (actually, Dijkstra felt the need to write a note on the subject [9]).
- We can define easily, and without off-by-ones, operators such as $\mathrm{count}_b[p \ldots q) = \mathrm{rank}(q) - \mathrm{rank}(p)$.

In any case, it is trivial to compute other variations of rank and select by suitably offsetting the arguments and the results.

We use $a \setminus b$ to denote integer division of a by b, $\gg$ and $\ll$ to denote right and left (zero-filled) shifting, &, | and $\oplus$ to denote bit-by-bit not, and, or, and xor; $\bar{x}$ denotes the bit-by-bit complement of x. We pervasively use precedence to avoid excessive parentheses, and we use the same precedence conventions of the C programming language: arithmetic operators come first, ordered in the standard way, followed by shifts, followed by logical operators; $\oplus$ sits between | and &.

We use L_k to denote the constant whose ones are in position $0, k, 2k, \ldots$ that is, the constant with the *lowest* bit of each k-bit subword set (e.g, $L_8 = 0x0101010101010101$). This constant is very useful both to spread values (e.g., $12 * L_8 = 0x1212121212121212$) and to sum them up, as it generates cumulative sums of k-bit subwords if the values contained in each k-bit subword, when added, do not exceed k bits. We use H_k to denote $L_k \ll k - 1$, that is, the constant with the *highest* bit of each k-bit subword set (e.g, $L_8 = 0x8080808080808080$).

Our model is a RAM machine with d-bit words that performs logic operations, additions, subtractions and multiplications in unit time using 2-complement arithmetic. We note that albeit multiplication can be proven to require $O(\log d)$ basic operations, modern processors have very fast multiplication (close to one cycle), so designing broadword algorithms without multiplications turns out to generate slower code.

3   rank9

We now introduce the layout of our data structure for ranking, which follows a traditional two-level approach but uses broadword sideways addition (Algorithm 1) for counting inside a word and interleaving to reduce cache misses. We assume the bit array b is represented as an array of words of 64 bits. The bit of position p is located in the word of index $p \setminus 64$ at position p mod 64, and we number bits inside each word in *little-endian* style.

To each subsequence of eight words starting at bit position p, called a *basic block*, we associate two words:

- the first word (first-level count) contains rank(p);
- the second contains the seven 9-bit values (second-level counts) rank($p + 64k$) − rank(p), for $1 \leq k \leq 7$, each shifted left by $9(k − 1)$ bits.

First and second level counts are stored in *interleaved form*—each first-level count is followed by its second-level counts. When we have to rank a position p living in the word $w = p \setminus 64$, we have just to sum the first-level count of the sequence starting at $w \setminus 8$, possibly a second-level count (if w mod $8 \neq 0$) and finally invoke sideways addition on the word containing p, suitably masked. Note that this apparently involves a test, but we can get around the problem as follow:

$$s \gg (t + (t \gg 60 \; \& \; 8)) * 9 \; \& \; 0x1FF,$$

where s is the second-level count and $t = w$ mod $8 − 1$. When w mod $8 = 0$, the expression $t \gg 60 \; \& \; 8$ has value 8, which implies that s is shifted by 63, obtaining zero (we are not using the most significant bit of s).

We call the resulting structure rank9 (the name, of course, is inspired by the fact that it stores 9-bit second-level counts). It requires just 25% additional space, and ranks are evaluated with at most *two* cache misses, as when the first-level count is loaded by the L1 cache, the second-level count is, too. No tests or precomputed tables are involved.[2]

The only dependence on the word length d is in the first cumulative phase of sideways addition. We need to cumulate at least b bits, where b is a power of two enough large to express d, that is, $b = \lceil \log d \rceil$. Thus, this phase requires $O(\log \log d)$ step. However, since Algorithm 1, with suitable constants, works up to $d = 256$, it can be considered constant time to all practical purposes (as we will never have 2^{256} bits).

Algorithm 1. The classical broadword algorithm for computing the sideways addition of x in $O(\log \log d)$ steps. The first step leaves in each pair of bits the number of ones originally contained in that pair. The following steps gather partial summations and, finally, the multiplication sums up them all.

---

0    $x = x - (x\ \&\ \text{0xAAAAAAAAAAAAAAAA}) \gg 1$
1    $x = (x\ \&\ \text{0x3333333333333333}) + ((x \gg 2)\ \&\ \text{0x3333333333333333})$
2    $x = (x + (x \gg 4))\ \&\ \text{0x0F0F0F0F0F0F0F0F}$
3    $x * L_8 \gg 56$

---

4 k-Bit Comparisons

Given x and y, consider them as sequences of $64 \setminus k$ (un)signed k-bit values. We would like to operate on them so that, in the end, each k-bit block contains 1 in the leftmost position iff the corresponding pair of k-bit values is ordered. At that point, it is easy to count how many ones are present using a multiplication. Knuth describes a broadword expression to this purpose, using the properties of the median (a.k.a. majority) ternary operator [4]. We just recall the operators we will be using in what follows (the subscript denotes the block size, while a superscript "u" denotes unsigned comparison):

$$x \leq_k^u y := \left(\left(((y \mid H_k) - (x\ \&\ \overline{H_k})) \mid x \oplus y \right) \oplus (x\ \&\ \overline{y}) \right) \&\ H_k$$

$$x \leq_k y := \left(((y \mid H_k) - (x\ \&\ \overline{H_k})) \oplus x \oplus y \right) \&\ H_k$$

$$x >_k^u 0 := \left(\left(((x \mid H_k) - L_k) \mid x \right) \right) \&\ H_k$$

5   select9

We would like to build upon rank9 selection capabilities. To this purpose, we work backwards, starting from selection in a word, moving to selection in a sequence of

---

[2] Of course, if more than 64 bits per word are available, more savings are possible: for instance, for 128-bit processors rank16, which uses 16-bit second-level counts, requires just 12.6% additional space.

words, and finally getting to selection over the bit array. In rank9 we conceded a shift-based access to non-aligned subwords, but in the case of select several accesses are needed (even in the optimal, non-aligned data structures), so we will limit ourselves to access only correctly aligned subwords of size $d/2^i$ (except, of course, for rank9 access).

The starting consideration for our select-in-a-word broadword algorithm is the observation that at the end of Algorithm 1 we use just the most significant byte of a multiplication that provides much more information—namely, the cumulative sums of the number of ones contained in each byte. If we compare each of these numbers with the desired index r, we can easily locate the byte containing the r-th one. With a typical broadword approach, we then solve the problem in the relevant byte in a similar manner.

We are now ready to introduce Algorithm 2. In the first lines we follow exactly Algorithm 1, building the bytewise cumulative sums s. Then, we compare in parallel each cumulative sum with r: the number of positive results is exactly the index of the byte containing the bit of rank r, so we extract it in b already multiplied by eight. To obtain the bytewise rank ℓ, we subtract from r the value found in the byte starting at bit $b - 8$ (if $b = 0$, $\ell = r$).

We now compute a word z that contains eight copies of the byte starting at position b (the one containing the bit of rank r); however, from the j-th copy we just keep bit j. We now compare each byte in parallel with zero, which make it possible to compute, with a multiplication by L_8, the rank of each bit. We compare the cumulative sums with eight copies of ℓ; again, the number of positive results is the index of the ℓ-th one, which we return, summed with b.

We note that, similarly to sideways addition, we need to compute the number of ones in subwords of size $\lceil \log d \rceil$. Now, however, we have another constraint: $\lceil \log d \rceil$ copies of each sum must fit into a word, that is, $\lceil \log d \rceil^2 \leq d$. This constraint cannot be satisfied with d a power of two unless $d \geq 64$.

Again, Algorithm 2 requires $O(\log \log d)$ operations in the initial phase, and up to $d = 256$ the only modifications required are suitable changes to the constants. Moreover, the constant operations significantly outnumber those of the initial phase. Finally, the algorithm contain several multiplications by L_8: they can be replaced by less than $\log d$ shifts and adds, as the number of ones in L_8 is very low.

Algorithm 2. Computes the index of the r-th one in x ($r < 2^{2^{\lceil \log \log d \rceil}}$). If no such bit exists, computes 72.

```
0   s = x − ((x & 0xAAAAAAAAAAAAAAAA) ≫ 1)
1   s = (s & 0x3333333333333333) + ((x ≫ 2) & 0x3333333333333333)
2   s = ((s + (s ≫ 4)) & 0x0F0F0F0F0F0F0F0F) ∗ L₈
3   b = ((s ≤₈ r ∗ L₈) ≫ 7) ∗ L₈ ≫ 53 & 7̄
4   ℓ = r − (((s ≪ 8) ≫ b) & 0xFF)
5   s = (((x ≫ b & 0xFF) ∗ L₈ & 0x8040201008040201 >₈ 0) ≫ 7) ∗ L₈
6   b + (((s ≤₈ ℓ ∗ L₈) ≫ 7) ∗ L₈ ≫ 56)
```

We now approach the problem of constant-time selection inside a block of `rank9`. The idea, by now familiar to the reader, is to locate the right word using parallel comparisons. More precisely, if s contains the subcount word and we have to locate the bit of rank r we can just compute

$$o = ((s \leq_9^u r * L_9) \gg 8) * L_9 \gg 54 \,\&\, 7$$

to know the offset in the block of the word containing the bit, and

$$r - (s \gg (o - 1 \,\&\, 7) * 9 \,\&\, 0x1FF)$$

to know the rank inside the word. Note that $o - 1 \,\&\, 7$ is 63 when $o = 0$, which implies that no correction is performed if the bit belongs to the first word in the block.

Binary-search selection. At this point, we could follow the steps of [3] and just perform a binary search over blocks, followed by the broadword block search we just described. Moreover, we could add a simple, one-level inventory that would help locating more quickly the region in which perform a binary search: we call this approach a *hinted bsearch*. In the experimental part, however, we will see that while (hinted) binary searches have excellent performances on evenly distributed arrays, they give worst results on uneven distributions.

Selecting in $d\sqrt{d}$ words. In general, the approach we described provides selection in $\sqrt{d}$ words. We are now going to use the broadword approach to provide selection in practical constant time inside $d\sqrt{d}$ words.

The idea is very simple: since by broadword comparison we can quickly locate, in a list of increasing integers, the first integer larger than a given integer x, given a sequence of $\sqrt{d}$ basic blocks, that we shall call an *intermediate block*, we can list the $\sqrt{d}$ first-level count of each block and perform selection by first locating the correct basic block, and then operating as we previously described. Note that since we need just to store the difference of each first-level count from the first one, we need very few bits ($2 \log d$), so a constant number of words will suffice. In our main example, we use two words to store eight 16-bit values containing the first-level counts.

To get to $d\sqrt{d}$ words (512, in our example) we repeat again the same trick, but now we consider a sequence of $\sqrt{d}$ intermediate blocks, called an *upper block*, and record the $\sqrt{d}$ first-level counts of the first basic block of each intermediate block. Using the parallel comparison operator as we did in the first part of this section, and using suitable constants (e.g., L_{16}) we can find in constant time the intermediate block and, again in constant time, the basic block containing the bit we are interested in.

We note that the cost of recording this information is very low: when $d = 64$ we need 16 bits for each basic block, which contains 512 bits.

Selecting over the whole bit array. Our interest in selecting over $d\sqrt{d}$ words stems from the fact that, by keeping track of the position of one each $d\sqrt{d}$ bits in a primary inventory space and allocating with care some secondary inventory, we can reduce in constant time our problem to selection in $d\sqrt{d}$ words.

More precisely, we record the position of each $d\sqrt{d}$-th bit. In our example, in the worst case (density close to 1) this information requires 12.5% additional space. Then, we allocate one word each α words for a secondary inventory. Consider two bits that appear consecutively in the primary inventory (in particular, their indices differ by $d\sqrt{d}$), and let p and q be their positions. For the $d\sqrt{d}$ bits inbetween we have at our disposal

$$q\backslash(\alpha d) - p\backslash(\alpha d)$$

words. If this number is at least $d\sqrt{d}$, we can record the position of each bit. Otherwise, we can describe the position of each bit in this range using

$$\log(\alpha d^2\sqrt{d}) = \log\alpha + \frac{5}{2}\log d$$

bits, so as long as

$$\log(\alpha d^2\sqrt{d}) = \log\alpha + \frac{5}{2}\log d \leq \frac{d}{2}$$

we can still describe the position of each bit using the upper and lower half of each word (note that, as we discussed, we are purposely avoiding to manipulate non-aligned subwords). The process can continue if there is enough space to describe the position of all $d\sqrt{d}$ bits: depending on α, more or less subword sizes can be used.

For the case $d = 64$, $\alpha = 4$ is a particularly good value because it generates an equality in the inequality

$$\log a + \frac{5}{2}\log d - 1 \leq \frac{d}{4},$$

which means that we can get to the point where we are recording the positions of all $d\sqrt{d} = 512$ bits using 128 words of secondary storage. Since these 128 words correspond to $512 = d\sqrt{d}$ words in our bit array, below this size we can use the broadword techniques described in the previous paragraph.

All in all, `select9` uses an underlying `rank9` structure, plus additional data occupying at most 37.5% of the original bit array. To rank a bit r, we first compute the positions p and q of the bit $r' = r - (r \bmod d\sqrt{d})$ and of the bit $r' + d\sqrt{d}$, respectively, using the primary inventory. Then, we compute the *span* associated to r'

$$s = (q\backslash d)\backslash\alpha - (p\backslash d)\backslash\alpha,$$

which represent the number of words from the secondary inventory we can use for the $d\sqrt{d}$ bits after r'. Finally, to locate the position of the bit of position r, we proceed as follows:

1. if $s < 2$, the bit can be located inside the basic block to which r' belongs;
2. if $s < 16$, the bit can be located using a two-word index collecting the first-level counts of an intermediate block;
3. if $s < 128$, the bit can be located using an eighteen-word two-level index collecting the first-level counts of an upper block, organised as we described above; note that by storing the two indices consecutively, we effectively interleave the data, generating a single cache miss for both reads;
4. if $s < 256$, we store explicitly the offset of each bit from r' (whose rank is known by first-level counting) in 16 bits;

5. if $s < 512$, we store explicitly the offset of each bit in 32 bits;
6. otherwise, we have enough space to store explicitly all bit positions.

It is easy to check that the choice $\alpha = 4$ makes it possible to store any of the alternative information required by the data structure.

In the worst case, `select9` will generate four cache misses: one to access the primary inventory, one to access the secondary inventory, one to locate the correct basic block, and one to select inside a basic block. The only test required when performing selection is comparing the value of s with the constants above.[3]

6 `simple`

The idea of broadword selection can be easily extended to a *bit search* algorithm that quickly locates a bit in a bit array. Assuming we want to locate the bit of rank r in a sequence of words, we simply have to load the first word into x and loop around the first three lines of Algorithm 2: if $r < s \gg 56$, we exit the loop and proceed as usual. Otherwise, we load x with the content of the next word, decrease r by $s \gg 56$ and iterate again.

Armed with this tool, we implement `simple`, an almost naive but surprisingly efficient select structure that does not depend on `rank9`. The structure is a two-level inventory similar to the `darray` dense select structure described in [10], but it has been suitably modified to have reduced access time an halved space occupancy in spite of 64-bit addressing.

We keep an inventory of ones at position multiples of $\lceil Lm/n \rceil$, where L is a constant limiting the size of the inventory ($L = 8192$ in our implementation). For each bit in the inventory, we allocate a number of words (again, upper bounded by a constant M) depending on the density. Inside, we record a 16-bit subinventory (if 16 bits are not enough, we use the space to point at a spill buffer where we record each bit position individually). We use the inventory and the subinventory to locate a position that is near the bit we intend to select, and then we perform a linear broadword bit search. The experimental results about this algorithm show that, in fact, it is the fastest, even in the presence of uneven bit distribution. It also has the advantage of providing just selection with a very limited space usage.

The memory occupancy depend mainly by the bound M. Due to the speed of broadword bit search, we have been able to halve it with respect to the value used in [10], without a noticeable effect on performance. As a result, we have almost halved the space occupancy.

7 Elias–Fano Representation of Monotone Sequences

For sparse arrays, we provide a 64-bit implementation of the Elias–Fano representation of monotone sequences [11,12], which is one of the earliest examples of a fully indexable dictionary. We briefly recall the main idea, translated into the bit array scenario: we

---

[3] We remark that we claimed in the introduction that our broadword algorithms contain no branching; but there is no contradiction, as this part of `select9` is not broadword.

record all bits positions, but while the lower $\ell = \lfloor \log(n/m) \rfloor$ bits are recorded explicitly, the $u = \lceil \log n \rceil - \lfloor \log(n/m) \rfloor$ upper bits are recorded in an array U of $m + u \setminus 2^{\ell}$ bits as follows: if the value of the upper u bits of the position of the i-th one is k, we set the bit in position $i + k$. It is easy to recover the original value by selecting the i-th bit in U and subtracting i. The space occupancy is bounded by $2m + m \log(n/m)$ bits [11], which is almost optimal as specifying a set of m elements out of n requires $\approx m \log(n/m)$ bits when $m \ll n$.

The only component we can improve is actually selection in U, which however is a very well-behaved dense array, so we use a version of `simple` that is wired to density $1/2$.

8   Experiments

We performed a number of experiments on a Linux-based system sporting a 64-bit Opteron processor running at 2814.501 MHz with 1 MiB of first-level cache. The tests show that on 64-bit architectures broadword programming provides significant performance improvements. We compiled using `gcc` 4.1.2 and options `-O9`.

Table 1. Percentage of space occupied by various select structures in a densely (50%) populated bit array. Note that the percentage shown for `select9` and hinted bsearch includes 25% for `rank9`. The preposterous values shown for Clark's structure are due to the very large lookup table.

Size	select9	Hinted bsearch	simple	darray	Kim	Clark
1 Ki	62.50%	37.50%	25.00%	67.19%	86.72%	544073.24%
16 Ki	56.25%	37.50%	14.45%	28.81%	72.80%	34074.85%
256 Ki	56.13%	37.23%	13.79%	27.73%	71.57%	2184.99%
4 Mi	56.12%	37.25%	13.78%	27.56%	71.67%	192.81%
64 Mi	56.12%	37.25%	13.78%	27.56%	71.67%	68.30%
1 Gi	56.13%	37.25%	13.78%	27.56%	71.67%	60.52%

The experimental setting for benchmarking operations that require few nanoseconds must be set up carefully. We generate random bit arrays and store a million test positions. During the tests, the positions are read with a linear scan, producing a minimal interference; generating random positions during the tests causes instead a significant perturbation of the results, mainly due to the slowness of the modulo operator. The tests are repeated ten times and averaged. We measure user time using the system function `getrusage()`.

We provide results for dense (50%) and sparse (1%) arrays of different sizes[4]. In the first case, however, we take care of experimenting over a highly uneven bit array (almost empty in the first half, almost full in the second half). Test positions are generated so to fall approximately half of the time in the dense part, and half of the time in the sparse part. The results obtained using this method highlight serious limitations of some approaches (e.g., binary search) which are not evident in experiments involving

---

[4] Note that we use the NIST-endorsed prefixes: Ki=2^{10}, Mi=2^{20}, etc.

Table 2. Nanoseconds per rank and select operations in densely populated (50%) bit arrays of increasing size. The space usage of rank structures is shown on their label; the space shown for Jacobson's structure is for the 1 Gi array (for smaller sizes, it grows significantly, as it happens for Clark's structure in Table 1: at size 2^{64}, it is still 37.5%; it becomes space-competitive with rank9 beyond 2^{100} bits). As noted in [3], once out of the cache access time increase linearly due to the memory-address resolution process.

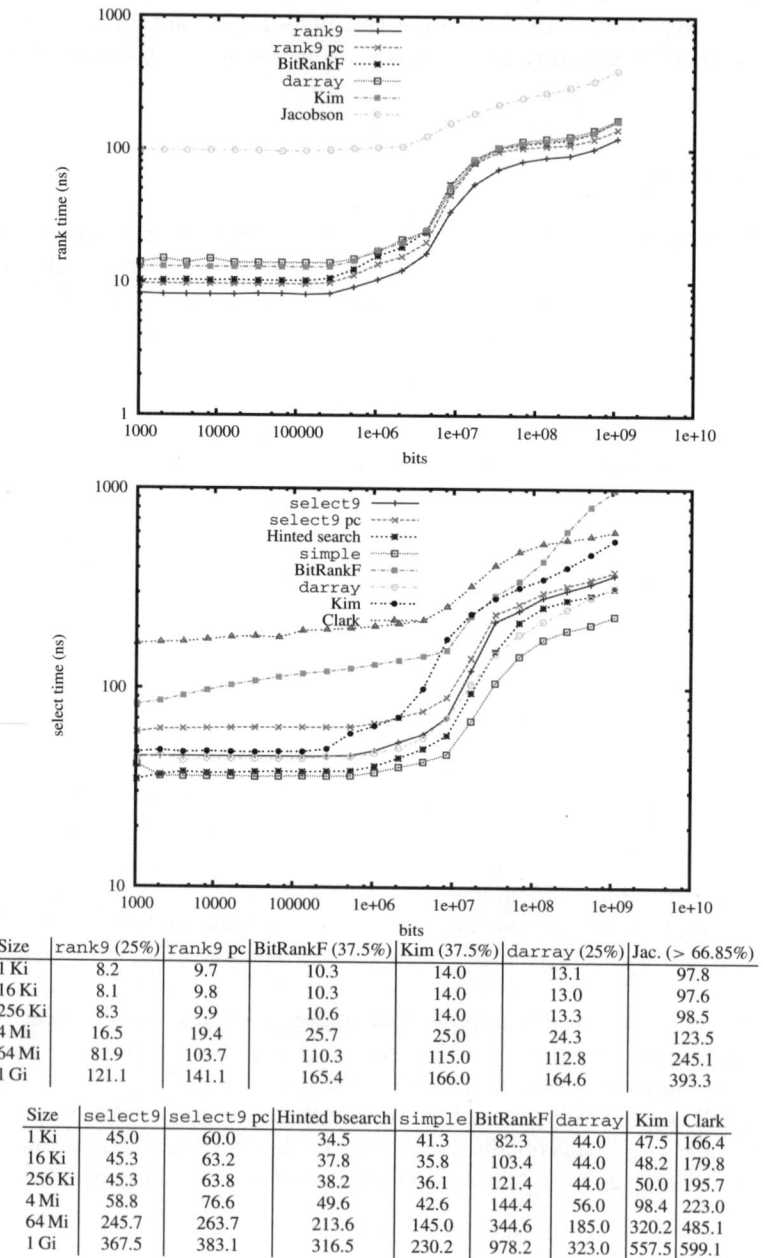

Size	rank9 (25%)	rank9 pc	BitRankF (37.5%)	Kim (37.5%)	darray (25%)	Jac. (> 66.85%)
1 Ki	8.2	9.7	10.3	14.0	13.1	97.8
16 Ki	8.1	9.8	10.3	14.0	13.0	97.6
256 Ki	8.3	9.9	10.6	14.0	13.3	98.5
4 Mi	16.5	19.4	25.7	25.0	24.3	123.5
64 Mi	81.9	103.7	110.3	115.0	112.8	245.1
1 Gi	121.1	141.1	165.4	166.0	164.6	393.3

Size	select9	select9 pc	Hinted bsearch	simple	BitRankF	darray	Kim	Clark
1 Ki	45.0	60.0	34.5	41.3	82.3	44.0	47.5	166.4
16 Ki	45.3	63.2	37.8	35.8	103.4	44.0	48.2	179.8
256 Ki	45.3	63.8	38.2	36.1	121.4	44.0	50.0	195.7
4 Mi	58.8	76.6	49.6	42.6	144.4	56.0	98.4	223.0
64 Mi	245.7	263.7	213.6	145.0	344.6	185.0	320.2	485.1
1 Gi	367.5	383.1	316.5	230.2	978.2	323.0	557.5	599.1

Table 3. Nanoseconds per select operation in a densely (50%) populated bit array of increasing size with uneven bit distribution: almost all bits in the first half are zeroes, and almost all bits in the second half are ones. The "switch" effect typical of structures that change their strategy depending on the density is very visible. Note the poor performance on large arrays of methods based on binary search.

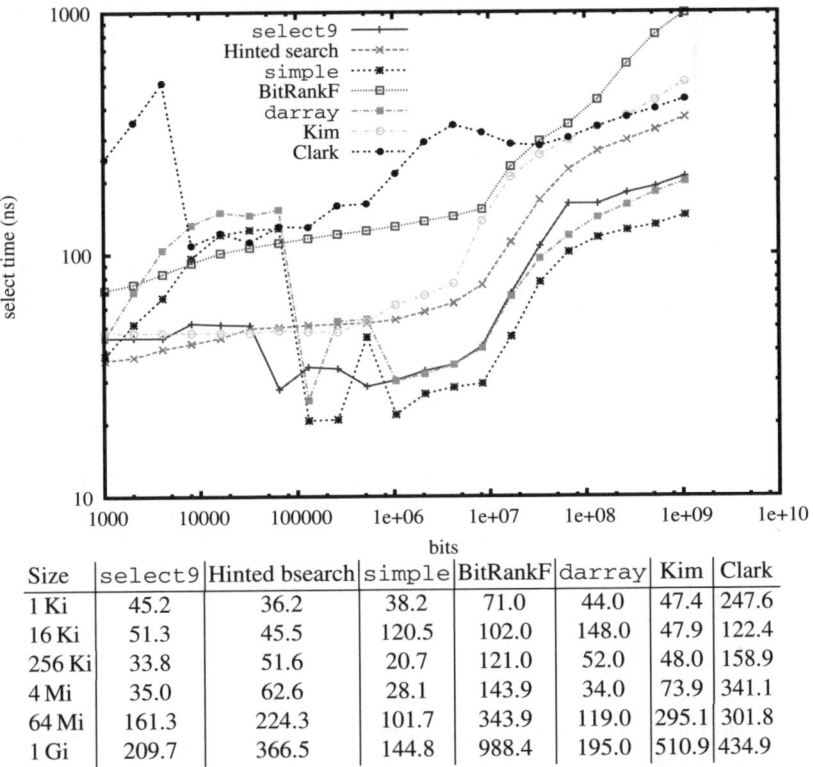

Size	select9	Hinted bsearch	simple	BitRankF	darray	Kim	Clark
1 Ki	45.2	36.2	38.2	71.0	44.0	47.4	247.6
16 Ki	51.3	45.5	120.5	102.0	148.0	47.9	122.4
256 Ki	33.8	51.6	20.7	121.0	52.0	48.0	158.9
4 Mi	35.0	62.6	28.1	143.9	34.0	73.9	341.1
64 Mi	161.3	224.3	101.7	343.9	119.0	295.1	301.8
1 Gi	209.7	366.5	144.8	988.4	195.0	510.9	434.9

Table 4. Percentage of space occupied by various select structures in a densely (50%) populated uneven bit array (see Table 3)

Size	select9	Hinted bsearch	simple	darray	Kim	Clark
1 Ki	56.25%	37.50%	25.00%	67.19%	94.53%	544073.24%
16 Ki	56.25%	37.50%	14.45%	28.81%	83.06%	34074.85%
256 Ki	56.20%	37.38%	63.96%	40.25%	80.93%	2184.99%
4 Mi	56.19%	37.37%	45.17%	43.23%	80.80%	192.81%
64 Mi	56.19%	37.38%	45.95%	43.61%	80.76%	68.30%
1 Gi	56.19%	37.38%	45.94%	43.60%	80.77%	60.52%

uniform bit arrays. Our results suggest that practical implementations of rank/select queries should be always tested against uneven bit arrays (and possibly even more adversarial settings).

We chose to compare our structures against practical ones: the code for the BitRankF structure proposed in [3] was provided by the authors. The authors of [10]

Table 5. Nanoseconds per select operation in bit arrays of increasing size with sparse (1%) bit population

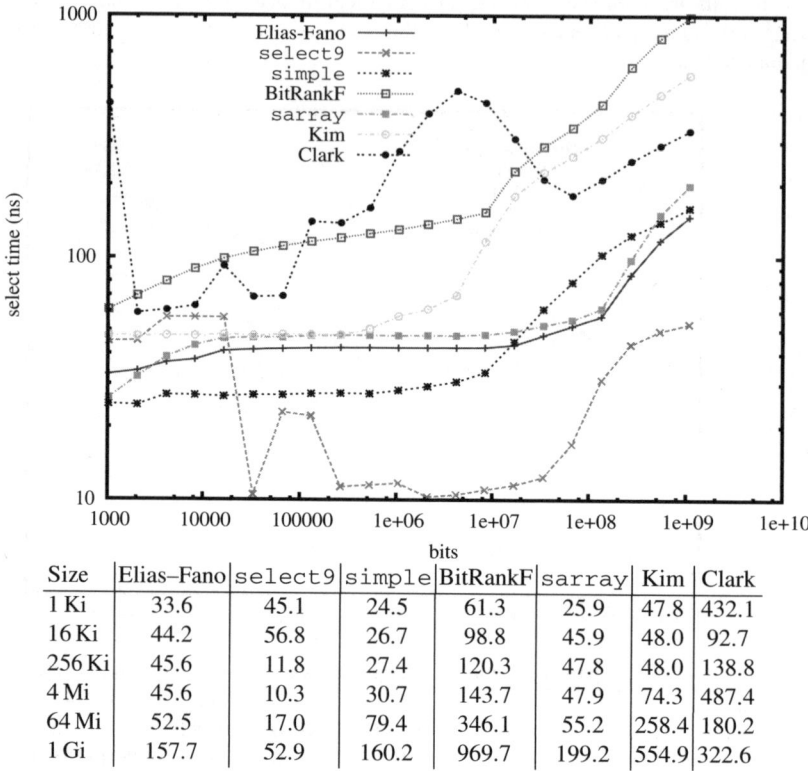

Size	Elias–Fano	select9	simple	BitRankF	sarray	Kim	Clark
1 Ki	33.6	45.1	24.5	61.3	25.9	47.8	432.1
16 Ki	44.2	56.8	26.7	98.8	45.9	48.0	92.7
256 Ki	45.6	11.8	27.4	120.3	47.8	48.0	138.8
4 Mi	45.6	10.3	30.7	143.7	47.9	74.3	487.4
64 Mi	52.5	17.0	79.4	346.1	55.2	258.4	180.2
1 Gi	157.7	52.9	160.2	969.7	199.2	554.9	322.6

provided code for their implementation of the Elias–Fano[5] representation (darray[6] and sarray), and for the byte-oriented select structure described by Kim et al. in [13].[7] All these structures exploit byte or word alignment to increase speed, as previous experiments have made clear [3] that non-aligned structures are extremely slow. Nonetheless, to let the reader have a feeling about what happens using $o(n)$-space constant-time structures we also provide results about Jacobson's [1] classic rank implementation and Clark's [14] select implementation.[8,9]

---

[5] It should be noted that in [10] no mention is made of the work of Elias and Fano. Moreover, their bit subdivision (using $\lceil \log(n/m) \rceil$ lower bits) causes a larger space occupation.

[6] We have decreased the bound M in darray to reduce further space occupancy; we can do so with an almost immaterial impact on performance due to the speed of broadword bit search.

[7] The authors of the latter paper, in spite of several communication attempts, did not provide code for their structures.

[8] The code for the latter was kindly provided by the authors of [3].

[9] We wish to thank one of the anonymous referees for pointing us at a series of papers about practical rank/select structures [8,15]. Unfortunately, at the time of this writing the authors distribute publicly just a few header files and two binary libraries for an unspecified operating system, without any source code or documentation.

Table 6. Percentage of space occupied by various select structures in bit arrays of increasing size with sparse (1%) bit population. Note that the percentage shown for `select9` includes 25% for `rank9`. Elias–Fano and `sarray` do not require the original bit array (which contributes an additional 100% to the other structures).

Select	Elias–Fano	select9	simple	sarray	Kim	Clark
1 Ki	84.77%	56.25%	25.00%	98.44%	45.31%	544073.24%
16 Ki	13.94%	50.39%	10.55%	15.33%	26.12%	34074.85%
256 Ki	9.45%	50.15%	9.01%	9.81%	22.65%	2184.99%
4 Mi	9.37%	50.13%	9.01%	9.64%	22.55%	192.81%
64 Mi	9.38%	50.13%	9.01%	9.64%	22.52%	68.30%
1 Gi	9.37%	50.13%	9.01%	9.63%	22.50%	60.52%

Looking at Table 2, `rank9` is the clear winner among ranking methods. For completeness, we provide results for a variant that trades broadword programming for *population counting* ("pc"), a standard table-based technique used in [3] that turns out to be slower.[10] The situation for select is more varied, and also Table 3 and 4 should be taken into account. Essentially, `simple` turns out to be the fastest and more space efficient data structure on evenly distributed arrays. If constant time is required in spite of adversarial distribution, `select9` is highly competitive if paired with `rank9`.

The results for selection on sparse arrays are reported in Table 5 and 6. Our implementation of the Elias–Fano representation provides support for very large (64-bit) arrays while keeping the excellent space occupancy of `sarray` (for lack of space we cannot report results on ranking, which are however in the same line). Among implementations requiring the original bit array, `select9` has excellent performance even on very large arrays. Its space occupancy is also very competitive if it used in conjunction with `rank9`, albeit `simple` has also very good timings, and the lowest space occupancy.

9 Conclusions

We have introduced some new ideas about the application of broadword programming [4] to bit-level manipulations typical of succinct static data structures. For densely populated arrays, `rank9` and `simple` are generally the best structures, both in term of time, space, and addressability. If a more robust performance guarantee is required, `select9` provide the fastest practical constant-time operations. For sparsely populated arrays, the Elias–Fano representation of monotone sequences, supported by dense broadword selection, provides good speed and nearly optimal space occupancy.

We wish to thank one of the anonymous referees for pointing us to Elias's paper [11], which in turn led us to Fano's *memorandum* [12].

---

[10] It is interesting to remark that testing in isolation broadword programming *vs.* popcounting for ranking or selecting in a word we obtained opposite results. This happens because when testing popcounting in isolation the whole processor cache and branch-prediction unit are servicing a single, small loop.

References

1. Jacobson, G.: Space-efficient static trees and graphs. In: 30th Annual Symposium on Foundations of Computer Science, Research Triangle Park, North Carolina, pp. 549–554. IEEE, Los Alamitos (1989)
2. Golynski, A.: Optimal lower bounds for rank and select indexes. In: Bugliesi, M., Preneel, B., Sassone, V., Wegener, I. (eds.) ICALP 2006. LNCS, vol. 4051, pp. 370–381. Springer, Heidelberg (2006)
3. Gonzàlez, R., Grabowski, S., Mäkinen, V., Navarro, G.: Practical implementation of rank and select queries. In: Poster Proceedings Volume of 4th Workshop on Efficient and Experimental Algorithms (WEA 2005), pp. 27–38. CTI Press, Ellinika Grammata (2005)
4. Knuth, D.E.: The Art of Computer Programming. Pre-Fascicle 1A. Draft of Section 7.1.3: Bitwise Tricks and Techniques (2007)
5. Fisher, R.J., Dietz, H.G.: Compiling for SIMD within a register. In: Carter, L., Ferrante, J., Sehr, D., Chatterjee, S., Prins, J.F., Li, Z., Yew, P.-C. (eds.) LCPC 1998. LNCS, vol. 1656, pp. 290–304. Springer, Heidelberg (1999)
6. Lamport, L.: Multiple byte processing with full-word instructions. Comm. ACM 18(8), 471–475 (1975)
7. Fredman, M.L., Willard, D.E.: Surpassing the information theoretic bound with fusion trees. J. Comput. System Sci. 47(3), 424–436 (1993)
8. Geary, R.F., Rahman, N., Raman, R., Raman, V.: A simple optimal representation for balanced parentheses. In: Sahinalp, S.C., Muthukrishnan, S.M., Dogrusoz, U. (eds.) CPM 2004. LNCS, vol. 3109, pp. 159–172. Springer, Heidelberg (2004)
9. Dijkstra, E.W.: Why numbering should start at zero. EWD, p. 831 (1982)
10. Okanohara, D., Sadakane, K.: Practical entropy-compressed rank/select dictionary. In: Proc. of the Workshop on Algorithm Engineering and Experiments, ALENEX 2007, SIAM, Philadelphia (2007)
11. Elias, P.: Efficient storage and retrieval by content and address of static files. J. Assoc. Comput. Mach. 21(2), 246–260 (1974)
12. Fano, R.M.: On the number of bits required to implement an associative memory. In: Memorandum 61, Computer Structures Group, Project MAC, MIT, Cambridge, Mass., n.d (1971)
13. Kim, D.K., Na, J.C., Kim, J.E., Park, K.: Efficient implementation of rank and select functions for succinct representation. In: Nikoletseas, S.E. (ed.) WEA 2005. LNCS, vol. 3503, pp. 315–327. Springer, Heidelberg (2005)
14. Clark, D.R.: Compact Pat Trees. PhD thesis, University of Waterloo, Waterloo, Ont., Canada (1998)
15. Delpratt, O., Rahman, N., Raman, R.: Compressed prefix sums. In: van Leeuwen, J., Italiano, G.F., van der Hoek, W., Meinel, C., Sack, H., Plášil, F. (eds.) SOFSEM 2007. LNCS, vol. 4362, pp. 235–247. Springer, Heidelberg (2007)

Efficient Implementations of Heuristics for Routing and Wavelength Assignment

Thiago F. Noronha[1], Mauricio G.C. Resende[2], and Celso C. Ribeiro[3]

[1] Department of Computer Science, Catholic University of Rio de Janeiro
Rua Marquês de São Vicente, 225, Rio de Janeiro, RJ 22453-900, Brazil
tfn@inf.puc-rio.br
[2] Algorithms and Optimization Research Department, AT&T Labs Research
Florham Park, NJ 07932-0971, United States
mgcr@research.att.com
[3] Department of Computer Science, Universidade Federal Fluminense
Rua Passo da Pátria, 156, Bloco E, Niterói, RJ 24210-240, Brazil
celso@ic.uff.br

Abstract. The problem of Routing and Wavelength Assignment in Wavelength Division Multiplexing (WDM) optical networks consists in routing a set of lightpaths and assigning a wavelength to each of them, such that lightpaths whose routes share a common fiber are assigned to different wavelengths. When the objective is to minimize the total number of wavelengths used, this problem is NP-hard. The current state-of-the-art heuristics were proposed in 2007 by Skorin-Kapov. The solutions provided by these heuristics were near-optimal. However, the associated running times reported were high. In this paper, we propose efficient implementations of these heuristics and reevaluate them on a broader set of testbed instances.

1 Introduction

Information in optical networks is transmitted through optical fibers as optical signals. Each link operates at a speed of the order of terabits per second, which is much faster than the currently available electronic devices for signal reception and transmission. *Wavelength Division Multiplexing* (WDM) technology allows more efficient use of the huge capacity of optical fibers, as far as it permits the simultaneous transmission of different channels along the same fiber, each of them using a different *wavelength*. An all-optical point-to-point connection between two nodes is called a *lightpath*. It is characterized by its route and the wavelength in which it is multiplexed. Two lightpaths may use the same wavelength, provided they do not share any common fiber. Such networks require a large number of available wavelengths, especially when wavelength conversion is not available.

Given an optical network and a set of lightpath requests, the problem of *Routing and Wavelength Assignment* (RWA) in WDM optical networks consists

C.C. McGeoch (Ed.): WEA 2008, LNCS 5038, pp. 169–180, 2008.

in routing the set of lightpaths and assigning a wavelength to each of them, such that lightpaths whose routes share a common fiber are assigned to different wavelengths. Variants of RWA are characterized by different optimization criteria and traffic patterns, see e.g. [3,13]. We consider the min-RWA offline variant, in which all lightpath requests are known beforehand. No wavelength conversion is available, i.e. a lightpath must be assigned the same wavelength on all fibers in its route. The objective is to minimize the total number of wavelengths used. This problem is also known as the *Path Coloring Problem*. Erlebach and Jansen [4] showed that min-RWA is NP-hard.

State-of-the-art heuristics for min-RWA are discussed in the next section. Implementation issues are discussed and new heuristics are proposed in Section 3. Computational experiments illustrating the efficiency of the new implementations on a broad set of test instances are reported in Section 4. Concluding remarks are drawn in the last section.

2   Related Work

Different heuristics have been proposed for solving min-RWA. Some approaches decompose the problem into two subproblems: the routing subproblem and the wavelength assignment subproblem [2,5,7,9], while others tackle the two subproblems simultaneously [8,12]. A functional classification of RWA heuristics can be found in [3].

The current state-of-art heuristics for min-RWA were proposed by Skorin-Kapov [12]. Each wavelength is represented by a different copy of a bidirected graph $G = (V, A)$ that represents the physical topology of the optical network. Vertices in V and arcs in A represent network nodes and fibers, respectively. Lightpaths arc-disjointly routed in the same copy of G are assigned the same wavelength. The copies of G are associated with the bins and the lightpaths with the items of a bin packing problem [1]. Problem min-RWA is reformulated as that of packing the lightpaths using a minimum number of bins.

The *size* of a lightpath is defined as the hop-count shortest path between its endnodes in G. We notice that lightpaths are not necessarily routed on shortest paths. Whenever a lightpath is placed in a bin (i.e., a copy of G), all arcs in its route are deleted from the corresponding copy of G to avoid that other lightpaths use them. Therefore, the next lightpaths packed in that bin might not be able to be routed on a shortest path.

Four min-RWA heuristics were developed based on classical bin packing heuristics: (i) FF-RWA, based on the First Fit heuristic, (ii) BF-RWA, based on the Best Fit heuristic, (iii) FFD-RWA, based on the First Fit Decreasing heuristic, and (iv) BFD-RWA, based on the Best Fit Decreasing heuristic. The first is equivalent to the Greedy-EDP-RWA [8] heuristic, except for the order in which some steps are executed [12].

The pseudo-codes of FF-RWA, BF-RWA, FFD-RWA, BFD-RWA are similar. They are summarized in Figure 1. The inputs are the graph G, the set τ of

```
begin heuristic(G, τ, d)
1.   Let t be a permutation of the lightpaths in τ;
2.   Set Ω ← ∅ and S ← ∅;
3.   for i = 1, ..., |t| do
4.       Find the bin ω ∈ Ω where the shortest path of tᵢ in ω has less than d arcs;
5.       if no such a bin exists then do
7.           ω ← new copy of G;
8.           Ω ← Ω ∪ {ω};
9.       end if
10.      Let pᵢ be the shortest path between the endnodes of tᵢ in ω;
11.      S ← S ∪ (pᵢ, ω);
12.      Delete edges in path pᵢ from ω;
13. end-for;
14. return S;
end
```

Fig. 1. Pseudo-code of heuristics FF-RWA, BF-RWA, FFD-RWA, and BFD-RWA

lightpath requests, and the value d of the maximum number of links in each route. As suggested in [12], d is set to be the maximum of the square root of the number of links in the network and the diameter of G (i.e., the maximum value of a shortest path between two nodes in the network). The output is a set S of tuples (p_i, ω_i), for $i = 1, \ldots, |\tau|$, where p_i is the route followed by lightpath t_i and ω_i is the wavelength with which it is multiplexed. A permutation t of lightpaths in τ is built in line 1. In FF-RWA and BF-RWA, lightpaths are randomly distributed in t, while in FFD-RWA and BFD-RWA, they are sorted in non-increasing order of their sizes. In line 2, the set S and the set Ω of copies of G are initialized. The lightpaths are routed and assigned a wavelength in lines 3 to 13, one at a time, according to their order in t. A bin $\omega \in \Omega$ in which lightpath t_i can be routed with less than d arcs is sought in line 4. FF-RWA and FFD-RWA stop at the first bin found, while BF-RWA and BFD-RWA scan all bins in Ω and select that in which t_i fits with the smallest number of arcs (since the arcs in each copy of G are not necessarily the same). Let p_i be the shortest path between the endnodes of t_i in ω. If there is no bin in Ω where p_i fits with less than d arcs, then t_i is routed on a new copy of G that is created in line 7 and added to set Ω in line 8. The tuple (p_i, ω) is added to the solution in line 11, and all arcs in p_i are deleted from ω in line 12 to avoid that other lightpaths are routed on those arcs in this copy of G.

Numerical results in [12] showed that FFD-RWA and BFD-RWA outperformed Greedy-EDP-RWA [8], one of the best heuristic in the literature for min-RWA. However, the running times reported in [12] were very high. On the largest instances, running times of up to 8 minutes (Pentium IV 2.8 GHz) were reported. In the next section, we propose five different implementation strategies for FF-RWA, BF-RWA, FFD-RWA, and BFD-RWA and evaluate them in a broad set of test instances in Section 4.

3   Implementation Issues

Let $n = |V|$, $m = |A|$, and $l = |\tau|$. Furthermore, let $c^{sp}(m, n)$ be the compu-
tational complexity of a one-to-all shortest path algorithm applied to G, and
$c^{del}(m, n)$ be the complexity of deleting an arc from G. The worst case complexity $T(n, m, l)$ of FF-RWA, BF-RWA, FFD-RWA, and BFD-RWA is calculated as
follows. For sake of simplicity, we assume that $n < l$, which holds for all real and
artificial networks in the literature. First, all-to-all shortest paths are calculated
in time $O(n \cdot c^{sp}(m, n))$ in line 1, and sets Ω and S are initialized in constant time
in line 2. Next, lines 3 to 13 are repeated for each lightpath t_i, with $i = 1, \ldots, l$.
In line 4, a bin where t_i fits with less than d arcs is found in time $O(l \cdot c^{sp}(m, n))$.
A new copy of G is created in time $O(m)$ in line 7 and added to set Ω in constant
time in line 8. Finally, the set S is updated in constant time in line 11, while the
arcs in p_i are deleted from w in time $O(d \cdot c^{del})$ in line 12. Therefore, the worst
case complexity of these heuristics is

$$T(n, m, l) = O(n \cdot c^{sp}(m, n)) + O(1) + O(l \cdot (l \cdot c^{sp}(m, n) + m + 1 + d \cdot c^{del}(m, n)))$$
$$= O(n \cdot c^{sp}(m, n) + l^2 \cdot c^{sp}(m, n) + l \cdot m + l \cdot d \cdot c^{del}(m, n)).$$

The efficiency of the heuristics depends on how fast a shortest path (SP, for
short) query is performed and how fast an arc is removed from G. The traditional
implementations using Dijkstra's algorithm and single linked lists to represent
the adjacency matrix of G lead to $c^{sp}(m, n) = O(n \cdot \log n + m)$ and $c^{del}(m, n) = O(n)$. Therefore,

$$T(n, m, l) = O(n \cdot (n \cdot \log n + m) + l^2 \cdot (n \cdot \log n + m) + l \cdot m + l \cdot d \cdot n)$$
$$= O(l^2 \cdot (n \cdot \log n + m)).$$

However, the hop-count shortest paths can be calculated using Breadth First
Search (BFS) in time $O(m)$. In addition, any arc can be deleted in time $O(1)$
using the representation of G by adjacency lists M as follows. For each node
$i \in V$, we keep a doubly linked list whose cells correspond to the arcs having
i as their origin. Furthermore, we keep an array P pointing to the address of
each cell in M. Whenever an arc (i, j) is to be removed, we use P to obtain
the address of its corresponding cell in constant time. Since the adjacency list
of node i is doubly linked, the cell corresponding to arc (i, j) can be deleted in
time $O(1)$. This data structure and the BFS algorithm were used in our standard
implementation STD of the four heuristics. Therefore, the complexity of the min-
RWA heuristics using STD is

$$T(n, m, l) = O(n \cdot m + l^2 \cdot m + l \cdot m + l \cdot d) = O(l^2 \cdot m).$$

The most expensive operation of the min-RWA heuristics appears in line 4
of Figure 1, where a SP query is performed in at most l bins for each of the l
lightpaths in τ. At this point, only the value of the shortest path is required.
Therefore, we propose another implementation based on an $n \times n$ distance matrix
in which each entry is the value of the shortest path between the two corresponding nodes in G. It is initialized in $O(n \cdot m)$ in line 1 and instantiated for each

new bin created in $O(n^2)$ in line 7. As long as arcs are deleted from a bin, the shortest paths on that bin may change and the corresponding distance matrix must be updated.

The new data structure allows SP queries to be performed in constant time. However, the efficiency of the heuristics depends on how fast the updates are performed. Given the graph $G = (V, A)$ and a node $v \in V$, the shortest path (SP, for short) graph of v is a subgraph $G^v = (V, A^v)$ of G, with $A^v \subseteq A$, such that the path from any vertex i to v in G^v corresponds to the shortest path from i to v in G. If the graph is acyclic, it is called a shortest path tree. We experimented with four algorithms for updating the distance matrix: RRg, RRt, NRRg, and NRRt. The first two are based on the work of Ramalingam and Reps [10] for dynamically updating SP graphs and SP trees, respectively, while the last two are adaptations of the former two algorithms.

Given a node $v \in V$ and the SP graph G^v, the algorithm of Ramalingam and Reps [10] for dynamically updating SP graphs is based on the observation that when the weight of an arc $a \in A$ increases, the shortest paths from v to many vertices in V might not change and do not need to be recalculated. Arcs are deleted by increasing their weights to infinity. If $a \notin A^v$, no update is necessary; otherwise a is removed from G^v. Next, if G^v remains connected, the algorithm stops. Otherwise, the set U of vertices whose shortest paths to v have changed is identified and removed from G^v. Then, each vertex $u \in U$ is inserted into a priority queue Q with a key equal to the weight of the least cost arc from u to one of the vertices that remained in G^v. Finally, the algorithm proceeds in a Dijkstra-like fashion.

A variant of this algorithm is that of Ramalingam and Reps [10] for dynamically updating SP trees. The algorithm is similar to the one described above. However, the identification of the vertices in U and the shortest path updates are performed more efficiently in SP trees. Every time an arc $a \in A^v$ is deleted, the data structure has to be updated. Using a Fibonacci heap to implement Q, the worst case time complexity of both algorithms is $O(n \cdot \log n + m)$. However, since only deletions are performed and the arcs have unit cost, it can be implemented in time $O(m)$ by using a bucket to implement Q.

Algorithm RRg keeps one SP graph for each vertex in V. The SP graphs are initialized in $O(n \cdot m)$ in line 1 of Figure 1 and instantiated for each new bin created in $O(n \cdot m)$ in line 7. After each arc in p_i is deleted from ω in line 12, RRg checks if any SP graph of ω must be updated. If so, the algorithm of Ramalingam and Reps [10] for SP graphs is used, and the distance matrix is updated. In the worst case scenario, n SP graphs are updated in $O(n \cdot m)$. Therefore, the complexity of the min-RWA heuristics using RRg is

$$T(n, m, l) = O(n \cdot m + l^2 + l \cdot n \cdot m + l \cdot d \cdot (n \cdot m)) = O(l^2 + l \cdot d \cdot n \cdot m).$$

Algorithm RRt keeps one SP tree for each vertex in V. The SP trees are initialized in $O(n \cdot m)$ in line 1 of Figure 1 and instantiated for each new bin created in time $O(n^2)$ in line 7. As before, after each arc deletion, RRt checks if any SP tree of ω must be updated. If so, the algorithm of Ramalingam and

Reps [10] for trees of shortest paths is used, and the distance matrix is updated. Therefore, the complexity of the min-RWA heuristics using RRt is

$$T(n, m, l) = O(n \cdot m + l^2 + l \cdot n^2 + l \cdot d \cdot (n \cdot m)) = O(l^2 + l \cdot d \cdot n \cdot m).$$

Algorithm RRg (resp. RRt) might not be efficient, since the number of SP graphs (resp. SP trees) to be updated after each lightpath is assigned a bin may be very high. To remedy this, we propose a compromise implementation. Algorithm NRRg (resp. NRRt) uses the same data structure as algorithm RRg (resp. NRRt), but without updating the latter as soon as an arc is deleted. Therefore, the distance matrix gives a lower bound to the shortest path between any two nodes in time $O(1)$, since the shortest paths can only increase after an arc deletion. If the lower bound is larger than d, the correct distance is not needed. Otherwise, it can be calculated in time $O(d)$ by retrieving the shortest path in the SP graph (resp. SP tree) of the sink node of the lightpath. If no arc along the shortest path has been deleted, the value stored in the distance matrix is exact. Otherwise, algorithm NRRg (resp. NRRt) updates only the corresponding SP graph (resp. SP tree) from scratch and the entries of the distance matrix that have changed. Therefore, the worst case complexity of an SP query is $O(m)$. However, no SP graph (resp. SP tree) update is necessary in line 12 of Figure 1 and the arc deletion can be done in $O(d)$ for each lightpath. Therefore, the complexity of the min-RWA heuristics using algorithm NRRg is

$$T(n, m, l) = O(n \cdot m + l^2 \cdot m + l \cdot n \cdot m + l \cdot d) = O(l^2 \cdot m),$$

while using algorithm NRRt it is

$$T(n, m, l) = O(n \cdot m + l^2 \cdot m + l \cdot n^2 + l \cdot d) = O(l^2 \cdot m).$$

4   Computational Experiments

Four sets of testbed instances were used in the computational experiments. Set X was randomly generated as in [12]. Sets Y and Z are proposed in this paper. Finally, set W is a collection of the most studied realistic instances in the literature, together with two new instances introduced in this paper. All network topologies are connected and each link corresponds to a pair of bidirected fibers. The traffic matrices are asymmetric, i.e. there might be a lightpath request from a node i to a node j and not from j to i. A description of each set is presented below.

The set X of instances was randomly generated exactly as in [12]. The instances have 100 nodes, the probability P_e that there is a link between a pair of nodes is equal to 0.03, 0.04, and 0.05, and the probability P_l that there is a connection between a pair of nodes is equal to 0.2, 0.4, 0.6, 0.8, and 1.0. The networks were randomly generated and only connected networks were considered. Fifteen groups of five instances each were created, combining each possible pair of values for P_e and P_l.

We observed that set X is mostly made up of easy instances. This is due to two structural characteristics that are present in most of its instances. First, there are nodes incident to only one link whose connections are all routed through the same link. Second, there are weakly connected components, i.e. disjoint subsets of nodes that are connected by only one link. Therefore, all connections whose endnodes are in different weakly connected components must be routed through the same link. These characteristics may imply high lower bounds on the number of wavelengths necessary to establish the set of lightpath requests. For most of the instances in set X, a solution with this number of wavelengths can be easily found.

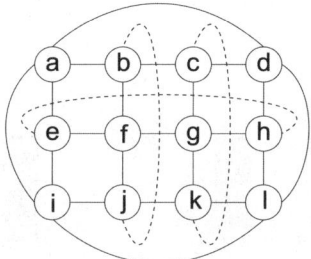

Fig. 2. Example of a 3×4 grid topology

Table 1. Description of test set W

Instance	Nodes	Links	Lightpaths	Maximum
Finland	31	51	930	1
EON	20	39	374	2
ATT	90	137	359	5
ATT2	71	175	4456	34
NSF.1	14	21	284	3
NSF.3	14	21	258	3
NSF.12	14	21	551	6
NSF.48	14	21	547	6
NSF2.1	14	22	284	3
NSF2.3	14	22	258	3
NSF2.12	14	22	551	6
NSF2.48	14	22	547	6

As an attempt to generate harder random instances, we propose the set Y. Networks in this set were randomly generated with the same number of nodes and the same values of P_e and P_l used for the instances in set X. However, we considered only networks whose node degrees are greater than or equal to 2 when P_e is equal to 0.04 and 0.05, and we restricted the diameter of the networks to 5, 6, and 7 for instances with P_e equal to 0.05, 0.04, and 0.03, respectively. As before, fifteen groups of five instances each were randomly generated, combining

Table 2. Average gaps and CPU times for implementations FF-RWA^{STD}, BF-RWA^{STD}, FFD-RWA^{STD}, and BFD-RWA^{STD} for sets X, Y, Z, and W

	FF-RWA		FFD-RWA		BF-RWA		BFD-RWA	
Set	Gap	T(s)	Gap	T(s)	Gap	T(s)	Gap	T(s)
X	4.7%	0.60	1.9%	0.71	3.0%	1.33	1.2%	2.10
Y	23.1%	0.52	17.0%	0.58	13.9%	0.76	8.4%	1.07
Z	13.3%	0.89	9.7%	1.12	10.8%	1.10	7.0%	1.41
W	7.3%	0.10	6.3%	0.10	7.5%	0.11	7.1%	0.12

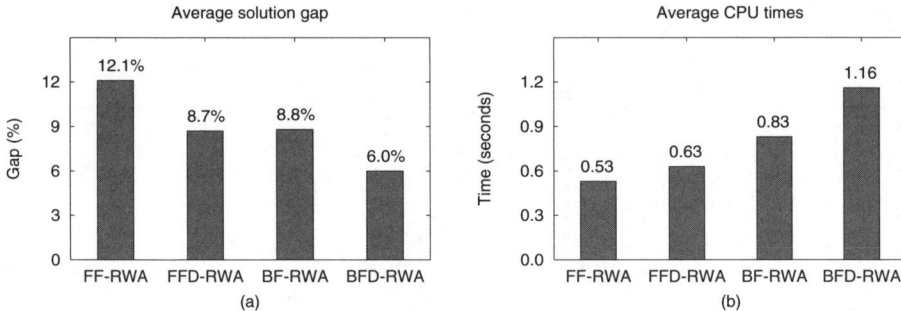

Fig. 3. (a) Average gaps and (b) CPU times for implementations FF-RWA^{STD}, BF-RWA^{STD}, FFD-RWA^{STD}, and BFD-RWA^{STD} over all the 187 instances

each possible pair of values for P_e and P_l. The traffic matrices are the same used for the instances in set X.

Instances in test set Z are built on $n \times m$ grids embedded on the torus. Each node is connected only to its nearest four nodes. Figure 2 gives an example of a 3×4 grid. Five grid networks with approximately 100 nodes (10×10, 8×13, 6×17, 5×20, 4×25) were generated. For each of them, five traffic matrices are randomly generated with the probability P_l that there is a connection between a pair of nodes being equal to 0.2, 0.4, 0.6, 0.8, and 1.0.

Finally, set W is a collection of the most studied realistic instances in the literature, together with two new instances ATT and ATT2 whose topologies and traffic matrices resemble those of real telecommunication networks. The topology of the Finland network was obtained from [5] and its traffic matrix was the same used in [9]. Networks EON, NSF, and NSF2 and their respective traffic matrices were downloaded from [6]. The first three columns of Table 1 display the name, the number of nodes, and the number of links in each instance of set W, respectively. The total number of lightpaths and the maximum number of lightpaths starting from the same node are given in the fourth and fifth columns, respectively.

We denote by FF-RWA^{STD}, FF-RWA^{RRg}, FF-RWA^{RRt}, FF-RWA^{NRRg}, and FF-RWA^{NRRt} the implementations of the heuristic FF-RWA using algorithms STD, RRg, RRt, NRRg, and NRRt, respectively. The same notation is extended to the

Table 3. Average CPU times for each heuristic using NRRt, NRRg, RRt, and RRg. Times are displayed as a percent deviation of the times using STD.

Heuristic	Set	NRRt (%)	NRRg (%)	RRt (%)	RRg (%)
FF-RWA	X	88.3	115.6	671.8	817.4
	Y	83.9	107.6	699.9	824.9
	Z	73.4	95.8	745.7	884.4
	W	98.4	119.0	423.8	497.6
FFD-RWA	X	84.8	112.4	562.2	690.1
	Y	81.1	103.9	626.0	739.1
	Z	62.0	79.7	595.2	699.1
	W	93.4	111.8	406.6	471.3
BF-RWA	X	50.3	67.5	254.2	333.2
	Y	67.8	88.1	400.7	496.2
	Z	63.4	83.4	568.1	685.7
	W	91.8	114.4	363.7	431.5
BFD-RWA	X	37.1	50.0	163.8	209.2
	Y	52.8	70.4	277.3	349.0
	Z	53.2	67.9	440.4	528.8
	W	88.7	107.5	344.0	407.5

other heuristics. The algorithms were coded in C++ and compiled with the GNU GCC version 4.0.3 with no compiler code optimization. The experiments were performed on a 2.8 GHz Pentium IV with 1 Gb of RAM memory using Linux Ubuntu 6.10. CPU times are reported in seconds. The quality of the heuristics is displayed as the gap (UB-LB)/LB between the cost UB of the solution provided by the heuristic and a lower bound LB for the cost of the optimal solution, which is calculated as suggested in [2].

The first experiments evaluate and compare the performance of FF-RWA[STD], FFD-RWA[STD], BF-RWA[STD], and BFD-RWA[STD] for the 187 instances in sets X, Y, Z, and W. Each heuristic was run five times with different seeds for the random number generator algorithm [11]. For each instance set, Table 2 displays the average gaps and CPU times for implementations FF-RWA[STD], FFD-RWA[STD], BF-RWA[STD], and BFD-RWA[STD], respectively. The average gaps for each heuristic over all instances are plotted in Figure 3a, while the average CPU times are plotted in Figure 3b.

The best of the five runs of BFD-RWA[STD] was optimal for 62 out of the 75 instances in set X and the average gap was 1.2%, which confirms the hypothesis that this set is mostly made up of easy instances. The average solution gaps for the instances in sets Y and Z proposed in this paper were greater than or equal to those in the other sets for all the heuristics, which indicates that the instances in sets Y and Z are harder than those in the literature.

Algorithm BFD-RWA[STD] found on average better results than the other heuristics for most of the instances tested. We notice in Figure 3a that the average gap observed for FFD-RWA (8.7%) is almost 50% larger than that corresponding to BFD-RWA (6.0%). The average gap observed for algorithm FFD-RWA[STD] was smaller than that for BFD-RWA[STD] exclusively for set W. However, this

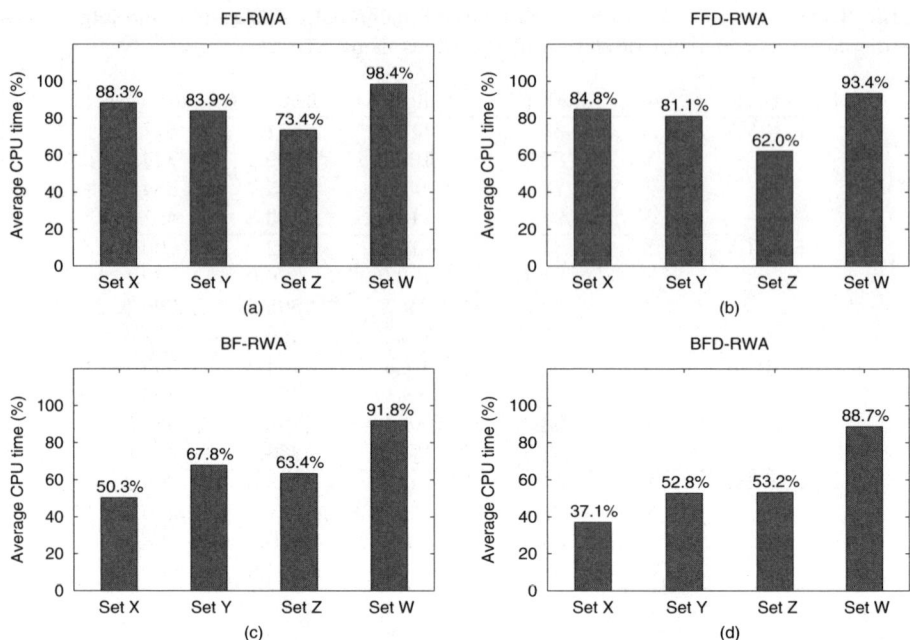

Fig. 4. Average CPU times of (a) FF-RWA$^{\texttt{NRRt}}$, (b) FFD-RWA$^{\texttt{NRRt}}$, (c) BF-RWA$^{\texttt{NRRt}}$, and (d) BFD-RWA$^{\texttt{NRRt}}$ for instance sets X, Y, Z, and W. Times are displayed as a percent deviation of the times using STD.

occurs because of the huge difference observed for instance ATT, where the FFD-RWA$^{\texttt{STD}}$ gap was 20.0% and the BFD-RWA$^{\texttt{STD}}$ gap was 32.0%. If we exclude this instance from set W, the average gap of FF-RWA$^{\texttt{STD}}$ would be 5.0% and that of BFD-RWA$^{\texttt{STD}}$ would be 4.9%.

As expected, CPU times of the best fit heuristics were greater than those of the first fit heuristics, because each iteration of FF-RWA$^{\texttt{STD}}$ and FFD-RWA$^{\texttt{STD}}$ stops at the first bin in which the lightpath can be routed with less than d arcs, while each iteration of BF-RWA$^{\texttt{STD}}$ and BFD-RWA$^{\texttt{STD}}$ scans all the bins looking for bins where the lightpath fits with the smallest number of arcs. Although solution gaps of BFD-RWA$^{\texttt{STD}}$ were on average smaller than those of the other heuristics, its running times were the longest. However, the maximum CPU times of BFD-RWA$^{\texttt{STD}}$ in set X was 10 seconds, which is much less than the 8 minutes reported for the implementation of [12] in the same set of instances and the same machine. The CPU times of BFD-RWA$^{\texttt{STD}}$ were always less than five seconds for instances in sets Y and Z and never greater than one second for those in set W.

The next experiments evaluate the performance of the heuristics FF-RWA, FFD-RWA, BF-RWA, and BFD-RWA using NRRt, NRRg, RRt, and RRg. The running times are compared with those using the respective standard implementation (STD). For each heuristic and each set of instances, Table 3 displays the average CPU times using NRRt, NRRg, RRt, and RRg as a percent deviation of the times using STD (i.e. the times of NRRt, NRRg, RRt, and RRg are divided by the times of STD). Each version of each heuristic was run five times with different seeds for the random number generator algorithm [11]. For all heuristics and instance sets, the implementations using RRt and NRRt were faster than those using RRg and NRRg, respectively. This is due to the fact that updating of SP graphs is more expensive than updating SP trees, and the SP graphs were not dense enough to compensate the trade off. Due to the number of SP graphs that must be updated after an arc deletion, the implementations of FF-RWA, FFD-RWA, BF-RWA, and BFD-RWA using RRt and RRg were slower than their respective implementations using STD.

NRRt was the best algorithm for updating the distance matrix. The numerical results for heuristics FF-RWA^{NRRt}, FFD-RWA^{NRRt}, BF-RWA^{NRRt}, and BFD-RWA^{NRRt} displayed in the third column of Table 3 are also plotted in Figure 4. The improvements observed in BF-RWA^{NRRt} and BFD-RWA^{NRRt} are greater than those observed in FF-RWA^{NRRt} and FFD-RWA^{NRRt}, when compared with their respective standard implementations. This is due to the fact that the number of SP queries in BF-RWA and BFD-RWA is greater than in FF-RWA and FFD-RWA, while the number of updates is approximately the same. The heuristic that took more advantage of NRRt was BFD-RWA, whose times were shortened, on average, to almost one half of those of BFD-RWA^{STD}. The maximum running time of the BFD-RWA^{NRRt} over all the 187 instances tested (including set X) was only 2.2 seconds, which is one quarter of the maximum running time of BFD-RWA^{STD}.

5   Concluding Remarks

This paper tackled the problem of routing and wavelength assignment in WDM optical networks. We proposed five different implementations of the best heuristics in the literature, as well as new testbed instances that allowed a precise comparison of the heuristics.

Computational experiments showed that BFD-RWA was the best heuristic for the instances tested. The new algorithms proposed in this paper shortened the average and maximum running times of BFD-RWA by 57% and 25%, respectively, with respect to those of the standard implementation. The maximum computation times of the best implementation of BFD-RWA was less than three seconds, while the times reported for the same heuristic in [12] were up to eight minutes on the same instances and the same Pentium IV 2.8 GHz computer.

References

1. Alvim, A.C.F., Ribeiro, C.C., Glover, F., Aloise, D.J.: A hybrid improvement heuristic for the one-dimensional bin packing problem. Journal of Heuristics 10, 205–229 (2004)
2. Bannerjee, D., Mukherjee, B.: Practical approach for routing and wavelength assignment in large wavelength routed optical networks. IEEE Journal on Selected Areas in Communications 14, 903–908 (1995)

3. Choi, J.S., Golmie, N., Lapeyrere, F., Mouveaux, F., Su, D.: A functional classification of routing and wavelength assignment schemes in DWDM networks: Static case. In: Proceedings of the 7th International Conference on Optical Communication and Networks, Paris, pp. 1109–1115 (2000)
4. Erlebach, T., Jansen, K.: The complexity of path coloring and call scheduling. Theoretical Computer Science 255, 33–50 (2001)
5. Hyytiä, E., Virtamo, J.: Wavelength assignment and routing in WDM networks. In: Fourteenth Nordic Teletraffic Seminar, Copenhagen, pp. 31–40 (1998)
6. Jaumard, B.: Network and traffic data sets for optical network optimization (last visited on January 3th, 2008), http://users.encs.concordia.ca/~bjaumard
7. Li, G., Simha, R.: The partition coloring problem and its application to wavelength routing and assignment. In: Proceedings of the First Workshop on Optical Networks, Dallas (2000)
8. Manohar, P., Manjunath, D., Shevgaonkar, R.K.: Routing and wavelength assignment in optical networks from edge disjoint path algorithms. IEEE Communications Letters 5, 211–213 (2002)
9. Noronha, T.F., Ribeiro, C.C.: Routing and wavelength assignment by partition coloring. European Journal of Operational Research 171, 797–810 (2006)
10. Ramalingam, G., Reps, T.W.: An incremental algorithm for a generalization of the shortest-path problem. Journal of Algorithms 21, 267–305 (1996)
11. Schrage, L.: A more portable Fortran random number generator. ACM Transactions on Mathematical Software 5, 132–138 (1979)
12. Skorin-Kapov, N.: Routing and wavelength assignment in optical networks using bin packing based algorithms. European Journal of Operational Research 177, 1167–1179 (2007)
13. Zang, H., Jue, J.P., Mukherjee, B.: A review of routing and wavelength assignment approaches for wavelength-routed optical WDM networks. Optical Networks Magazine 1, 47–60 (2000)

Myopic Distributed Protocols for Singleton and Independent-Resource Congestion Games[*]

D. Kalles[1], A.C. Kaporis[2], and P.G. Spirakis[2,3]

[1] Hellenic Open University, Patras, Greece & Open University of Cyprus
[2] Department of Computer Engineering and Informatics, University of Patras, Greece
[3] Research Academic Computer Technology Institute, University of Patras Campus, Greece

Abstract. Let n atomic players be routing their unsplitable flow on m resources. When each player has the option to drop her current resource and select a better one, and this option is exercised sequentially and unilaterally, then a Nash Equilibrium (NE) will be eventually reached. Acting sequentially, however, is unrealistic in large systems. But, allowing concurrency, with an arbitrary number of players updating their resources at each time point, leads to an oscillation away from NE, due to big groups of players moving simultaneously and due to non-smooth resource cost functions. In this work, we validate experimentally simple concurrent protocols that are realistic, distributed and myopic yet are scalable, require only information local at each resource and, still, are experimentally shown to quickly reach a NE for a range of arbitrary cost functions.

1 Introductory Motivation

Alice enters a large University library at the evening determined to copy some pages from a friend's notes. Miraculously, she finds a quite peaceful environment where no student opts to shift from her copier at hand. All students know that no copier will decrease their waiting time. This operating point is a *Nash equilibrium* (NE) over copiers and it is quite straightforward to think of other library's facilities also being operated at a NE. Suppose, however, that Alice observes groups of students rushing to copiers when she enters the library in the early morning. She observes student S, currently pending on copier C, contemplating to move to copier C' which seems more appealing, either because it is faster, or less crowded or both. At this critical decision-making point, there are two issues for S. The first issue is that, if a group of students shift to C' alongside S, then S's waiting time is likely to increase. The second issue is that, if printer C''s speed decreases abruptly even due to the slightest increase in demand, then it is even more likely that S's waiting time will increase, and it may do so beyond any anticipation on the part of S. These obstacles naturally give rise to oscillations. No oscillations occur if all students shift to copiers sequentially, one at a time. But, Alice is old enough to know that only Wonderland's disciplined students are determined to shift sequentially. Of course, even in Wonderland, certain side effects do persist: acting sequentially may last long until a NE is reached.

---

[*] The 2nd and 3rd author were partially supported by the IST Program of the European Union under contract number IST-015964 (AEOLUS).

Back in the real world, however, imposing global synchronization is unrealistic. On the other hand, it seems realistic that students pending on copier C will briefly discuss their options before deciding how some of them might wisely move to C'. Their on-the-fly discussions are independent and not affected by decisions taken within any other group of students currently pending on any other copier. Moreover, it is also unrealistic that their local speculations will improve by any global (thus, expensive) information supplied (such information might consist of all copier's congestion and average waiting time). It is also realistic to assume that a group of students migrating from one copier to another will most likely not collide with some other group (also) on a migrating path.

The central question is thus framed as:

> *Question:* Can we model such concurrent migrations as a simple distributed protocol within available resources, based on local speculations and greedy decisions taken on the fly? Can we show that such a distributed protocol, despite its simplicity, is powerful enough so as to quickly reach a NE?

Apparently, as soon as all massive and concurrent migrations to copiers have taken place, it may turn out that many students feel tempted to subsequently shift to newly appealing copiers. This may lead to an endless copier-oriented migration process, oscillating eternally away from a NE, and presenting formidable obstacles in our attempt at analyzing concurrent selfish play. Note that our example is modeled as a *singleton* congestion game, where each player selects one resource over the available ones. It can also be generalized (and become more severe) if not all players' tasks are of the same value, i.e. if weights are introduced. The general problem identified in the above example regards *all* situations where selfish actors must compete for a set of resources and can make decisions based only on information about where they are and where they might want to go (i.e. they do not have access to what other distant actors do). Not surprisingly, there are numerous other fields of computer science that deal with similar situations, most notably in load balancing, scheduling and (most generally) distributed computing.

In this paper we focus on the development of decision making protocols to be used by actors who want to decide to which resource they should migrate. We want these protocols to withstand selfish behaviour and only use information that is available to actors in their current resource (and the one they want to move to). We experimentally show that the protocols we develop lead towards a Nash Equilibrium for a wide range of resource cost functions, including pessimistic mixtures of such cost functions, allowing both weighted and unweighted versions. Moreover, they do so in a number of steps that scales similarly with a baseline protocol that makes quite strong assumptions on the type of actors and the resources.

The rest of this paper is structured as follows: we first review the work on sequential and concurrent congestion games, as well as work on realistic assumptions for experimental settings. We then move on to describe in detail the (widely used) congestion game model we employ and, based on that, we describe two selfish protocols for making migration decisions. Then, we validate the protocols and finally we discuss the implications of our findings.

2 Related Work

2.1 *Sequential* and *Concurrent* Congestion Games

Congestion games (CG) provide a natural model for non-cooperative resource allocation and have been the subject of intensive research in algorithmic game theory. A *congestion game* is a non-cooperative game where selfish players compete over a set of resources. The players' strategies are subsets of resources. The cost of each player for selecting a particular resource is given by a non-negative and non-decreasing latency function of the load (or congestion) of the resource. The individual cost of a player is equal to the total cost for the resources in her strategy. A natural solution concept is that of a pure Nash equilibrium (NE), a state where no player can decrease her individual cost by unilaterally changing her strategy. On a *singleton* CG each player can select only one amongst m resources [16].

In a classical paper, Rosenthal [20] showed that pure Nash equilibria on atomic congestion games correspond to local minima of a natural potential function, which decreases every time a single player changes her strategy and improves her individual cost. Hence every sequence of improving moves will eventually converge to a pure Nash equilibrium. However, this may require an exponential number of steps, since computing a pure Nash equilibrium of a congestion game is *PLS-complete* [10].

There are strong reasons why sequential protocols are subject to critique. Using the Elementary Step System hypothesis, under which at most one user performs an improving move in each round, greatly facilitates the analysis [6,8,13,14,17,18,19], but is quite unrealistic.

Unlike sequential play, concurrent play may eternally oscillate away from NE, due mainly to two reasons: first, players have limited global info on making decisions and, second, the cost of resources may increase unboundedly on new demand. Exceedingly big group of players can cause bottleneck phenomena on their destination resources. This can be avoided by allowing each user to sample *uniformly* and independently, with appropriately small probability, for a new resource. If the resources have (nearly) identical cost functions, this migration probability usually depends only on the departure-destination pair of resources, eliminating any requirement for global information [3]. However, if cost functions are arbitrary, we need more information to better tune this migration probability. Concerning the second obstacle, suppose that user i finds appealing to migrate to resource e and that e is associated with a smooth cost function. Then, moving to e will likely not substantially affect i's *a priori* estimated migration profit, even if many other users opt for e. However, that anticipated profit may deteriorate abysmally if e's cost function is not smooth. A common way out is to consider cost functions that satisfy an α-bounded jump condition [5].

A typical approach [9] considers n players concurrently probing for a better link amongst m parallel links per round (singleton CG). However, this migration protocol, though concurrent, is not completely decentralized, since it uses global information in order to allow only appropriate groups of users to migrate. More precisely, only users with latency exceeding the overall average link latency are allowed to sample (on parallel) for a new link j. This boosts convergence time, requiring expectedly $O(\log \log n + \log m)$ rounds.

The analysis of a concurrent protocol on identical links and players was presented in [3]. Therein, the important aspect of the analysis is that no global information was given to the migrants. On parallel during round t, each user b on resource i_b with load $X_{i_b}(t)$ selects a random resource j_b and if $X_{i_b}(t) > X_{j_b}(t)$ then b migrates to j_b with probability $1 - X_{j_b}(t)/X_{i_b}(t)$. Despite users performing only uniform sampling, this protocol reaches an ε-NE in $O(\log \log n)$, or an exact NE in $O(\log \log n + m^4)$ rounds, in expectation. The reason that proportional sampling turns out to not be so crucial here, is the fact that all links are identical, so there is no need to reroute many users to any particular speedy link. Thus, an important question is to what extent such myopic distributed protocols can cope with links that have large differences amongst their latency functions.

2.2   Insights from Distributed Computing and Traffic Distributions

In our work, there are quite a few points where our research draws from advances in other fields of computing, beyond that of algorithmic game theory.

A key such point is the distribution of weights of players in the weighted version of a CG. Therein, the problem of estimating the typical workload distribution over servers of the Web has attracted a lot of research. Knowledge of this distribution helps evaluate the performance of proxies, servers, virtual networks and other Web related applications. The work in [2] has influenced a lot of subsequent research. It presents experimental evidence that up to a critical file size (the *cutoff* value) the distribution behaves as *Log-normal*, while for larger sizes as *Power Law*. The second such key point concerns the nature of protocols that decide who migrates between resources and how, as well as the extent to which such migrations effectively and efficiently achieve some notion of optimality. The field that has been most influential in that respect is that of load balancing. Drawing from [4] we note that key results from that field recommend that migration protocols are realistic when they assume that (now, we switch to the game nomenclature) a number of players moving from one resource at a given time point actually move to the *same* target, and are not distributed amongst more than one target [7]. This differentiation is described as the contrast between *diffusion* and *dimension exchange* methods, where the latter impose that a resource will only communicate (sample) with *one* potential target resource, to determine where to allow some of its migrants to move to (if at all). It is important to note that this assumption improves the robustness of the migration protocol since, when considering which players to move out of a resource, we do not need to collect expensive information (as is the case, for example, in [9]) from *all* available resources but we just focus on sampling one potential target. To appreciate the robustness potential consider what would happen in a network where we might need to sample many resources, yet find that many of the links seem to be broken, as is quite likely of course.

The justification for our protocols can be further seen in [12], where load balancing between processors is examined and the recommendations therein suggest that it is reasonable to expect more than one migrant per time slot from the same resource, though all migrants from that source resource move to the same target resource. Indeed, therein it is argued that the standard way of moving one migrant per time is an unwarranted pessimism and that it is more realistic to assume that more than one player may move at a time out of source resource and towards the (same) destination one. Therein, it is also argued why a resource cannot be expected to communicate in parallel with other resources, leading to the observation that sequential communication means that all migrants from a source will all go to the same target. Note that the above points have been also stressed in [7].

Morever, also according to [12], we note that our protocols indeed realistically assume that only local information is made available to the migrating candidates; note that, in stark contrast to this recommendation, [9] assume that players have access to accurate global statistics (like average load) to compute their next move.

A further justification for our protocols is the design pattern discussion in [1], where analogues are drawn to several biological processes that have influenced the design of distributed computing protocols and algorithms, and where a central recurring theme is the identification of processes that rely on strictly local information yet achieve some notion of effective global behavior.

3 An Efficient Selfish Distributed Protocol

The basic idea of our protocol is that, per round, each player independently and concurrently can selfishly move on the basis of her corresponding costs, as measured for the current and destination resources. In essence, a player could decide to migrate if the anticipated cost, *after* moving to a target resource, is favourably compared to her current cost.

There are finite sets of n players $N = \{1, \ldots, n\}$ and m resources $E = \{e_1, \ldots, e_m\}$, respectively. The strategy space of player i is $S_i = \{X \subseteq E : |X| = 1\}$; player i selects as her strategy $s_i(t)$ a *single* edge at round t. The game consists of a sequence of rounds $t = 0, \ldots, t^*$. Initially player i selects a random recourse $s_i(0) \in S_i$. Next, per round $t = 1, \ldots, t^*$, each player i updates *concurrently and independently* his current strategy $s_i(t)$ to $s_i(t + 1)$ according to an appropriate protocol. The number $f_e(t)$ of players on resource e is $f_e(t) = |\{j : e \in s_j(t)\}|$. On an unweighted CG, resource e has a cost $\ell_e(f_e(t))$, which is a function of the number of players on e, $f_e(t)$. On a weighted CG, each player j has weight w_j and the weight $w_e(t)$ of players on resource e is $w_e(t) = \sum_{\{j:e \in s_j(t)\}} w_j$, which is the corresponding sum-weight of the players on e. On an weighted CG, that cost is $\ell_e(w_e(t))$, a function of the sum of weights of the players on e. The cost $c_i(t)$ of player i is the cost of the resource where this player resides, l_e. A given state is a NE, if it is not beneficial for any player to change unilaterally her strategy at hand.

The above description is, actually, a protocol that has been shown to possess an interesting property of converging to an *almost*-NE in a logarithmic number of steps [11]. In this paper, drawing on the understanding that it is realistic to assume group migrations, we extend the above protocol to allow exactly that and then proceed to experimentally show that an *exact* NE can be efficiently reached, for a variety of cost functions.

First, we present our protocol for the unweighted case.

B2B: During round t, do in parallel on each resource $e \in E$:

1. $\forall$ **player i on e, sample a random resource e_i'.**
   /* Each player samples myopically a new destination resource. This requires no global information. */
2. $\forall$ **player i on e, let** $OUT_{e_i'} = \max\{0, \ldots, f_e(t)\}$ **sufficient to hold** $\ell_{e_i'}(f_{e_i'}(t) + OUT_{e_i'}) < \ell_e(f_e(t) - OUT_{e_i'} + 1)$.
   /* Each player on a given resource estimates the maximum number of her binmates that can follow her to her new sample destination, in a way that the destination will remain appealing *after* migration. Again, no global information is required. */
3. **Select at RANDOM a LEADER player i,** amongst those that have sampled appealing resources ($OUT_{e_i'} > 0$) and allow her with all her estimated binmates to migrate to e_i'.

Note that all players selfishly sample for a potential target resource. Of those players who have found such a resource, our protocol selectes one at random and drags a number of mates along. That selection is inexpensive. Moreover, since dragged mates expect a better cost compared to what they now have, one may realistically assume that there is no considerable "negotiation" (communication) cost between the migrants.

We now present the weighted case.

W-B2B: During round t, do in parallel on each resource $e \in E$:

1. $\forall$ player i on e, sample a random resource e_i'.
2. $\forall$ player i on e, select an arbitrary subset S_i of i's mates, with their *maximum* corresponding weight sum $WOUT_{e_i'}$, sufficient to hold: $\ell_{e_i'}(w_{e_i'}(t) + WOUT_{e_i'}) < \ell_e(w_e(t) - WOUT_{e_i'} + w_i)$.
3. Select a random LEADER player i, amongst those that have sampled appealing resources ($WOUT_{e_i'} > 0$) and allow her with all her estimated binmates to migrate to e_i'.

The above descriptions can be easily extended to consider more than one leader per resource and to more sophisticated techniques for selecting leaders; hereafter, we shall be using the notations **B2B** and **B2B(1)** interchangeably.

4 Experimental Validation of Singleton Congestion Games

At the initialization of an unweighted CG instance I, each one of n players randomly selects one of m resources. For each resource $e \in I$, a random cost is assigned. Following that, we experiment with PURE and MIX classes of cost functions as detailed below.

The m resources of each random CG instance I are associated with one of the following **PURE** classes of cost functions: **LIN**: $\ell_e(x) = a_e x + b_e$, **LOG**: $\ell_e(x) = b_e \log_{a_e} x$, **EXP**: $\ell_e(x) = b_e a_e^x$, **MM1**: $\ell_e(x) = \frac{b_e}{a_e - x}$, where x is load and a_e, b_e coefficients characterizing resource e. Within a given PURE cost class, coefficients are independently drawn per cost function, with each coefficient drawn uniformly from $[0 + \epsilon, A]$, $\epsilon = 1.05$, with parameter $A = 10$, in order to minimize concentration of coefficients's around any point $y^* \in [0 + \epsilon, A]$, and to avoid similarity amongst the m resource's

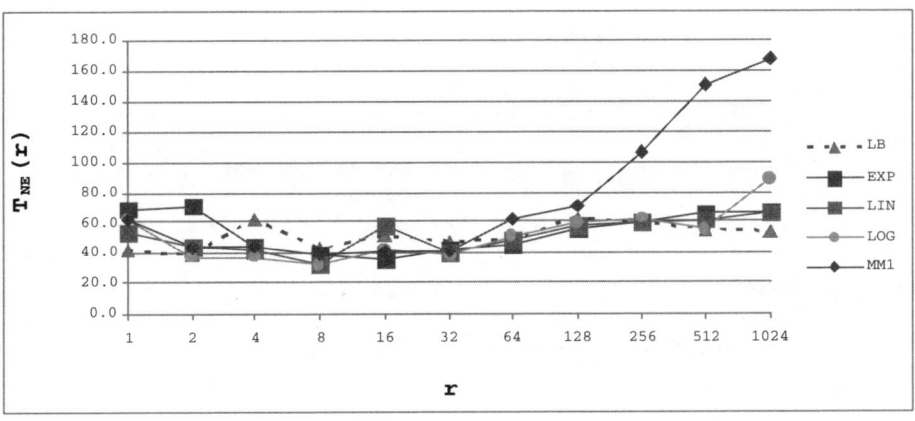

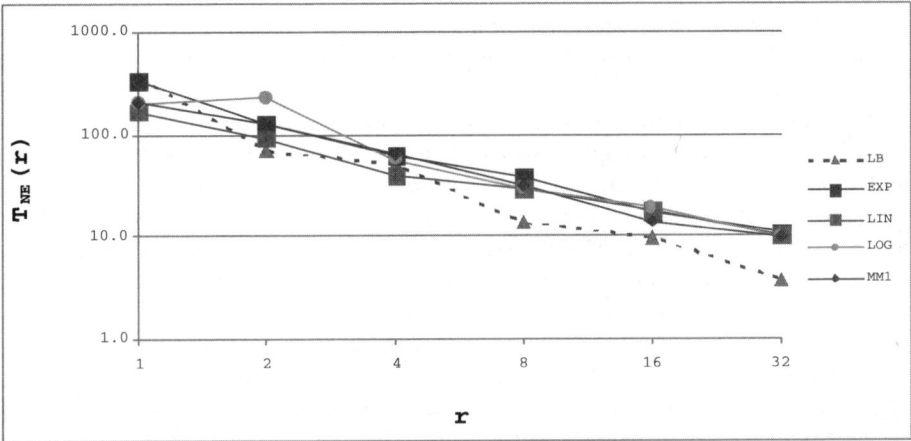

Fig. 1. Unweighted CGs and PURE **cost classes. Top: B2B(1)**'s speed scaling with n players. Resources are fixed at $m_0 = 2^6$. Players are $n = n(r) = rm_0$ with r increasing *exponentially* as $r = 2^t, t = 0, 1, \ldots, 10$. **Bottom: B2B(1)**'s speed scaling with m resources. Players are fixed at $n_0 = 2^8$ with $m = n_0/r$ decreasing exponentially as $m = n_0/2^y$ when $y = 0, 1, \ldots, 5$.

functions [21]. Additionally, to avoid similar random costs within a game instance, we introduce **MIX** classes (see details below) containing mixtures of the above PURE classes.

In Figure 1 (Top) we plot our protocols' speed scaling with n players, when the cost functions are drawn from one of the above PURE classes of cost functions. The *density* r is the ratio of n players to m resources. $T_{NE}(r)$ is the number of rounds until a NE for a given protocol (averaged over 10 random instances). As the x-axis suggests, we examine the state of things in increasingly sparse intervals (the numbers of rounds as powers of 2) and declare a NE whenever no player can select a better resource. Deciding whether a NE is reached is $\Theta(m^2)$ and is carried out in two passes. In the first pass, for

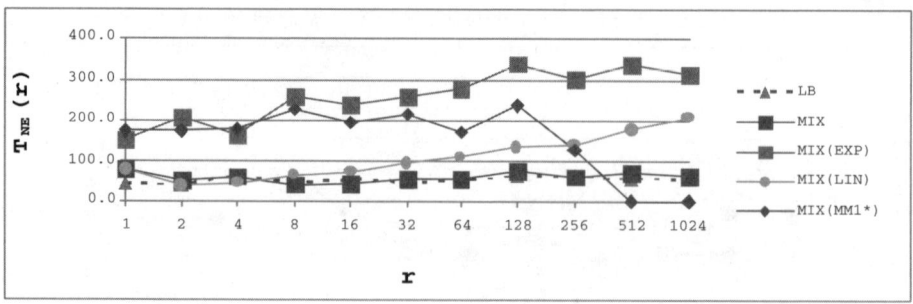

Fig. 2. Unweighted CGs & MIX cost classes: **B2B(1)**'s speed scaling with n players

each resource we compute its "future" cost, should only one newcomer arrive. In the second pass, for each resource, we examine whether its "future" cost is less than the current cost of some other resource; if the answer is yes, then a move is possible and we are not at a NE.

The lower LB-plot serves for comparison and shows the running time of the protocol in [3, Fig. 2] (see last paragraph of section 2.1), on an input of m_0 identical resources and $n = rm_0$ identical players. Each plot labeled "$\mathcal{C}$" shows the running time $T_{NE}(r)$ of our protocol **B2B(1)** on an input of m_0 resources, each drawn from a fixed PURE class $\mathcal{C} \in \{$LIN, LOG, EXP, MM1$\}$ and on $n = rm_0$ identical players. Since each $\mathcal{C}$ plot (with m_0 fixed) is almost parallel to the LB-plot, **B2B(1)** scales similarly to the protocol in [3, Fig. 2] with respect to n. The rather erratic behaviour of **MM1** instances is due to the overall capacity $\sum_e a_e$ being exactly equal to $m + n$ (with all resources having integer costs), which means that there is only one solution that allows non-congested links (in that solution, each resource is populated by its capacity minus 1). This means that costs increases are not α-bounded (so, they will peak abruptly) on almost all resources and for almost all rounds except the final one. Larger densities magnify this problem. If we relax the above very strong constraint and allow the overall capacity $\sum_e a_e$ to be at least 10% more than the number of balls, we have found out that the $\mathcal{MM}\infty$ plot reverts back to being parallel to the baseline LB plot throughout. Again, we remind the reader that we compare our protocols to LB not because we have found a deficiency but, simply, because it is so straighforward and powerful on identical resources and, so, serves as a good baseline. In Figure 1 (Bottom) we illustrate how **B2B(1)** scales with m resources. Again, all plots are almost parallel to the LB-plot of the protocol in [3, Fig. 2].

A main concern of ours was to use a quite wide window of rate of cost growth with respect to the cost classes within instance I, ranging from **DLOG** ($\ell_e(x) = b_e \log_{a_e} \ln x$), with a very smooth rate, to **EXP** and **MM1** classes, the most peaky ones. Towards this, we tried to illustrate more accurately the ability of our protocols to handle resources with fierce behavior on the slightest change of load, by considering classes **MIX, MIX(C)**. A random CG instance I belongs to class **MIX**, if each resource $e \in I$ has a cost function formed randomly according to a random PURE class. On such instances, see how **B2B(1)** scales with n in Figure 2. A CG instance I belongs to class

MIX(C)[1] if exactly 1 resource belongs to **DLOG**, while the remaining $m-1$ resources belong to a fixed PURE class C. In doing so, we essentially tried to mislead our protocols, by hiding 1 precious resource amongst $m-1$ costly ones. The corresponding scaling, as regards m, of **B2B(1)** is omitted due to space limitations. Note that **MIX(EXP)** instances seem to be the most difficult for protocol **B2B(1)**.

As far as weighted singleton CGs are concerned, we assign a random weight X to each player, according to the distribution in Section 2.2. So, for $x < 133000$ (cutoff value) a random weight X has Lognormal density $f(x) = (\sqrt{2\pi}\sigma x)^{-1} e^{-(\ln x-\mu)^2/(2\sigma^2)}$ with parameters $\mu = 9.357, \sigma = 1.318$. If $x \geq 133000$ a random weight X obeys Pareto $g(x) = ak^a x^{-a-1}$ with parameters $k = 133000$ and $a = 1.1$.

Detecting a NE for the weighted case is somewhat subtler. In the first pass, we calculate the lightest player (feather) per resource in $\Theta(n)$. In the second pass, we try to see whether there exists any feather that can move to a less expensive resource; that costs $\Theta(m^2)$; if the answer is yes, we are not at a NE. The results for fixed resources are shown in Figure 3 and in Figure 4.

5   On the Validity and the Implications of the Results

Besides being competitive, our protocols avoid oscillations for α-*bounded jump* cost functions, with experimentally tested $\alpha \leq 10$. Such cost functions include polynomials of bounded degree and exponentials that scale up to 10^x. Remarkably, our protocols remain fast for MM1 cost functions (widely met in real word applications) that do not satisfy any α-bounded condition.

Our protocols's speed is compared to $O(\log \log n + m^4)$ achieved in [3] and to $O(\log \log n + \log m)$ in [9]. Both protocols [3,9] scale (as $\log \log n$) with n players. Using global information one may scale as $\log m$, by performing *proportional sampling* amongst m resources, guiding the migration of players towards appealing resources [9]. When just *uniform sampling* amongst m resources is employed, however, scaling deteriorates to m^4 [3].

Briefly reviewing the properties of our protocols, we note that:

- Our protocols are as simple and myopic as in [3], requiring no tuning of migration probability and exhibiting a similar speed scaling with n players as in [3,9]. They, also, scale with m resources as in [3], but they are not as fast as $O(\log m)$ in [9] (therein, proportional sampling amplifies fast resources).
- Our protocols employ a realistic amount of parallelism. Specifically, they assume that during a migration step, players moving out of one resource may only go to the same target resource [12,7].
- On any symmetric CG with α-bounded latencies the *sequential* protocol in [5, Th. 1.3] reaches an[2] ε-NE in $\lceil n\alpha\varepsilon^{-1} \log(nC) \rceil$ rounds, which is $\geq poly(n)$ on

---

[1] In class **MIX(MM1*)** 1 precious resource has an MM1 cost function of high capacity and $m-1$ resources belong to **MM1**. In very dense settings, this lead to game instances where, at the outset, all links were congested (infinite latency cost); this is also a NE state since no one wants to move to a congested state.

[2] At an ε-NE bicriteria state, no player unilaterally changing her strategy can decrease the cost at hand by more than an ε-portion.

the number of players. Our concurrent protocols reach a NE in $O(\log \Phi(0))$, with $\Phi(0)$ being Rosenthal's potential value at round $t = 0$, where it is well known that $\Phi(0) \leq nC$ (see the open problem in [5, Sec 7: Case 4]).

- Our protocols apply to a wide class of costs. The protocol in [3] balances load (number of players) over identical resources, with each cost being equal to load. The protocol in [9] is limited to linear cost functions with no constant term.
- Our protocols can be extended to apply to independent-resource CGs.
- Our protocols can also handle weighted players, unlike [3]. This property is also shared by [9], however, therein this is done with the use of global information and *just* for linear costs with no constant terms. Note that, with weighted players, an arbitrary weight assignment requires $\Omega(\sqrt{n})$ rounds till a NE [9]. We experimentally improve this lower bound using a realistic weight distribution (see Section 2.2).

We now summarize some experiments to estimate the sensitivity of our protocols to some experimentation parameters (figures were omitted due to space limitations).

First, we experimented with the running time of **B2B** within a particular cost class with coefficients $a_e, b_e \in [1.05, A]$ for $A = 10, 100, 1000, 10000$. These experiments were carried out with $m_0 = 64$ resources and density $r_0 = 32$. We observed that the corresponding average running time of **B2B** was almost the same or even better for values $A > 10$, which suggests that all our previous results are pessimistic and even better speed-ups should be expected.

For the second type of experiments, we have developed the **Hint** variant, where each leader player j migrating out of e additionally transmits her $OUT_{e'_j}$-value as a token to exactly 1 other random player amongst all available players. Essentially, this is as if she drops a hint and the first passer-by picks it up. Then, the LEADER player takes the best choice amongst her own-sampled one and the most recent OUT-token she may have received. So, motivated by the hardness of class **MIX(EXP)** for protocol **B2B**, as illustrated in Figure 2 (Top), we compared **B2B(1)** and **Hint** on this class. **Hint** was just slightly better throughout, suggesting that this problem class is tough indeed.

Generalizing the introductory example, suppose our student also has to consider her best choice amongst *many* groups of University resources (for example, the fasted public bus from town to campus, the most efficient pc in the lab, the least crowded studying room) and that the University has k such groups of resources. Then, this amounts to a concurrent congestion game, where the strategy of player i is a k_i-tuple of resources, each drawn from a group of similar resources. Such CGs are *Independent Resource* ones. We now note (using the nomenclature introduced in Section 3) that the strategy space of player i is $S_i = \{X \subseteq E : |X| = k_i, 1 \leq k_i \leq m\}$; player i selects as her strategy $s_i(t)$, a k_i-tuple of edges at round t. Essentially, the set E of all resources is partitioned into k parts (or colors) $E_1, \ldots, E_k$, each part E_j containing all resources of the same kind (color). The cost of player i is the sum of the corresponding costs of the resources in her strategy.

There is a convenient transparency amongst independent-resource and singleton congestion games [15], with each player i competing over k_i kinds of resources interpreted as k_i *clones*, each acting independently and selfishly in her corresponding group of resources. Now, let n_j the number of clones in the subset E_j, containing all resources of a given kind. Then, it is convenient to view the *overall game* G as k independent

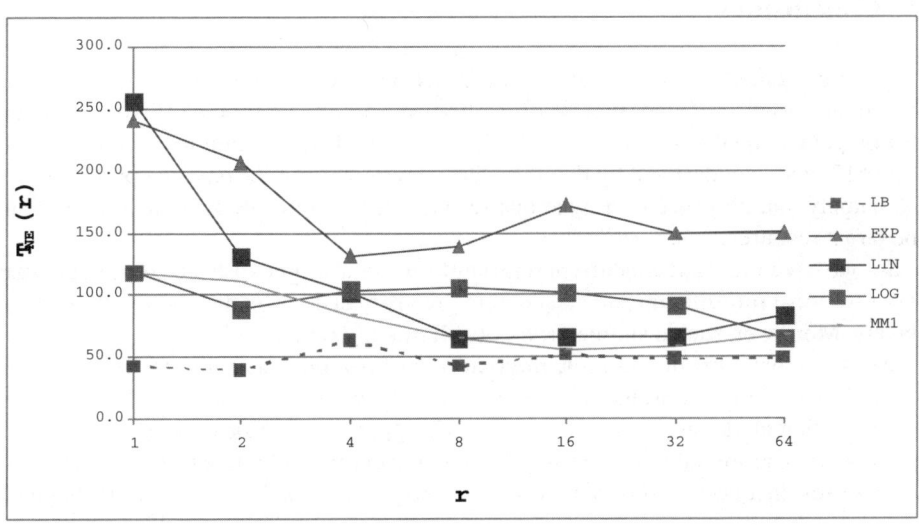

Fig. 3. Weighted CGs & PURE cost classes, as scaling for various densities, for fixed resources

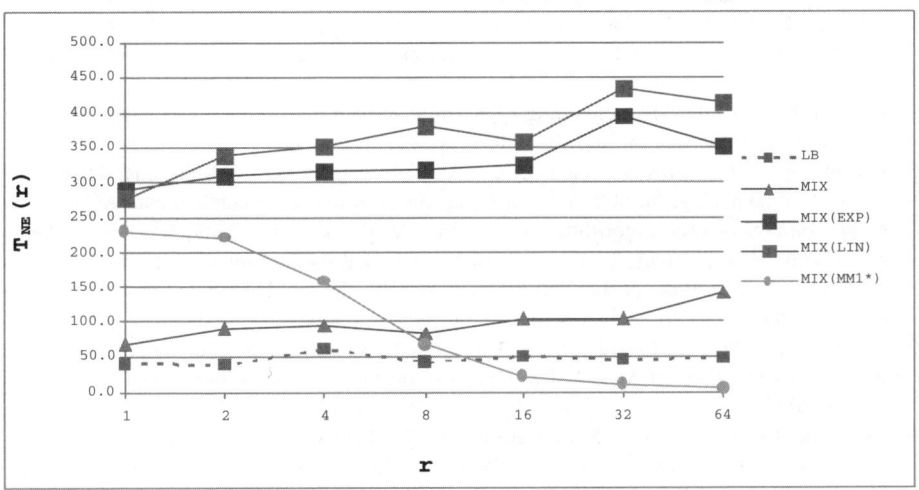

Fig. 4. Weighted CGs & MIX cost classes, as scaling for various densities, for fixed resources

congestion sub-games $G_1, \ldots, G_k$, with sub-game G_j is a singleton CG on n_j players over $|E_j|$ resources, $j = 1, \ldots, k$. This method will raise up to at most k times the corresponding singleton protocol's running time (actually, the experimental results for the independent-resource CGs are even better than k times the corresponding ones for singleton CGs, as shown in Section 4, but we omit them due to space limitations.)

6   Conclusions

We have presented a protocol for leading concurrent congestion games to Nash Equilibrium in a number of steps that is competitive to a baseline protocol [3]. We stress that the competitive quality of our protocol follows the "realistic assumption" recommendation of [7] and is underlined by the unrealistic assumption of the baseline protocol [12] [7], namely that players currently at one resource may arbitrarily migrate to more than one target resource.

So, we have modeled concurrent migrations using a simple distributed protocol that only uses local information and can contain greedy decisions on the part of all involved players. Moreover, such a simple protocol still quickly reaches a NE.

We stress that allowing multiple migrants out of resource per time step is a straightforward extension of a protocol developed in [11], which was theoretically shown to possess efficient (logarithmic) interesting NE approximation properties. Our experimental demonstration that the *extended* protocol can efficiently reach an *exact* NE naturally raises the question whether a relevant result could also be theoretically obtained.

References

1. Babaoglu, O., Canright, G., Deutsch, A., Di Caro, G.A., Ducatelle, F., Gambardella, L.M., Ganguly, N., Jelasity, M., Montemani, R., Urnes, T.: Design Patterns from Biology for Distributed Computing. ACM Transactions on Autonomous and Adaptive Systems 1(1), 26–66 (2006)
2. Barford, P., Crovella, M.: Generating representative web workloads for network and server performance evaluation. In: SIGMETRICS, pp. 151–160 (1998)
3. Berenbrink, P., Friedetzky, T., Goldberg, L.A., Goldberg, P., Hu, Z., Martin, R.: Distributed selfish load balancing. In: SODA 2006: Proceedings of the seventeenth annual ACM-SIAM symposium on Discrete algorithm, pp. 354–363. ACM Press, New York, NY, USA (2006)
4. Berenbrink, P., Friedetzky, T., Hu, Z.: A new analytical method for parallel, diffusion-type load balancing. In: Proc. of the 20th International Parallel and Distributed Processing Symposium (IPDPS) (2006)
5. Chien, S., Sinclair, A.: Convergece to Approximate Nash Equilibria in Congestion Games. In: Proc. of the 18th ACM-SIAM Symposium on Discrete Algorithms (SODA 2007) (to appear, 2007)
6. Christodoulou, G., Mirrokni, V.S., Sidiropoulos, A.: Convergence and approximation in potential games. In: Durand, B., Thomas, W. (eds.) STACS 2006. LNCS, vol. 3884, pp. 349–360. Springer, Heidelberg (2006)
7. Cybenko, G.: Dynamic Load Balancing for Distributed Memory Multiprocessors. Journal of Parallel Distributed Computing 7(2), 279–301 (1989)
8. Even-Dar, E., Kesselman, A., Mansour, Y.: Convergence Time to Nash Equilibria. In: Baeten, J.C.M., Lenstra, J.K., Parrow, J., Woeginger, G.J. (eds.) ICALP 2003. LNCS, vol. 2719, pp. 502–513. Springer, Heidelberg (2003)
9. Even-Dar, E., Mansour, Y.: Fast convergence of selfish rerouting. In: Proceedings of the Sixteenth Annual ACM-SIAM Symposium on Discrete Algorithms (SODA 2005), pp. 772–781 (2005)
10. Fabrikant, A., Papadimitriou, C., Talwar, K.: The Complexity of Pure Nash Equilibria. In: Proc. of the 36th ACM Symp. on Theory of Computing (STOC 2004), pp. 604–612 (2004)

11. Fotakis, D., Kaporis, A.C., Spirakis, P.G.: Atomic congestion games: Fast, myopic and concurrent. In: Proceedings of the 1st International Symposium on Algorithmic Game Theory (SAGT) (to appear, 2008)
12. Ghosh, B., Muthukrishnan, S.: Dynamic load balancing in parallel and distributed networks by random matchings. In: Proc. of the 6th Annual ACM Symposium on Parallel Algorithms and Architectures (SPAA), pp. 220–225 (1994)
13. Goemans, M.X., Mirrokni, V.S., Vetta, A.: Sink equilibria and convergence. In: FOCS, pp. 142–154 (2005)
14. Goldberg, P.W.: Bounds for the convergence rate of randomized local search in a multiplayer load-balancing game. In: Proc. of the twenty-third annual ACM symposium on Principles of distributed computing (PODC 2004), pp. 131–140. ACM Press, New York, NY, USA (2004)
15. Ieong, S., McGrew, R., Nudelman, E., Shoham, Y., Sun, Q.: Fast and compact: A simple class of congestion games. In: AAAI, pp. 489–494 (2005)
16. Koutsoupias, E., Papadimitriou, C.: Worst-case Equilibria. In: Meinel, C., Tison, S. (eds.) STACS 1999. LNCS, vol. 1563, pp. 404–413. Springer, Heidelberg (1999)
17. Libman, L., Orda, A.: Atomic resource sharing in noncooperative networks. Telecommunication Systems 17(4), 385–409 (2001)
18. Mirrokni, V.S., Vetta, A.: Convergence issues in competitive games. In: APPROX-RANDOM, pp. 183–194 (2004)
19. Orda, A., Rom, R., Shimkin, N.: Competitive routing in multiuser communication networks. IEEE/ACM Transactions on Networking 1(5), 510–521 (1993)
20. Rosenthal, R.W.: A Class of Games Possessing Pure-Strategy Nash Equilibria. International Journal of Game Theory 2, 65–67 (1973)
21. Valiant, G., Roughgarden, T.: Braess's paradox in large random graphs. In: EC 2006: Proceedings of the 7th ACM conference on Electronic commerce, pp. 296–305. ACM Press, New York, NY, USA (2006)

When to Reap and When to Sow –
Lowering Peak Usage with Realistic Batteries

Amotz Bar-Noy, Yi Feng, Matthew P. Johnson, and Ou Liu

Department of Computer Science
The Graduate Center of the City University of New York

Abstract. In some energy markets, large clients are charged for both total energy usage and peak energy usage, which is based on the maximum
single energy request over the billing period. The problem of minimizing peak charges was recently introduced as an online problem in [4],
which gave optimally competitive algorithms. In this problem, a battery
(previously assumed to be perfectly efficient) is used to store energy for
later use. In this paper, we extend the problem to the more realistic setting of *lossy* batteries, which lose to conversion inefficiency a constant
fraction of any amount charged (e.g. 33%). For this setting, we provide
efficient and optimal offline algorithms as well as possibly competitive
online algorithms. Second, we give *factor-revealing* LPs, which provide
some quasi-empirical evidence for competitiveness. Finally, we evaluate
these and other, heuristic algorithms on real and synthetic data.

1   Introduction

Power companies charge some high-consumption clients not just for the total
amount of power consumed, but also for how quickly they consume it. Within
the billing period (typically a month), the client is charged for the amount of
energy used (*usage charge*, in kWh) and for the maximum request (*peak charge*,
in kW). If demands are given as a sequence $(d_1, d_2, \ldots, d_n)$, then the total bill is
of the form $c_1 \sum_i d_i + c_2 \max_i\{d_i\}$, i.e., a weighted sum of the total usage and the
maximum usage. This means a client who powers a 100kW piece of machinery
for one hour and then uses no more energy for the rest of the month would be
charged more than a client who uses a total of 100kWh spread evenly over the
course of the month. Since the per-unit cost for peak charge may be on the order
of 100 times the per-unit cost for total usage[1], this difference can be significant.
Indeed, this is borne out in our experiments.

At least one start-up company [1] is currently marketing battery-based systems intended to reduce peak energy charges. In such a system, a battery is
placed between the power company and a high-consumption client site, in order to smooth power requests and shave the peak. The client site will charge
to the battery when demand is low and discharge when demand is high. Spikes

---

[1] The Orlando Utilities Commission website [3], e.g. quotes rates of 6.388¢/kWh (*"energy charge"*) and $6.50/kW (*"demand charge"*).

C.C. McGeoch (Ed.): WEA 2008, LNCS 5038, pp. 194–207, 2008.
© Springer-Verlag Berlin Heidelberg 2008

in the demand curve can thus be made consistent with a relatively flat level of supplied power, yielding lower cost for the client and more tractable requests for the provider. It is interesting to note that a battery system may actually *raise* energy usage, since there may be energy loss due to inefficiency in AC/DC conversion. This loss may be as much as 33% of the amount charged. Serving peak requests during periods of high demand is a difficult and expensive task for the power company, however, and the event of a black-out inflicts high societal costs. While a battery system may involve higher total energy requests, it may still benefit the system as a whole by easing the strain of peak demands.

In the online setting, the essential choice faced at each time is whether (and by how much) to invest in the future or to cash in on a prior investment. The investment in our setting is a request for more energy than is needed at the time. If the algorithm only asks for the minimum needed, then it is vulnerable to spikes in demand; if it asks for much more energy than it needs, then this request could itself become a new, higher peak. The strictness of the problem lies in the fact that we want *every* request to be low, not just to minimize a total.

In [4], we gave H_n-competitive algorithms for the online lossless setting and matching lower bounds on competitiveness. (These algorithms are in fact only partially online since they depend on having the maximum demand D revealed in advance. Lacking this information, no non-trivial competitiveness is possible.) Those algorithms assume perfectly efficient batteries, however, and will fail if run on realistic, lossy batteries. In the present paper, we adapt these algorithms to the lossy setting, testing them on both synthetic and actual customer usage data. Moreover, we test more aggressive, heuristic algorithms, as well as algorithms that accept predictions, with error, of future demands. We also consider new settings and objective functions, such as total cost. Finally, we provide factor-revealing LPs, which we use to provide quasi-empirical evidence of the competitiveness of the lossy algorithms.

Background. As noted above, the problem we study was introduced in [4]. There are many related problems in commodity production, storage, warehousing, etc. More specifically, there are many inventory problems based on the Economic Lot Sizing model [6], in which demand levels for a product vary over a discrete finite time-horizon and are known in advance. See [4] for a full discussion.

The goal in the minimax work-scheduling problem [7] is to minimize the maximum amount of work done in any timeslot over a finite time-horizon. Our online problem is related to a previously studied special case of this in which jobs with deadlines are assigned online. In that problem, all work must be done by deadline but cannot be begun until assigned. While the problems differ in important respects (see [4]), the objectives are similar. Indeed, while the α-policy of [7] performs α *times the maximum per-unit-timeslot amount of work that OPT would have done, when running on the partial input received so far*, many of our algorithms ensure that the savings at each point is a multiple of the optimal savings so far.

2   Model and Algorithms

Definition 1. *The* demand curve *is the timeslot-indexed sequence of energy demands* $(d_1, ..., d_n)$. *The* request curve *is the timeslot-indexed sequence of energy requests* r_i. *Battery charge level* b_i *indicates the (non-negative) amount of energy present in the battery* at the start *of timeslot* i. D *is the revealed maximum demand* $\max_i\{d_i\}$, *and* R *is the maximum request* $\max_i\{r_i\}$.

The demand curve (combined with battery information) is the problem instance; the request curve is the problem solution. In the absence of battery loss and overflow/underflow, the battery level at timeslot i is simply $b_i = b_{i-1} + r_{i-1} - d_{i-1}$. It is forbidden for b_i to ever fall below 0 (underflow). That is, the request r_i and the battery level b_i must sum to at least the demand d_i at each time i.

By discretizing we assume wlog that battery level, demand, and request values are all expressed in the same units ("kWh"). Peak charges are based linearly on the max request. We optimize for the peak charge, not for total energy usage, since the bulk of this is a fixed cost. There are several independent optional extensions, leading to many problem variants. The battery can have *maximum capacity* B or be unbounded; with some batteries, there is as already noted an automatic percentage *loss* $0 \leq \ell \leq 1$ in all charged energy, due to AC/DC conversion; the problem may be online, offline, or in between; we consider the settings in which the peak demand D is revealed in advance, or estimates of the individual demands are known, perhaps based on historical data. The loss model works as follows. For each unit of energy charged, only $r = 1 - \ell$ units will be available for discharge, due to the combined inefficiencies of charging and discharging. This loss could be broken into separate components ℓ_1, ℓ_2 for charge and discharge, but since the loss does not depend on time, doing so would have essentially no effect. For simplicity, we merge these losses into a single loss that occurs instantly at the time of charging.

Threshold algorithms. For a particular snapshot (d_i, r_i, b_i), demand d_i must be supplied through a combination of the request r_i and a change in battery $b_i - b_{i-1}$. This means that there are only three possible modes for each timestep: request exactly the demand, request more than the demand and *charge* the difference, or request less than the demand and *discharge* the difference. Our online and offline algorithms are *threshold algorithms*. If $(T_1, T_2, ..., T_n)$ are the chosen thresholds, then the algorithms behave as follows:

```
for each timeslot i
    if d_i < T_i
        charge min(B − b_i, T_i − d_i)
    else
        discharge min(d_i − T_i, b_i)
        if d_i − T_i < b_i
            T_i ← T_i + (d_i − T_i − b_i)
```

The algorithm schema amounts to the rule: *at each timeslot* i, *request an amount as near to* T_i *as the battery constraints will allow.* Our offline algorithms use a constant T (though in practice an offline algorithm could naturally lower its requests to avoid overflow); our online algorithms compute T_i dynamically for each timeslot i.

Definition 2. *Let* overflow *be the situation in which* $T_i - d_i > B - b_i$, *i.e., there is not enough room in the battery for the amount we want to charge. Let* underflow *be the situation in which* $d_i - T_i > b_i$, *i.e., there is not enough energy in the battery for the amount we want to discharge. Call a threshold algorithm* feasible *if underflow never occurs (overflow merely indicates a lower effective request).*

The second *if* statement of the algorithm schemas is executed only if underflow occurs. The competitiveness guarantee of Algorithm 2.b for the lossless setting was achieved in [4] by showing that such underflow would never occur. The factor-revealing LP below provides evidence that such underflow also never occurs in the lossy setting. Our heuristic algorithms choose lower, more aggressive thresholds, with the result that such underflow does (or rather would) occur. Since meeting demand is a strict requirement, in the event of underflow, the request rises accordingly to keep the battery level non-negative, which is what the *if* statement does.

Although there is no constant-ratio competitive algorithm for unbounded n, our intended application in fact presumes a fixed time-horizon. If the billing period is one month, and peak charges are computed as 30-minute averages, then for this setting H_n is approximately 7.84. If we assume that the battery can fully recharge at night, so that each day can be treated as a separate time period, then for a 12-hour daytime time-horizon H_n is approximately 3.76.

3   GA and Lossy Battery Algorithms

The algorithms for lossy batteries are structurally similar to those for lossless, except that computations of *average* are replaced with *generalized average* (GA). In all cases, the average computed over an interval will correspond to the best possible maximum request over that interval, which can be found by examining all subintervals. The algorithms used are shown below:

Alg.	online	threshold T_i	running-time
1	no	$\hat{\mu}(1,n)$	$O(n^2 \log n)$
2.a	yes	$D - \frac{D - \hat{\mu}(1,i)}{H_n}$	$O(n^2 \log n)$
2.b	yes	$D - \frac{D - \mu(s_i,i)}{H_{(n-s_i+1)}}$	$O(n \log n)$

Definition 3. *Given n real values $(y_1, y_2, \ldots, y_n)$ and constants $0 < r \leq 1$ and $B \geq 0$, let the* **generalized average** *$GA(y_1, y_2, \ldots, y_n)$ be the value μ satisfying $U(a) = B + r \cdot L(a)$, where: $U(a) = \sum_{i=1}^{n} \max(y_i - a, 0)$ and $L(a) = \sum_{i=1}^{n} \max(a - y_i, 0)$. We call $U(a)$ and $L(a)$ μ's L/U values or μ's upper and*

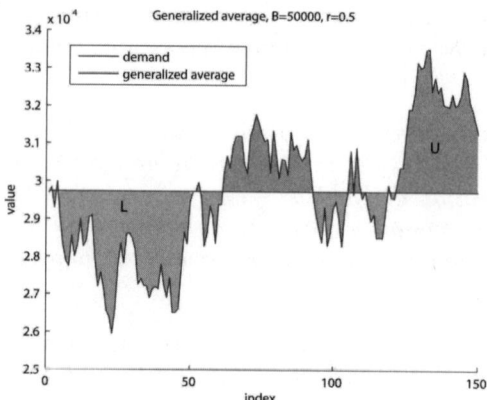

Fig. 1. Generalized Average, with U shaded darker and L shaded lighter

lower. *Treating the input values $y_1, ..., y_n$ as a step function $y = y(x)$, they correspond to the area above μ and below y (upper) and the area below μ and above y (lower).*

Some intuition may be provided by considering what is likely the simplest way, in terms of coding, of computing a GA: binary search in the range $[-B, \max_i\{d_i\}]$. For each candidate value μ, if $B + r \cdot L(\mu) > U(\mu)$ then shift downward and otherwise shift upward, until the two values are sufficiently close.

Note that μ need not be one of the y_i values. When $B = 0$ and $r = 1$, the generalized average is simply the mean of the values y_i; when $B = 0$ and r approaches 0, the generalized average approaches the maximum.

Computing a GA in $O(n \log n)$ is not difficult. First sort the values y_i, and let the values' subscripts now reflect their new positions. Next, set $L(y_1) = 0$, since y_1 is the smallest y_i. For each subsequent i up to n, $L(y_i)$ can be computed in constant time (we omit details due to space constraints). Similarly, compute each $U(y_i)$, starting with $U(y_n)$. Once all the U/Ls are computed, μ's *neighbors* (y_i, y_{i+1}), i.e., the two nearest input values that μ lies between, can be found by inspection, and given these μ can be computed in constant time. Unlike the ordinary arithmetic mean, however, computing a GA in $O(n)$ requires more effort.

Our recursive algorithm, whose behavior we sketch in words, both is inspired by the well-known linear-time deterministic Selection Algorithm [5], and calls it as a subroutine. The bulk of the work is in finding μ's *neighbors*. Given these data points (and their L/U values), we can solve for the correct value μ in constant time. (The cases–*not shown in the pseudocode*–when the solution μ is among the data points, and when μ is less than *all* the points can be checked as special cases.) The algorithm for finding the neighboring data points to μ takes the set of points y_i as input. Let $0 \leq r < 1$ and B be the parameters to the GA. The first parameter to the algorithm is the set of values to be averaged; all other parameters to the first (non-recursive) call are set to 0.

```
GenAvgNbrs(A[], X_U, X_L, W_U, W_L) :
    if length(A) == 2
        return A;
    else p = Select-Median(A);                           (a)
         (A_L, A_U) = Pivot(A,p);                         (b)
         U_p = Upper(A,p); L_p = Lower(A,p);              (c)
         U = U_p + X_U + W_U · (max(A) − p); L = L_p + X_L + W_L · (p − min(A));
         if U < r · L + B
             return GenAvgNbrs(A_U ∪ p, L, X_U, W_L + |A_L|, W_U);
         else if U > r · L + B
             return GenAvgNbrs(A_L ∪ p, X_L, U, W_L, W_U + |A_U|);
         else
             return p;
```

Theorem 1. $GA(y_1, ..., y_n)$ *can be computed in* $O(n)$ *time.*

Proof. (sketch) With $|A| = n$, lines a,b,c each take time $O(n)$ since *Select-Median* uses the Selection Algorithm, *Pivot* is the usual Quicksort pivoting algorithm, and Upper and Lower are computed directly. (Min and max can be passed in separately, but we omit them for simplicity.) The function makes one recursive call, whose input size is by construction half the original input size. Hence the total running time is $O(n)$.

The bulk of the work done by our algorithms for lossy batteries is to compute the GA for a series of ranges $[i, j]$, as i stays fixed (as e.g. 1) and j increases iteratively (e.g. from 1 to n). It is straightforward to do this in $O(n^2)$ time, by maintaining a sorted sublist of the previous elements, inserting each new y_j and computing the new GA in linear time. Unlike ordinary averages, $GA[i, j]$ and the value y_{j+1} do not together determine $GA[i, j + 1]$.[2] (The GA could also be computed separately for each region $[1, j]$.) This yields offline algorithms for the lossy unbounded and bounded settings, with running times $O(n^2)$ and $O(n^3)$. Through careful use of data structures, we obtain faster algorithms, with running times $O(n \log n)$ and $O(n^2 \log n)$, respectively.

Theorem 2. *The values* $GA[1, j]$, *as j ranges from 1 to n can be computed in* $O(n \log n)$.

Proof. (sketch) A balanced BST is used to store previous work so that going from $GA[i, j]$ to $GA[i, j + 1]$ is done in $O(\log n)$. Each tree node stores a y_i value plus other data (its L/U, etc.) used by GenAvgNbrs to run in $O(\log n)$. Each time a new data point y_i is inserted into the tree, its data must be computed (and the tree must be rebalanced). Unfortunately, each insertion *partly* corrupts *all other nodes' data*. Using a lazy evaluation strategy, we initially update only $O(\log n)$ values. After the insert, GenAvgNbrs is run on the tree's current set of data

---

[2] For example, when $B = 10$ and $r = .5$, $GA(5, 10, 15) = GA(3, 21, 3) = 7$, but $GA(5, 10, 15, 20) = 10.83 \neq GA(3, 21, 3, 20) = 11.33$.

points, in $O(\log n)$ time, relying only on the nodes' data that is guaranteed to be correct. Running on the BST, GENAVGNBRS's subroutines (Select-Median, Pivot, and selection of the subset to recurse on) now complete in $O(\log n)$, for a total of $O(n \log n)$.

Definition 4. *Let $\mu(i, j)$ be the GA of the demands over region $[i, j]$. Let $\hat{\mu}(h, k) = \max_{h \le i \le j \le k} \mu(i, j)$. At time i, let s_i be the most recent time when the battery was full.*

Theorem 3. *For the offline/lossy setting, Algorithm 1 ($T_i = \hat{\mu}(1, n)$) is optimal, feasible, and runs in time $O(n^2 \log n)$.*

Proof. Within any region $[i, j]$, the battery may help in two ways. First, the battery may be able to lower the local peak by sometimes charging and sometimes discharging. Second, the battery in the best case would start with charge B at timestep i. With battery loss percentage ℓ, the total amount discharged from the battery over this period can be at most B plus $(1 - \ell)$ times the total amount charged. The optimal threshold over this region cannot be less than $GA(d_i, ..., d_j)$ with $(1 - \ell, B)$ chosen as its parameters (r, B).

The threshold used is $T = \hat{\mu}(1, n)$. It suffices to show that the battery will be nonnegative after each time j. Suppose j is the first time underflow occurs. Let $i - 1$ be the last timestep prior j with a full battery (or 0 if this has never occurred). Then there is no underflow *or* overflow in $[i, j)$, so the total charged in region $[i, j]$ is exactly $U(T) = \sum_{t=i}^{j} \max(T - d_t, 0)$ and the total discharged will be $L(T) = \sum_{t=i}^{j} \max(d_t - T, 0)$. The amount of energy available for discharge over the entire period is $B + r \cdot L(T)$. Overflow at time j means $U(T) > B + r \cdot L(T)$, but this contradicts the definition of T.

To compute the thresholds, compute $GA[i, j]$ iteratively (varying j from i to n) for each value i. Each i value takes $O(n \log n)$, for a total of $O(n^2 \log n)$.

Corollary 1. *If the battery is effectively unbounded, then a similar optimal algorithm can be obtained, which runs in time $O(n \log n)$.*

4   Factor-Revealing LPs

If no underflow occurs, then algorithms 2.a and 2.b are H_n-competitive by construction. (Recall that the objective function is the peak reduction amount.) In this section, we use the factor-revealing LP technique of Jain et al. [8] to provide some quasi-empirical evidence that no such underflow can ever occur.

A factor-revealing LP is defined based on a particular algorithm for a problem. The LP variables correspond to possible instances, of a certain size n, of the optimization problem. (We therefore have an indexed family of linear programs.) The optimal solution value of the linear program reveals something about the algorithm it is based on. In the original Facility Location application, the objective function was the ratio of the cost incurred by the approximation algorithm in covering the facility and the optimal cost (assumed wlog to be 1) of doing so,

$$\begin{array}{l}
\textbf{\textit{min:}} \ b_{n+1} \\
\textbf{\textit{s.t.:}} \ b_{i+1} = b_i + T_i - d_i, \text{ for all } i \\
\qquad b_i \leq B \\
\qquad T_i = D - (D - opt_i)/H_n, \text{ for all } i \\
\qquad opt_i \geq (1/i)(-B + \sum_{j=1}^{i} d_j), \text{ for all } i \\
\\
\qquad b_1 = B \\
\qquad d_i \leq D, \text{ for all } i \\
\qquad D \geq 0, B = 1
\end{array}$$

Fig. 2. Factor-revealing linear program for lossless batteries (LP1)

so the maximum possible value of this ratio provided an upper bound on the algorithm's approximation guarantee.

The size index of our LPs is the number of timesteps n. The objective function is the final battery level b_{n+1}. The constraints are properties describing the behavior of the algorithm; some of the constraints perform book-keeping, including keeping track of the battery level over time. We first provide the factor-revealing LP for the lossless setting (Fig. 2), which is simpler than the lossy.

We now explain this program. The battery is initialized to B and can never supersede this level. As we argue below, we can limit ourselves without loss of generality to demand sequences in which the algorithm never wishes to charge to a level greater than B, i.e. no overflow occurs. For such inputs, the threshold scheme's first min has no effect and we always have that $b_{i+1} = b_i + T_i - d_i$. Threshold T_i is constrained in the LP to equal the expression for Algorithm 2.b's threshold, with opt_i lower-bounded by the closed-form expression for the analog of GA for lossless batteries [4]. Moreover, this value is less than or equal to the corresponding value used in Algorithm 2.a, with the effect that T_i is less than or equal to the corresponding threshold of Algorithm 2.a at every time i. This in turn means that feasibility of Algorithm 2.b implies feasibility of Algorithm 2.a. D is included in the program for clarity.

We solved this LP, written in AMLP, with LP solvers on the NEOS server [2], for several values $n \leq 100$. The solution value found was 0, consistent with the known result [4]. We now state the following lemma, which allows us to limit our attention to a certain family of demand sequences.

Lemma 1 ([4]). *If there is a demand sequence $d_1, d_2, ..., d_n$ in which underflow occurs for Algorithm 2.a, then there is also a demand sequence in which underflow continues to the end (i.e., $b_n < 0$) and no overflow ever occurs.*

Theorem 4. *If the optimal solution value LP1, for parameter size n, is at least 0, then Algorithms 2.a is H_n-competitive in the lossless setting, for problem size n.*

Proof. Suppose underflow were to occur at some time t, and let s be the most recent time prior to t when the battery was full. Then by the lemma, $[s, t]$ can be assumed wlog to be $[1, n]$. The assumptions that the battery starts at level B and never reaches this level again (though it may rise and fall non-monotonically) are implemented by the constraints stating that $b_1 = B$ and $b_i \geq B$. Since no overflow occurs, the first min in the threshold algorithm definition has no effect, and the battery level changes based only on T_i and d_i, i.e., it rises by amount $T_i - d_i$ (which may be negative), which is stated in constraints. Since $opt_i = \hat{\mu}(1, i)$ is the max of expressions for all sequences of $[1, i]$, in particular we have that $opt_i \geq (-B + \sum_{j=1}^{i} d_i)/i$. The optimal solution value to such an LP equals the lowest possible battery level which can occur, given any possible problem instance of size n, when using any algorithm consistent the these constraints. Since in particular the behavior of Algorithms 2.a is consistent with these constraints, the result follows.

Corollary 2. *If the optimal solution value LP1, for all parameter sizes $\leq n$, is at least 0, then Algorithm 2.b is H_n-competitive in the lossless setting, for problem sizes $\leq n$.*

Proof. In Algorithm 2.b, threshold T_i is defined based on the region beginning after the last overflow at position $s = s_i$, shifting $[s, t]$ to $[1, t'] = [1, t-s+1]$ has *no effect*. If 2.b *instead* always used H_n, then the lemma above would directly apply, since the algorithm would then perform identically on $[s, t]$ to $[1, t']$, and then the underflow could again be extended to the end. The result that LP1's optimal solution value is ≥ 0 would then imply that this *modified* Algorithm 2.b is feasible for inputs of size n.

In fact, the actual Algorithm 2.b uses H_{n-s+1}, with $s = s_i$, at time i. In effect, Algorithm 2.b treats the demand suffix $d_{s+1}, ..., d_n$ as an independent problem instance of size $n - s + 1$. Each time the battery overflows, the harmonic number subscript is modified, and there is a new subregion and a new possibility for underflow. If all sizes of overflow-free subregions are underflow-free, then the algorithm is feasible. Therefore, if LP1's optimal solution is non-negative, Algorithm 2.b is feasible.

The more complicated Factor-revealing program for the lossy setting is shown in Fig. 3. The additional difficulty here is that the program has quadratic constraints.

We omit a full description of this program and only remark that the two main difficulties are 1) that there is an essential asymmetry in the battery behavior, which complicates the first constraints; and 2) that since we have no closed formula for GA, (a lower bound on) the optimal threshold must be *described* rather than *computed*.

There are three sets of quadratic constraints, indicated by stars. In fact, it is possible to remove these and convert LP2 to a linearly-constrained quadratic program, defined for a fixed constant efficiency L. Unfortunately, the resulting QP is not convex.

We then have the following results.

Theorem 5. *If the optimal solution value LP2, for parameter size n, is at least 0, then Algorithms 2.a is H_n-competitive in the lossy setting, for problem size n.*

$$\begin{aligned}
&\textbf{\textit{min:}}\ b_{n+1}\\
&\textbf{\textit{s.t.:}}\ b_{i+1} = b_i + L \cdot ch_i - dis_i,\ \text{for all } i\\
&\qquad ch_i \cdot dis_i = 0 \qquad\qquad\qquad\qquad\quad (*)\\
&\qquad b_i \leq B,\ \text{for all } i\\
&\qquad ch_i, dis_i \geq 0,\ \text{for all } i\\[6pt]
&\qquad b_1 = B\\
&\qquad D \geq d_i,\ \text{for all } i\\
&\qquad B = 1, D \geq 0\\[6pt]
&\qquad T_i = D - (D - opt_i)/H_n\\
&\qquad T_i = d_i - dis_i + ch_i\\
&\qquad opt_i \geq ga_i,\ \text{for all } i\\[6pt]
&\qquad B + L \cdot \left(\textstyle\sum_{j=1}^{i} cho_{i,j}\right) = \sum_{j=1}^{i} diso_{i,j},\ \text{for all } i \quad (*)\\
&\qquad cho_{i,j} \cdot diso_{i,j} = 0,\ \text{for all } (j,i) : j \leq i \qquad (*)\\
&\qquad ga_i = d_j - diso_{i,j} + cho_{i,j},\ \text{for all } (j,i) : j \leq i\\
&\qquad cho_{i,j}, diso_{i,j} \geq 0,\ \text{for all } (j,i) : j \leq i
\end{aligned}$$

Fig. 3. Factor-revealing quadratically-constrained LP for lossy batteries (LP2)

Corollary 3. *If the optimal solution value LP2, for all parameter sizes $\leq n$, is at least 0, then Algorithm 2.b is H_n-competitive in the lossy setting, for problem sizes $\leq n$.*

We solved the second program, implemented in AMLP, using several solvers (MINLP, MINOS, SNOPT) on the NEOS server [2], for several values $n \leq 100$. (The number of variables is quadratic in n, and there are limits to the amount of memory NEOS provides.) In all case, we found the solution value found was (within a negligible distance of) non-negative. Although these solvers do not guarantee a globally optimal solution (at least not for non-convex quadratically-constrained LPs), we believe this performance provides some "quasi-empirical" evidence for the correctness of Algorithms 2.a and 2.b.

5   Performance Evaluation

5.1   Experiment Setup

We performed experiments on three datasets: a regular business day's demand from an actual Starbucks store, a simulated weekday demand of a residential user, and a randomly generated demand sequence. Each dataset is of length $n = 200$. The demand curves are shown in Fig. 4. The parameters in our simulation 1) battery size B (typically $B = 500K$), 2) battery charging loss factor L (typically $L = 0.33$), 3) aggressiveness c (with $c = 0$ for 2.b and $c = 1$ for 2.b-opt).

Since the objective is peak minimization, we modify the algorithms so that the requests are monotonically increasing (except when prevented by overflow).

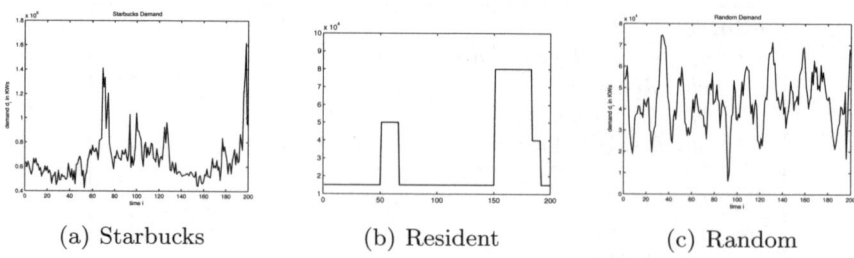

(a) Starbucks   (b) Resident   (c) Random

Fig. 4. Input data: demand versus time

Since the peak must be at least $D - B$, we similarly force this to be the minimum request (again barring overflow). Although the underlying offline algorithm assumes that $b_1 = B$, other lower initial battery levels can be simulated by artificially increasing the initial demand(s). In the next subsection, we discuss a sample of the experiments performed.

5.2   Simulation Results

We know that in the lossless setting, our algorithms are H_n-competitive in terms of *peak reduction*, since no underflow occurs. We first wish to test the performance of the corresponding lossy algorithms, as well as other, heuristic algorithms.

Test 1 - battery size. In this test, we measured the peaks produced by the different algorithms running with various battery sizes, for settings including lossy and lossless and initial battery levels of $b_1 = 0$ and $b_1 = B$. We observe the same general patterns throughout. For random input, performance is averaged over 50 runs. We observe (Fig. 5) that increasing the battery size reduces the peak in our optimal algorithm; we also see that Algorithm *2.b* constantly outperforms Algorithm *2.a*, and that they both are within H_n of opt, as expected. We include the heuristic Algorithm *2.b-opt*, for comparison, which at each point attempts threshold $opt_i = \hat{\mu}(1, i)$, i.e., to have, at all times, the same peak as opt would have had so far. This is a very bold algorithm. We see that it can perform badly with too large a battery since its aggressiveness can then have greater effect, increasing the likelihood of underflow.

In our next test, we seek a middle-ground between the conservativeness of *2.b* and the boldness of *2.b-opt*.

Test 2 - aggressiveness. We vary the boldness in an algorithm based on *2.b* by using a threshold $T_i = D - \frac{D - \hat{\mu}(1,i)}{1 + (H_{n-s+1} - 1)c}$, with parameter c. When $c = 1$, the algorithm is *2.b* (most conservative); when $c = 1$, it is *2.b-opt* (most aggressive). In this test, we measure the performance as c varies from 0 to 1 with increment of 0.1. We compare the performance of Algorithm *2.b* as a reference. We used two battery sizes on the scale of battery size in the Starbucks installation. We observe (Fig. 5) that increased aggressiveness improves performance, but only up to a point, for reasons indicated above. We note that the best aggressive factor c can depend on both battery size B and the input data.

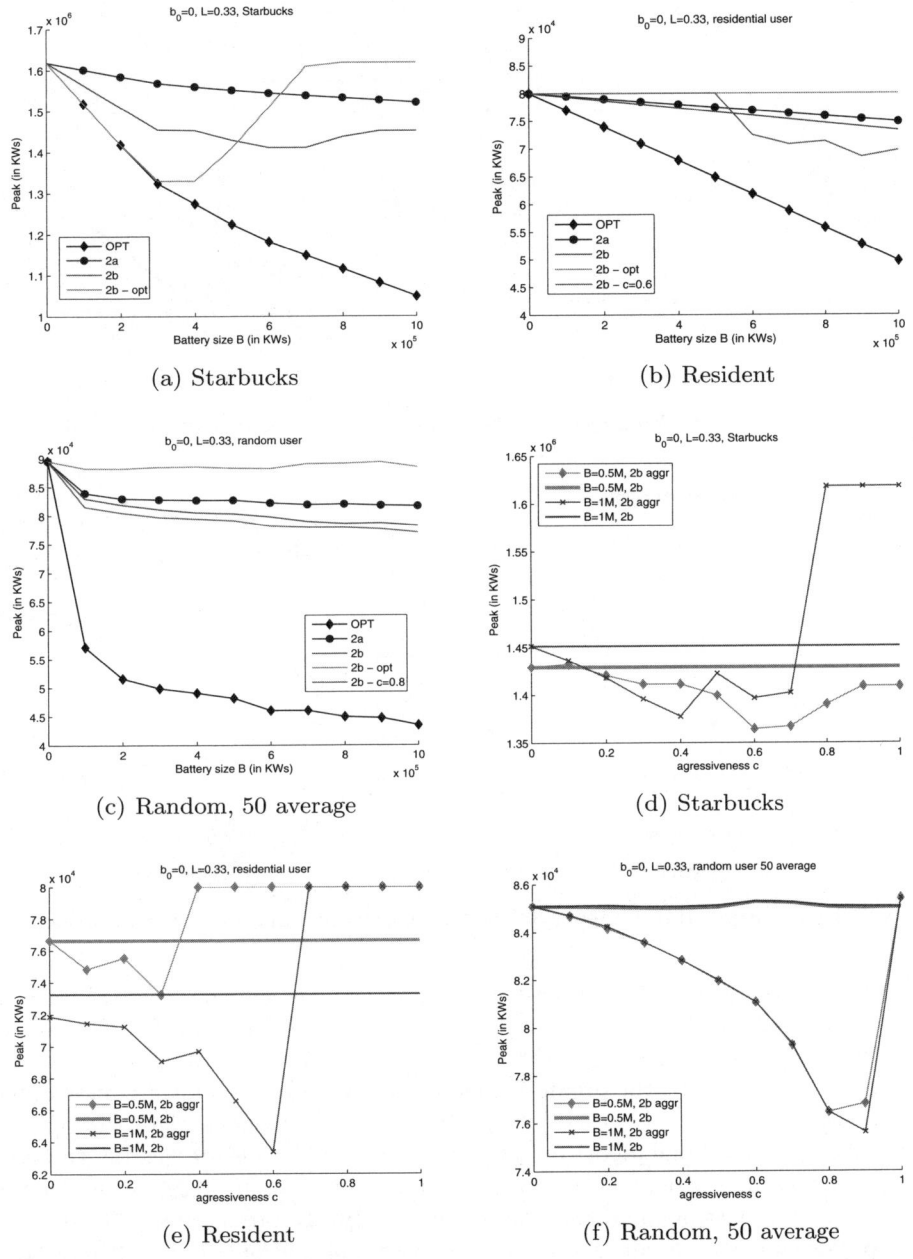

Fig. 5. Test 1: Peak versus B, $b_1 = 0$, $L = 0.33$ (a) (b) (c); Test 2: Peak versus aggressiveness c, $b_1 = 0$, $L = 0.33$ (d) (e) (f)

Although we naturally find that too much *unmotivated* boldness can be damaging, there are potential situations in which significant boldness can be justified.

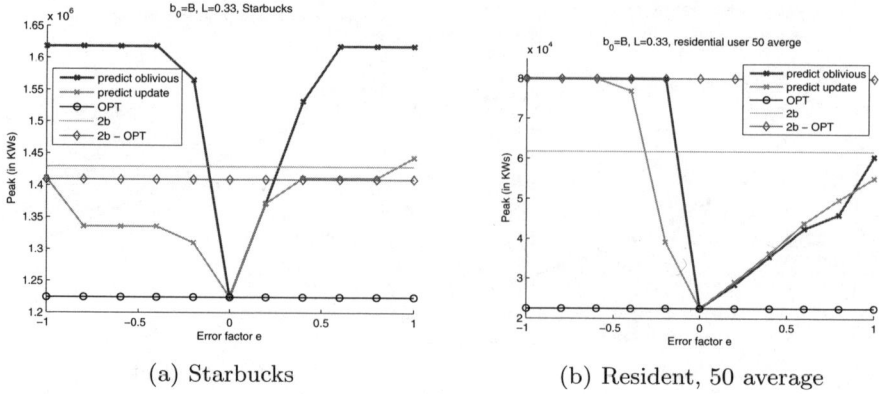

(a) Starbucks   (b) Resident, 50 average

Fig. 6. Test 3: Peak versus prediction error e, $b_1 = B = 500k$, $L = 0.33$

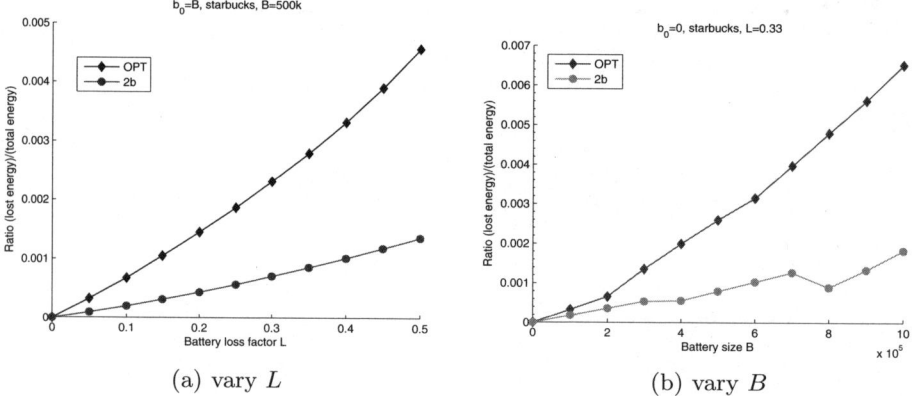

(a) vary L (b) vary B

Fig. 7. Test 4: Ratio of energy loss and energy demand, $b_1 = 0$, Starbucks

Test 3 - predictions. Suppose we are give error-prone but reasonably accurate predictions of future demands, based e.g. on historical data. In this test, we test two prediction-based algorithms. Let p_i be the predicted demand sequence. Let error $e \in [-1, 1]$ be the prediction error level, with p_i is uniformly distributed in $[d_i, d_i(1 + e)]$ or $[d_i(1 + e), d_i]$ if $e < 0$. First, the *oblivious* algorithm simply runs the optimal offline algorithm on the p_i values. The *update* version runs the offline algorithm on demand sequence $< d_1, \ldots, d_i, p_{i+1}, p_n >$, in which the d_i are the actual past demands and the p_i are the future predictions. We compare the performances of prediction algorithms with optimal offline algorithm, Algorithms *2.b* and *2.b-opt* as references. We vary the prediction values from most optimistic $e = -1$ to most conservative $e = 1$.

We see (Fig. 6) that the performance of the prediction algorithm varies in roughly inversely proportion to the error level. If the prediction error is less than 20%, both prediction algorithms outperform the two online algorithms. As e approaches zero, the performance naturally converges to the optimal.

Test 4 - lost energy. As noted in the introduction, the use of lossy batteries increases the total energy used. In this test, we compare the lost energy during charging process with the total energy demand (Fig. 7). We verify that the amount of lost energy is negligible compared with the total energy demand. We naturally find, however, that larger B and larger loss factor L increase energy loss. We believe that the facts that the fraction of lost energy is small and that the per-unit energy charge is significantly lower than the per-unit peak charge vindicate our choice to focus on peak charge.

6   Conclusion

In this paper, we presented optimal offline algorithms and both heuristic and possibly competitive online algorithms for the peak reduction problem with lossy batteries. The factor-revealing LPs for the lossy setting presently provide only quasi-empirical evidence for competitiveness. The potential future availability of global quadratically-constrained LP solvers, however, could provide computer-aided proof of such competitiveness, at least for instances of bounded size. Several additional future extensions suggest themselves:

- additional free but unreliable energy sources (e.g. solar power)
- limited battery charging/discharging speed
- battery loss over time ("self-discharge')
- multi-dimensional demands and resulting complex objective functions.

Acknowledgements. This work was supported by grants from the NSF (grant number 0332596) and the New York State Office of Science, Technology and Academic Research. We thank Deniz Sariöz and Ted Brown for useful discussions. We also thank Ib Olsen of Gaia for posing the problem and for providing the Starbucks dataset.

References

1. Gaia Power Technologies, gaiapowertech.com
2. NEOS server, Argonne National Lab, www-neos.mcs.anl.gov/neos/solvers/
3. Orlando Utilities Commission, www.ouc.com/account/rates/electric-comm.htm
4. Bar-Noy, A., Johnson, M., Liu, O.: Peak shaving through resource buffering. Technical Report TR-2007018, CUNY Graduate Center, Dept. of Computer Science (November 2007)
5. Cormen, T.H., Leiserson, C.E., Rivest, R.L., Stein, C.: Introduction to Algorithms, 2nd edn. MIT Press, McGraw-Hill (2001)
6. Florian, M., Lenstra, J., Kan, A.R.: Deterministic production planning: algorithms and complexity. Management Science 26 (1980)
7. Hunsaker, B., Kleywegt, A.J., Savelsbergh, M.W.P., Tovey, C.A.: Optimal online algorithms for minimax resource scheduling. SIAM J. Discrete Math. (2003)
8. Jain, K., Mahdian, M., Markakis, E., Saberi, A., Vazirani, V.V.: Greedy facility location algorithms analyzed using dual fitting with factor-revealing lp. J. ACM 50(6), 795–824 (2003)

Characterizing the Performance of Flash Memory Storage Devices and Its Impact on Algorithm Design*

Deepak Ajwani[1], Itay Malinger[2], Ulrich Meyer[3], and Sivan Toledo[4]

[1] Max Planck Institut für Informatik, Saarbrücken, Germany
[2] Tel Aviv University, Tel Aviv, Israel
[3] Johann Wolfgang Goethe Universität, Frankfurt a.M., Germany
[4] Massachusetts Institute of Technology, Massachusetts, USA

Abstract. Initially used in digital audio players, digital cameras, mobile phones, and USB memory sticks, flash memory may become the dominant form of end-user storage in mobile computing, either completely replacing the magnetic hard disks or being an additional secondary storage. We study the design of algorithms and data structures that can exploit the flash memory devices better. For this, we characterize the performance of NAND flash based storage devices, including many solid state disks. We show that these devices have better random read performance than hard disks, but much worse random write performance. We also analyze the effect of misalignments, aging and past I/O patterns etc. on the performance obtained on these devices. We show that despite the similarities between flash memory and RAM (fast random reads) and between flash disk and hard disk (both are block based devices), the algorithms designed in the RAM model or the external memory model do not realize the full potential of the flash memory devices. We later give some broad guidelines for designing algorithms which can exploit the comparative advantages of both a flash memory device and a hard disk, when used together.

1 Introduction

Flash memory is a form of non-volatile computer memory that can be electrically erased and reprogrammed. Flash memory devices are lighter, more shock resistant, consume less power and hence are particularly suited for mobile computing. Initially used in digital audio players, digital cameras, mobile phones, and USB memory sticks, flash memory may become the dominant form of end-user storage in mobile computing: Some producers of notebook computers have already launched models (Apple MacBook Air, Sony Vaio UX90, Samsung Q1-SSD and Q30-SSD) that completely abandon traditional hard disks in favor of flash memory (also called solid state disks). Market research company In-Stat

* Supported in part by the DFG grant ME 3250/1-1, and by MADALGO - Center for Massive Data Algorithmics, a Center of the Danish National Research Foundation.

C.C. McGeoch (Ed.): WEA 2008, LNCS 5038, pp. 208–219, 2008.

predicted in July 2006 that 50% of all mobile computers would use flash (instead of hard disks) by 2013.

Frequently, the storage devices (be it hard disks or flash) are not only used to store data but also to actually compute on it if the problem at hand does not completely fit into main memory (RAM); this happens on both very small devices (like PDAs used for online route planning) and high-performance compute servers (for example when dealing with huge graphs like the web). Thus, it is important to understand the characteristics of the underlying storage devices in order to predict the real running time of algorithms, even if these devices are used as an external memory. Traditionally, algorithm designers have been assuming a uniform cost access to any location in the storage devices. Unfortunately, real architectures are becoming more and more sophisticated, and will become even more so with the advent of flash devices. In case of hard disks, the access cost depends on the current position of the disk-head and the location that needs to be read/written. This has been well researched; and there are good computation models such as the external memory model [1] or the cache-oblivious model [6] that can help in realistic analysis of algorithms that run on hard disks. This paper attempts to characterize the performance (read/writes; sequential/random) of flash memory devices; to see the effects of random writes, misalignment and aging etc. on the access cost and its implications on the real running time of basic algorithms.

External memory model. The external memory model (or the I/O model) proposed by Aggarwal and Vitter [1] is one of the most commonly used model when analyzing the performance of algorithms that do not fit in the main memory and have to use the hard disk. It assumes a single central processing unit and two levels of memory hierarchy. The internal memory is fast, but has a limited size of M words. In addition, we have an external memory which can only be accessed using I/Os that move B contiguous words between internal and external memory. At any particular time stamp, the computation can only use the data already present in the internal memory. The measure of performance of an algorithm is the number of I/Os it performs.

State of the art for flash memories. Recently, there has been growing interest in using flash memories to improve the performance of computer systems [4,9,11]. This trend includes the experimental use of flash memories in database systems [9,11], in Windows Vista's use of USB flash memories as a cache (a feature called ReadyBoost), in the use of flash memory caches in hard disks (e.g., Seagate's Momentus 5400 PSD hybrid drives, which include 256 MB on the drive's controller), and in proposals to integrate flash memories into motherboards or I/O busses (e.g., Intel's Turbo Memory technology).

Most previous algorithmic work on flash memory concerns *operating system* algorithms and data structures that were designed to efficiently deal with flash memory cells wearing out, e.g., block-mapping techniques and flash-specific file systems. A comprehensive overview on these topics was recently published by Gal and Toledo [7]. The development of application algorithms tuned to flash memory is in its absolute infancy. We are only aware of very few published results beyond file systems and wear leveling:

Wu et al. [12,13] proposed flash-aware implementations of *B*-trees and *R*-trees without file system support by explicitly handling block-mapping within the application data structures.

Goldberg and Werneck [8] considered point-to-point shortest-path computations on pocket PCs where preprocessed input graphs (road networks) are stored on flash-memory; due to space-efficient internal-memory data-structures and locality in the inputs, data manipulation remains restricted to internal memory, thus avoiding difficulties with unstructured flash memory write accesses.

Goals. Our first goal is to see how standard algorithms and data structures for basic algorithms like scanning, sorting and searching designed in the RAM model or the external memory model perform on flash storage devices. An important question here is whether these algorithms can effectively use the advantages of the flash devices (such as faster random read accesses) or there is a need for a fundamentally different model for realizing the full potential of these devices.

Our next goal is to investigate why these algorithms behave the way they behave by characterizing the performance of more than 20 different low-end and high-end flash devices under typical access patterns presented by basic algorithms. Such a characterization can also be looked upon as a first step towards obtaining a model for designing and analyzing algorithms and data structures that can best exploit flash memory. Previous attempts [9,11] at characterizing the performance of these devices reported measurements on a small number of devices (1 and 2, respectively), so it is not yet clear whether the observed behavior reflects the flash devices, in general. Also, these papers didn't study if these devices exhibit any second-order effects that may be relevant.

Our next goal is to produce a benchmarking tool that would allow its users to measure and compare the relative performance of flash devices. Such a tool should not only allow users to estimate the performance of a device under a given workload in order to find a device with an appropriate cost-effectiveness for a particular application, but also allow quick measurements of relevant parameters of a device that can affect the performance of algorithms running on it.

These goals may seem easy to achieve, but they are not. These devices employ complex logical-to-physical mapping algorithms and complex mechanisms to decide which blocks to erase. The complexity of these mechanisms and the fact that they are proprietary mean that it is impossible to tell exactly what factors affect the performance of a device. A flash device can be used by an algorithm designer like a hard disk (under the external memory or the cache-oblivious model), but its performance may be far more complex.

It is also possible that the flash memory becomes an additional secondary storage device, rather than replacing the hard disk. Our last, but not least, goal is to find out how one can exploit the comparative advantages of both in the design of application algorithms, when they are used together.

Outline. The rest of the paper is organized as follows. In Section 2, we show how the basic algorithms perform on flash memory devices and how appropriate the standard computation models are in predicting these performances. In Section 3, we present our experimental methodology, and our benchmarking program, which we use to measure and characterize the performance of many different flash devices. We also show the effect of random writes, misalignment and aging on the performance of these devices. In Section 4, we provide an algorithm design framework for the case when flash devices are used together with a hard disk.

2 Implications of Flash Devices for Algorithm Design

In this section, we look at how the RAM model and external memory model algorithms behave when running on flash memory devices. In the process, we try to ascertain whether the analysis of algorithms in either of the two models also carry over to the performance of these algorithms obtained on flash devices.

In order to compare the flash memory with DRAM memory (used as main memory), we ran a basic RAM model list ranking algorithm on two architectures - one with 4GB RAM memory and the other with 2GB RAM, but 32 GB flash memory. The list ranking problem is that given a list with individual elements randomly stored on disk, find the distance of each element from the head of the list. The sequential RAM model algorithm consists of just hoping from one element to its next, and thereby keeping track of the distances of node from the head of the list. Here, we do not consider the cost of writing the distance labels of each node.

We stored a 2^{30}-element list of long integers (8 Bytes) in a random order, i.e. the elements were kept in the order of a random permutation generated beforehand. While ranking such a list took minutes in RAM, it took days with flash. This is because even though the random reads are faster on flash disks than the hard disk, they are still much slower than RAM. Thus, we conclude that RAM model is not useful for predicting the performance (or even relative performance) of algorithms running on flash memory devices and that standard RAM model algorithms leave a lot to be desired if they are to be used on flash devices.

Table 1. Runtime of basic algorithms when running on Seagate Barracuda 7200.11 hard disk as compared to 32 GB Hama Solid State disk

Algorithm	Hard Disk	Flash
Generating a random double and writing it	0.2 μs	0.37 μs
Scanning (per double)	0.3 μs	0.28 μs
External memory Merge-Sort (per double)	1.06 μs	1.5 μs
Random read	11.3 ms	0.56 ms
Binary Search	25.5 ms	3.36 ms

As Table 1 shows, the performance of basic algorithms when running on hard disks and when running on flash disks can be quite different, particularly when it comes to algorithms involving random read I/Os such as binary search on a sorted array. While such algorithms are extremely slow on hard disks necessitating B-trees and other I/O-efficient data structures, they are acceptably fast on flash devices. On the other hand, algorithms involving write I/Os such as merge sort (with two read and write passes over the entire data) run much faster on hard disk than on flash.

It seems that the algorithms that run on flash have to achieve a different tradeoff between reads and writes and between sequential and random accesses than hard disks. Since the cost of accesses don't drop or rise proportionally over the entire spectrum, the algorithms running on flash devices need to be qualitatively different from the one on hard disk. In particular, they should be able to tradeoff write I/Os at the cost of extra read I/Os. Standard external memory algorithms that assume same cost for reading and writing fail to take advantage of fast random reads offered by flash devices. Thus, there is a need for a fundamentally different model for realistically predicting the performance of algorithms running on flash devices.

3   Characterization of Flash Memory Devices

In order to see why the standard algorithms behave as mentioned before, we characterize more than 20 flash storage devices. This characterization can also be looked at as a first step towards a model for designing and analyzing algorithms and data structures running on flash memory.

3.1   Flash Memory

Large-capacity flash memory devices use NAND flash chips. All NAND flash chips have common characteristics, although different chips differ in performance and in some minor details. The memory space of the chip is partitioned into blocks called *erase blocks*. The only way to change a bit from 0 to 1 is to erase the entire unit containing the bit. Each block is further partitioned into *pages*, which usually store 2048 bytes of data and 64 bytes of meta-data (smaller chips have pages containing only 512+16 bytes). Erase blocks typically contain 32 or 64 pages. Bits are changed from 1 (the erased state) to 0 by *programming* (writing) data onto a page. An erased page can be programmed only a small number of times (one to three) before it must be erased again. Reading data takes tens of microseconds for the first access to a page, plus tens of nanoseconds per byte. Writing a page takes hundreds of microseconds, plus tens of nanoseconds per byte. Erasing a block takes several milliseconds. Finally, erased blocks wear out; each block can sustain only a limited number of erasures. The guaranteed numbers of erasures range from 10,000 to 1,000,000. To extend the life of the chip as much as possible, erasures should therefore be spread out roughly evenly over the entire chip; this is called *wear leveling*.

Because of the inability to overwrite data in a page without first erasing the entire block containing the page, and because erasures should be spread out over the chip, flash memory subsystems map *logical block addresses* (LBA) to physical addresses in complex ways [7]. This allows them to accept new data for a given logical address without necessarily erasing an entire block, and it allows them to avoid early wear even if some logical addresses are written to more often than others. This mapping is usually a non-trivial algorithm that uses complex data structures, some of which are stored in RAM (usually inside the memory device) and some on the flash itself.

The use of a mapping algorithm within LBA flash devices means that their performance characteristics can be worse and more complex than the performance of the raw flash chips. In particular, the state of the on-flash mapping and the volatile state of the mapping algorithm can influence the performance of reads and writes. Also, the small amount of RAM can cause the mapping mechanism to perform more physical I/O operations than would be necessary with more RAM.

3.2   Configuration

The tests were performed on many different machines – a 1.5GHz Celeron-M with 512M RAM, a 3.0GHz Pentium 4 with 2GB OF RAM, a 2.0Ghz Intel dual core T7200 with 2GB OF RAM, and a 2 x Dual-core 2.6 GHz AMD Opteron with 2.5 GB OF RAM. All of these machines were running a 2.6 Linux kernel.

The devices include USB sticks, compact-flash and SD memory cards and solid state disks (of capacities 16GB and 32GB). They include both high-end and low-end devices. The USB sticks were connected via a USB 2.0 interface, memory cards were connected through a USB 2.0 card reader (made by Hama) or PCMCIA interface, and solid state disks with IDE interface were installed in the machines using a 2.5 inch to 3.5 inch IDE adapter and a PATA serial bus.

Our benchmarking tool and methodology. Standard disk benchmarking tools like zcav fail to measure things that are important in flash devices (e.g., write speeds, since they are similar to read speeds on hard disks, or sequential-after-random writes); and commercial benchmarks tend to focus on end-to-end file-system performance, which does not characterize the performance of the flash device in a way that is useful to algorithm designers. Therefore, we decided to implement our own benchmarking program that is specialized (designed mainly for LBA flash devices), but highly flexible and can easily measure the performance of a variety of access patterns, including random and sequential reads and writes, with given block sizes and alignments, and with operation counts or time limits.

3.3   Result and Analysis

Performance of steady, aligned access patterns. Figure 1 shows the performance of two typical devices under the aligned access patterns. The other devices that we tested varied greatly in the absolute performance that they

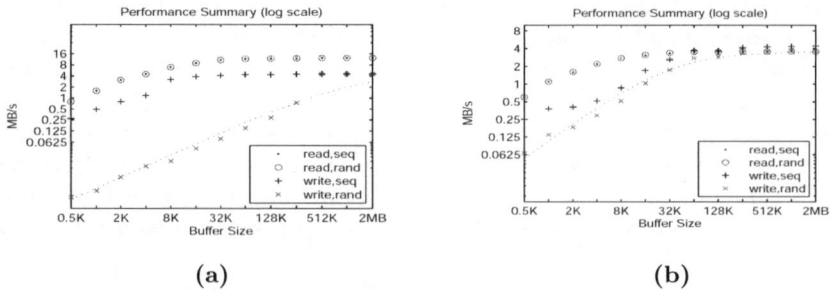

Fig. 1. Performance (in logarithmic scale) of the (a) 1 GB Toshiba TransMemory USB flash drive and the (b) 1 GB Kingston compact-flash card

Table 2. The tested devices and their performance (in MBps) under sequential and random reads and writes with block size of 512 Bytes and 2 MB

DEVICE		Buffer size 512 Bytes				BUFFER SIZE 2 MB			
NAME	SIZE	SR	RR	SW	RW	SR	RR	SW	RW
KINGSTON DT SECURE	512M	0.97	0.97	0.64	0.012	33.14	33.12	14.72	9.85
MEMOREX MINI TRAVELDRIVE	512M	0.79	0.79	0.37	0.002	13.15	13.15	5.0	5.0
TOSHIBA TRANSMEMORY	512M	0.78	0.78	0.075	0.003	12.69	12.69	4.19	4.14
SANDISK U3 CRUZER MICRO	512M	0.55	0.45	0.32	0.013	12.8	12.8	5.2	4.8
M-SYSTEMS MDRIVE	1G	0.8	0.8	0.24	0.005	26.4	26.4	15.97	15.97
M-SYSTEMS MDRIVE 100	1G	0.78	0.78	0.075	0.002	12.4	12.4	3.7	3.7
TOSHIBA TRANSMEMORY	1G	0.8	0.8	0.27	0.002	12.38	12.38	4.54	4.54
SMI FLASH DEVICE	1G	0.97	0.54	0.65	0.01	13.34	13.28	9.18	7.82
KINGSTON CF CARD	1G	0.60	0.60	0.25	0.066	3.55	3.55	4.42	3.67
KINGSTON DT ELITE HS 2.0	2G	0.8	0.8	0.22	0.004	24.9	24.8	12.79	6.2
KINGSTON DT ELITE HS 2.0	4G	0.8	0.8	0.22	0.003	25.14	25.14	12.79	6.2
MEMOREX TD CLASSIC 003C	4G	0.79	0.17	0.12	0.002	12.32	12.15	5.15	5.15
120X CF CARD	8G	0.68	0.44	0.96	0.004	19.7	19.5	18.16	16.15
SUPERTALENT SOLID STATE FLASH DRIVE	16G	1.4	0.45	0.82	0.028	12.65	12.60	9.84	9.61
HAMA SOLID STATE DISK 2.5" IDE	32G	2.9	2.18	4.89	0.012	28.03	28.02	24.5	12.6
IBM DESKSTAR HARD DRIVE	60G	5.9	0.03	4.1	0.03	29.2	22.0	24.2	16.2
SEAGATE BARRACUDA 7200.11 HARD DISK	500G	6.2	0.063	5.1	0.12	87.5	69.6	88.1	71.7

achieved, but not in the general patterns; all followed the patterns shown in Figures 1a and 1b.

In all the devices that we tested, small random writes were slower than all the other access patterns. The difference between random writes and other access patterns is particularly large at small buffer sizes, but it is usually still evident even on fairly large block sizes (e.g., 256KB in Figure 1a and 128KB in Figure 1b). In most devices, small-buffer random writes were at least 10 times slower than sequential writes with the same buffer size, and at least 100 times slower than sequential writes with large buffers. Table 2 shows the read/write access time with two different block sizes (512 Bytes and 2 MB) for sequential and random accesses on some of the devices that we tested.

We believe that the high cost for random writes of small blocks is because of the LBA mapping algorithm in these devices. These devices partition the virtual and physical address spaces into chunks larger than an erase block; in many cases 512KB. The LBA mapping maps areas of 512KB logical addresses to physical ranges of the same size. On encountering a write request, the system writes the new data into a new physical chunk and keeps on writing contiguously in this physical chunk till it switches to another logical chunk. The logical chunk is now mapped twice. Afterwards, when the writing switches to another logical chunk, the system copies over all the remaining pages in the old chunk and erases it. This way every chunk is mapped once, except for the active chunk, which is mapped twice. On devices that behave like this, the best random-write performance (in seconds) is on blocks of 512KB (or whatever is the chunk size). At that size, the new chunk is written without even reading the old chunk. At smaller sizes, the system still ends up writing 512KB, but it also needs to read stuff from the old location of this chunk, so it is slower. We even found that on some devices, writing randomly 256 or 128KB is slower than writing 512KB, in absolute time.

In most devices, reads were faster than writes in all block sizes. This typical behavior is shown in Figure 1a.

Another nearly-universal characteristic of the devices is the fact that sequential reads are not faster than random reads. The read performance does depend on block size, but usually not on whether the access pattern is random or sequential.

The performance in each access pattern usually increases monotonically with the block size, up to a certain saturation point. Reading and writing small blocks is always much slower than the same operation on large blocks.

The exceptions to these general rules are discussed in detail in [2].

Comparison to hard disks. Quantitatively, the only operation in which LBA flash devices are faster than hard disks is random reads of small buffers. Many of these devices can read a random page in less than a millisecond, sometimes less than 0.5ms. This is at least 10 times faster than current high-end hard disks, whose random-access time is 5-15ms. Even though the random-read performance of LBA flash devices varies, all the devices that we tested exhibited better random-read times than those of hard disks.

In all other aspects, most of the flash devices tested by us are inferior to hard disks. The random-write performance of LBA flash devices is particularly bad and particularly variable. A few devices performed random writes about as fast as hard disks, e.g., 6.2ms and 9.1ms. But many devices were more than 10 times slower, taking more than 100ms per random write, and some took more than 300ms.

Even under ideal access patterns, the flash devices we have tested provide smaller I/O bandwidths than hard disks. One flash device reached read throughput approaching 30MB/s and write throughput approaching 25MB/s. Hard disks can achieve well over 100MB/s for both reads and writes. Even disks designed

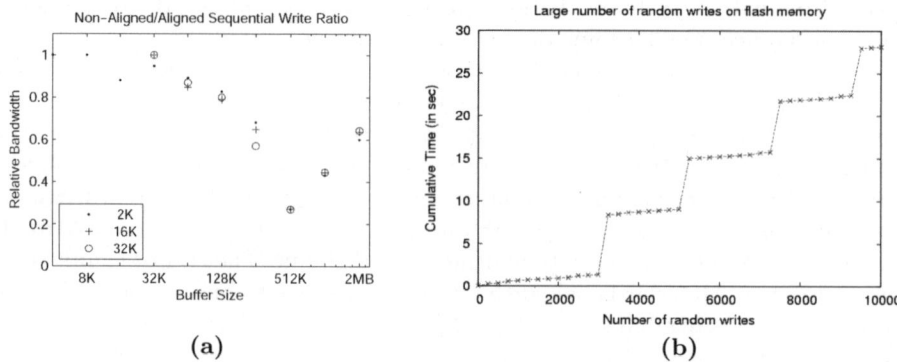

Fig. 2. (a) Effect of misalignment on the performance of flash devices (b) Total time taken by large number of random writes on a 32 GB Hama Solid state disk

for laptops can achieve throughput approaching 60MB/s. Flash devices would need to improve significantly before they outperform hard disks in this metric. The possible exception to this conclusion is large-capacity flash devices utilizing multiple flash chips, which should be able to achieve high throughput by writing in parallel to multiple chips.

Performance of large number of random writes. We observed an interesting phenomenon (Figure 2b) while performing large number of random writes on a 32 GB Hama (2.5" IDE) solid state disk. After the first 3000 random writes (where one random write is writing a 8-byte real number at a random location in a 8 GB file on flash), we see some spikes in the total running time. Afterwards, these spikes are repeated regularly after every 2000 random writes. This behavior is not restricted to Hama solid state disk and is observed in many other flash devices.

We believe that it is because the random writes make the page table more complex. After a while, the controller rearranges the pages in the blocks to simplify the LBA mapping. This process takes 5-8 seconds while really writing the data on the disk takes less than 0.8 seconds for 2000 random writes, causing the spikes in the total time.

Effects of misalignment. On many devices, misaligned random writes achieve much lower performance than aligned writes. In this setting, alignment means that the starting address of the write is a multiple of the block size. We have not observed similar issues with sequential access and with random reads.

Figure 2a shows the ratio between misaligned and aligned random writes. The misalignment is by 2KB, 16KB and 32KB. All of these sizes are at most as large as a single flash page. Many of the devices that we have tested showed some performance drop on misaligned addresses, but the precise effect varied from device to device. For example, the 128MB SuperTalent USB device is affected by misalignment by 2KB but not by misalignments of 16KB or 32KB.

Effect of random writes on subsequent operations. On some devices, a burst of random writes slows down subsequent sequential writes. The effect can last a minute or more, and in rare cases hours (of sustained writing). No such effect was observed on subsequent reads.

In these experiments, we performed t seconds of random writing, for $t = 5, 30$ and 60. We then measured the performance of sequential writes during each 4 second period for the next 120 seconds. For very small blocks, the median performance in the two minutes that follow the random writes can drop by more than a factor of two. Even on larger blocks, performance drops by more than 10%.

Effects of Aging. We were not able to detect a significant performance degradation as devices get older (in terms of the number of writes and erasures). On a (512MB KINGSTON DATATRAVELER II+) device, we observed that the performance of each access pattern remains essentially constant, even after 320,000 sequential writes on the entire device. The number of writes exceeded the rated endurance of the device by at least a factor of 3.

4   Designing Algorithms to Exploit Flash When Used Together with a Hard Disk

Till now, we discussed the characteristics of the flash memory devices and the performance of algorithms running on architectures where the flash disks replace the hard disks. Another likely scenario is that rather than replacing hard disk, flash disk may become an additional secondary storage, used together with hard disk. From the algorithm design point of view, it leads to many interesting questions. A fundamental question here is how can we best exploit the comparative advantages of the two devices while running an application algorithm.

The simple idea of directly using external memory algorithms with input and intermediate data randomly striped on the two disks treats both the disks as equal. Since the sequential throughput and the latency for random I/Os of the two devices is likely to be very different, the I/Os of the slower disk can easily become a bottleneck, even with asynchronous I/Os.

The key idea in designing efficient algorithms in such a setting is to restrict the random accesses to a static data-structure. This static data-structure is then kept on the flash disk, thereby exploiting the fast random reads of these devices and avoiding unnecessary writing. The sequential read and write I/Os are all limited to the hard disk.

We illustrate this basic framework with the help of external memory BFS algorithm of Mehlhorn and Meyer [10].

The BFS algorithm of Mehlhorn and Meyer [10] involves a preprocessing phase to restructure the adjacency lists of the graph representation. It groups the nodes of the input graph into disjoint clusters of small diameter and stores the adjacency lists of the nodes in a cluster contiguously on the disk. The key idea is that by spending only one random access (and possibly some sequential accesses depending on cluster size) in order to load the whole cluster and then keeping the cluster data in some efficiently accessible data structure (hot pool) until it is all used up, the total amount of I/Os can be reduced by a factor of up to $\sqrt{B}$ on sparse graphs. The neighboring nodes of a BFS level can be computed simply by scanning the hot pool and not the whole graph. Removing the nodes visited in previous two levels by parallel scanning gives the nodes in the next BFS level (a property true only for undirected graphs). Though some edges may be scanned more often in the pool, random I/Os to fetch adjacency lists is considerably reduced.

This algorithm is well suited for our framework as random I/Os are mostly restricted to the data structure keeping the graph clustering, while the hot pool accesses are mostly sequential. Also, the graph clustering is only stored once while the hot pool is modified (read and written) in every iteration. Thus, we keep the graph clustering data structure in the flash disk and the hot pool on the hard disk.

We ran a fast implementation [3] of this algorithm on a graph class that is considered difficult for the above mentioned algorithm. This graph class is a tree with $\sqrt{B}+1$ BFS levels. Level 0 contains only the source node which has an edge to all nodes in level 1. Levels $1 \ldots \sqrt{B}$ have $\frac{n}{\sqrt{B}}$ nodes each and the ith node in j^{th} level $(1 < j < \sqrt{B})$ has an edge to the ith node in levels $j-1$ and $j+1$.

As compared to striping the graph as well as pool randomly between the hard disk and the flash disk, the strategy of keeping the graph clustering data structure in flash disk and hot pool in hard disk performs around 25% better. Table 3 shows the running time for the second phase of the algorithm for a 2^{28}-node graph. Although the number of I/Os in the two cases are nearly the same, the time spent waiting for I/Os is much smaller for our disk allocation strategy, leading to better overall runtime.

The cluster size in the BFS algorithm was chosen in a way so as to balance the random reads and sequential I/Os on the hard disks, but now in this new setting, we can reduce the cluster size as the random I/Os are being done much faster by the flash memory. Our experiments suggest that this leads to even further improvements in the runtime of the BFS algorithm.

Table 3. Timing (in hours) for the second phase of Mehlhorn/Meyer's BFS algorithm on 2^{28}-node graph

Operation	Random striping		Our strategy	
	1 Flash + 1 Hard disk	2 Hard disks	Same cluster size	Smaller cluster size
I/O wait time	10.5	6.3	7.1	5.8
Total time	11.7	7.5	8.1	6.3

5   Discussion

Our results indicate that there is a need for more experimental analysis to find out how the existing external memory and cache-oblivious data structures like priority queues and search trees perform, when running on flash devices. Such experimental studies should eventually lead to a model for predicting realistic performance of algorithms and data structures running on flash devices, as well as on combinations of hard disks and flash devices. Coming up with a model that can capture the essence of flash devices and yet is simple enough to design and analyze algorithms and data structures, remains an important challenge.

As a first model, we may consider a natural extension of the standard external-memory model that will distinguish between block accesses for reading and writing. The I/O cost measure for an algorithm incurring x read I/Os and y write I/Os could be $x + c_W \cdot y$, where the parameter $c_W > 1$ is a penalty factor for write accesses.

An alternative approach might be to assume different block transfer sizes, B_R for reading and B_W for writing, where $B_R < B_W$ and $c_R \cdot x + c_W \cdot y$ (with $c_R, c_W > 1$) would be the modified cost measure.

References

1. Aggarwal, A., Vitter, J.S.: The input/output complexity of sorting and related problems. Communications of the ACM 31(9), 1116–1127 (1988)
2. Ajwani, D., Malinger, I., Meyer, U., Toledo, S.: Characterizing the performance of flash memory storage devices and its impact on algorithm design. Max Planck Institut für Informatik, Research report no. MPI-I-2008-1-001
3. Ajwani, D., Meyer, U., Osipov, V.: Improved external memory BFS implementations. In: ALENEX 2007, pp. 3–12 (2007)
4. Birrell, A., Isard, M., Thacker, C., Wobber, T.: A design for high-performance flash disks. SIGOPS Oper. Syst. Rev. 41(2), 88–93 (2007)
5. Chen, P.M., Patterson, D.A.: A new approach to I/O performance evaluation—self-scaling I/O benchmarks, predicted I/O performance. In: ACM SIGMETRICS Conference on Measurement and Modeling of Computer Systems, pp. 1–12, 10–14 (1993)
6. Frigo, M., Leiserson, C.E., Prokop, H., Ramachandran, S.: Cache-oblivious algorithms. In: FOCS, pp. 285–297. IEEE Computer Society Press, Los Alamitos (1999)
7. Gal, E., Toledo, S.: Algorithms and data structures for flash memories. ACM Computing Surveys 37, 138–163 (2005)
8. Goldberg, A., Werneck, R.: Computing point-to-point shortest paths from external memory. In: ALENEX 2005, SIAM, Philadelphia (2005)
9. Lee, S.-W., Moon, B.: Design of flash-based DBMS: An in-page logging approach. In: Chan, C.Y., Ooi, B.C., Zhou, A. (eds.) SIGMOD Conference, pp. 55–66. ACM, New York (2007)
10. Mehlhorn, K., Meyer, U.: External-memory breadth-first search with sublinear I/O. In: Möhring, R.H., Raman, R. (eds.) ESA 2002. LNCS, vol. 2461, pp. 723–735. Springer, Heidelberg (2002)
11. Myers, D., Madden, S.: On the use of NAND flash disks in high-performance relational databases. Manuscript (2007)
12. Wu, C.-H., Chang, L.-P., Kuo, T.-W.: An efficient B-tree layer for flash-memory storage systems. In: Chen, J., Hong, S. (eds.) RTCSA 2003. LNCS, vol. 2968, Springer, Heidelberg (2004)
13. Wu, C.-H., Chang, L.-P., Kuo, T.-W.: An efficient R-tree implementation over flash-memory storage systems. In: Proceedings of the eleventh ACM international symposium on Advances in geographic information systems, pp. 17–24. ACM Press, New York (2003)

Fast Local Search for the Maximum Independent Set Problem*

Diogo V. Andrade[1], Mauricio G.C. Resende[2], and Renato F. Werneck[3]

[1] Google Inc., 76 Ninth Avenue, New York, NY 10011, USA
diogo@google.com
[2] AT&T Labs Research, 180 Park Ave, Florham Park, NJ 07932, USA
mgcr@research.att.com
[3] Microsoft Research Silicon Valley, 1065 La Avenida, Mtn. View, CA 94043, USA
renatow@microsoft.com

Abstract. Given a graph $G = (V, E)$, the independent set problem is that of finding a maximum-cardinality subset S of V such that no two vertices in S are adjacent. We present a fast local search routine for this problem. Our algorithm can determine in linear time whether a maximal solution can be improved by replacing a single vertex with two others. We also show that an incremental version of this method can be useful within more elaborate heuristics. We test our algorithms on instances from the literature as well as on new ones proposed in this paper.

1   Introduction

The *maximum independent set problem* (MIS) takes a connected, undirected graph $G = (V, E)$ as input, and tries to find the largest subset S of V such that no two vertices in S have an edge between them. Besides having several direct applications [2], MIS is closely related to two other well-known optimization problems. To find the *maximum clique* (the largest complete subgraph) of a graph G, it suffices to find the maximum independent set of the complement of G. Similarly, to find the *minimum vertex cover* of $G = (V, E)$ (the smallest subset of vertices that contains at least one endpoint of each edge in the graph), one can find the maximum independent set S of V and return $V \setminus S$. Because these problems are NP-hard [11], for most instances one must resort to heuristics to obtain good solutions within reasonable time.

Most successful heuristics [1,7,8,9,12,14,15] maintain a single current solution that is slowly modified by very simple operations, such as individual insertions or deletions and swaps (replacing a vertex by one of its neighbors). In particular, many algorithms use the notion of *plateau search*, which consists in performing a randomized sequence of swaps. A swap does not improve the solution value by itself, but with luck it may cause a non-solution vertex to become free, at which point a simple insertion can be performed. Grosso et al. [8] have recently obtained

---

* Part of this work was done while the first author was at Rutgers University and the third author at Princeton University.

C.C. McGeoch (Ed.): WEA 2008, LNCS 5038, pp. 220–234, 2008.
© Springer-Verlag Berlin Heidelberg 2008

exceptional results in practice by performing plateau search almost exclusively. Their method (as well as several others) occasionally applies a more elaborate operation for diversification purposes, but spends most of its time performing basic operations (insertions, deletions, and swaps), often chosen at random.

This paper expands the set of tools that can be used effectively within meta-heuristics. We present a fast (in theory and practice) implementation of a natural local search algorithm. It is based on (1,2)-swaps, in which a single vertex is removed from the solution and replaced by two others. We show that one can find such a move (or prove that none exists) in linear time. In practice, an incremental version runs in sublinear time. The local search is more powerful than simple swaps, but still cheap enough for effective use within more elaborate heuristics. We also briefly discuss a generalization of this method to deal with (2,3)-swaps, i.e., two removals followed by three insertions.

Another contribution is a more elaborate heuristic that illustrates the effectiveness of our local search. Although the algorithm is particularly well-suited for large, sparse instances, it is competitive with previous algorithms on a wide range of instances from the literature. As an added contribution, we augmented the standard set of instances from the literature with new (and fundamentally different) instances, never previously studied in the context of the MIS problem.

This paper is organized as follows. Section 2 establishes the notation and terminology we use. Our local search algorithm is described in Section 3. Section 4 illustrates how it can be applied within a more elaborate heuristic. Experimental results are presented in Section 5, and final remarks are made in Section 6.

2 Basics

The input to the MIS problem is a graph $G = (V, E)$, with $|V| = n$ and $|E| = m$. We assume that vertices are labeled from 1 to n. We use the adjacency list representation: each vertex keeps a list of all adjacent vertices, sorted by label. One can enforce the ordering in linear time by applying radix sort to all edges.

A solution S is simply a subset of V in which no two vertices are adjacent. The *tightness* of a vertex $v \notin S$, denoted by $\tau(v)$, is the number of neighbors of v that belong to S. We say that a vertex is *k-tight* if it has tightness k. The tightnesses of all vertices can be computed in $O(m)$ time: initialize all values to zero, then traverse the adjacency list of each solution vertex v and increment $\tau(w)$ for every arc (v, w). Vertices that are 0-tight are called *free*. A solution is *maximal* if it has no free vertices.

Our algorithms represent a solution S as a permutation of all vertices, partitioned into three blocks: first the $|S|$ vertices in the solution, then the free vertices, and finally the nonfree vertices. The order among vertices within a block is irrelevant. The sizes of the first two blocks are stored explicitly. In addition, the data structure maintains, for each vertex, its tightness (which allows us to determine if the vertex is free) and its position in the permutation (which allows the vertex to be moved between blocks in constant time).

This structure must be updated whenever a vertex v is inserted into or removed from S. The only vertices that change are v itself and its neighbors,
so each such operation takes time proportional to the degree of v, denoted by
$\deg(v)$. This is more expensive than in simpler solution representations (such as
lists or incidence vectors), but the following operations can be easily performed
in constant time: (1) check if the solution is maximal (i.e., if the second block
is empty); (2) check if a vertex is in the solution (i.e., if it belongs to the first
block); (3) determine the tightness of a non-solution vertex; and (4) pick a vertex
within any of the three blocks uniformly at random.

3   Local Search

A (j, k)-*swap* consists of removing j vertices from a solution and inserting k
vertices into it. For simplicity, we refer to a (k, k)-swap as a k-*swap* (or simply
a *swap* when $k = 1$), and to a $(k - 1, k)$-swap as a k-*improvement*. We use the
term *move* to refer to a generic (j, k)-swap.

Our main local search algorithm is based on 2-improvements. These natural
operations have been studied before (see e.g. [6]); our contribution is a faster
implementation. Given a maximal solution S, we would like to replace some
vertex $x \in S$ with two vertices, v and w (both originally outside the solution),
thus increasing the total number of vertices in the solution by one. We test
each solution vertex $x \in S$ in turn. In any 2-improvement that removes x, both
vertices inserted must be neighbors of x (by maximality) that are 1-tight (or the
new solution would not be valid) and not adjacent to each other.

To process x efficiently, we first build a list $L(x)$ consisting of the 1-tight
neighbors of x, sorted by vertex identifier. If $L(x)$ has fewer than two elements,
we are done with x: it is not involved in any 2-improvement. Otherwise, we
must find, among all candidates in $L(x)$, a pair $\{v, w\}$ such that there is no
edge between v and w. We do this by processing each element $v \in L(x)$ in turn.
For a fixed candidate v, we check if there is a vertex $w \in L(x)$ (besides v) that
does not belong to $A(v)$, the adjacency list of v. Since both $L(x)$ and $A(v)$ are
sorted by vertex identifier, this can be done by traversing both lists in tandem.
All elements of $L(x)$ should appear in the same order within $A(v)$; if there is a
mismatch, the missing element is the vertex w we are looking for.

We claim that this algorithm finds a valid 2-improvement (or determines that
none exists) in $O(m)$ time. This is clearly a valid bound on the time spent
scanning all vertices (i.e., traversing their adjacency lists), since each vertex is
scanned at most once. Each solution vertex x is scanned to build $L(x)$ (the list
of 1-tight neighbors), and each 1-tight non-solution vertex v is scanned when its
only solution neighbor is processed. (Non-solution vertices that are not 1-tight
are not scanned at all.) We still need to bound the time spent traversing the
$L(x)$ lists. Each list $L(x)$ may be traversed several times, but each occurs in tandem with the traversal of the adjacency list $A(v)$ of a distinct 1-tight neighbor
v of x. Unless the traversal finds a valid swap (which occurs only once), traversing

$L(x)$ costs no more than $O(\deg(v))$, since each element of $L(x)$ (except v) also occurs in $A(v)$. This bounds the total cost of such traversals to $O(m)$.

An alternative linear-time implementation is as follows. As before, process each solution vertex x in turn. First, temporarily remove x from S. Then, for each newly-free neighbor v of x, insert v into S and check if the solution becomes maximal. If it does, simply remove v and process the next neighbor of x; if it does not, inserting any free vertex will yield a valid 2-improvement.

We have also considered more powerful local search algorithms. In particular, using generalized (and more complicated) versions of the techniques above, one can detect a 3-improvement (of prove that none exists) in $O(m\Delta)$ time, where Δ is the maximum vertex degree. Similarly, 2-swaps can be implemented in linear time. Due to space constraints, we omit a full description of these algorithms.

3.1   Incremental Version

A typical local search procedure does not restrict itself to a single iteration. If a valid 2-improvement is found, the algorithm will try to find another in the improved solution. This can of course be accomplished in linear time, but we can do better with an *incremental* version of the local search, which uses information gathered in one iteration to speed up later ones.

The algorithm maintains a set of *candidates*, which are solution vertices that might be involved in a 2-improvement. So far, we have assumed that all solution vertices are valid candidates, and we test them one by one. After a move, we would test all vertices again. Clearly, if we establish that a candidate x cannot be involved in a 2-improvement, we should not reexamine it unless we have good reason to do so. More precisely, when we "discard" a candidate vertex x, it is because it does not have two independent 1-tight neighbors. Unless at least one other neighbor of x becomes 1-tight, there is no reason to look at x again.

To accomplish this, we maintain a list of candidates that is updated whenever the solution changes. Any move (including a 2-improvement) can be expressed in terms of insertions and deletions of individual vertices. When we insert a vertex v into the solution, its neighbors are the only vertices that can become 1-tight, so we simply (and conservatively) add v to the list of candidates. When a vertex x is removed from the solution, the update is slightly more complicated. We must traverse the adjacency list of x and look for vertices that became 1-tight due to its removal. By definition, each such vertex v will have a single neighbor y in the solution; y must be inserted into the candidate list. We can find the solution vertex adjacent to each 1-tight neighbor v in constant time, as long as we maintain with each non-solution vertex the list of its solution neighbors.[1] Therefore, we could still update the candidate list after removing x in $O(\deg(x))$ time. For simplicity, however, we do not maintain the auxiliary data structures in our implementation, and explicitly scan each 1-tight neighbor of x.

---

[1] Standard doubly-linked lists will do, but updating them is nontrivial. In particular, when removing a vertex x from the solution, we must be able to remove in constant time the entry representing x in the list of each neighbor v. This can be accomplished by storing a pointer to that entry together with the arc (x, v) in x's adjacency list.

Although we have framed our discussion in terms of 2-improvements, these updates can of course be performed for any sequence of removals and/or insertions. As we will see, this means we can easily embed the incremental local search algorithm into more elaborate heuristics. Once invoked, the local search itself is quite simple: it processes the available candidates in random order, and stops when the list of candidates is empty.

3.2   Maximum Clique

Although our experiments focus mainly on the MIS problem, it is worth mentioning that one can also implement a linear-time 2-improvement algorithm for the maximum clique problem. Simply running the algorithm above on the complement of the input is not enough, since the complement may be much denser.

Given a maximal clique C, we must determine if there is a vertex $x \in C$ and two vertices $v, w \notin C$ such that the removal of x and the insertion of v and w would lead to a larger clique. Such a move only exists if the following holds: (1) v and w are neighbors; (2) both v and w are adjacent to all vertices in $C \setminus \{x\}$; and (3) at least one of v or w is not a neighbor of x (by maximality). For a vertex v with tightness $|C| - 1$, define its *missing neighbor* $\mu(v)$ as the only solution vertex to which v is not adjacent. There is a 2-improvement involving v if it has a neighbor w such that either (1) $\tau(w) = |C|$ or (2) $\tau(w) = |C| - 1$ and $\mu(w) = \mu(v)$. Knowing this, the local search procedure can be implemented in $O(m)$ time as follows. First, determine the tightness of all vertices, as well as the missing neighbors of those that are $(|C|-1)$-tight. Then, for each $(|C|-1)$-tight vertex v, determine in $O(\deg(v))$ time if it has a valid neighbor w.

4   Metaheuristics

4.1   Iterated Local Search

To test our local search, we use it within a heuristic based on the *iterated local search* (ILS) metaheuristic [13]. We start from a random solution S, apply local search to it, then repeatedly execute the following steps: (1) $S' \leftarrow$ perturb(S); (2) $S' \leftarrow$ localsearch(S'); (3) set $S \leftarrow S'$ if certain conditions are met. Any reasonable stopping criterion can be used, and the algorithm returns the best solution found. The remainder of this section details our implementation of each step of this generic algorithm.

Perturbations are performed with the *force(k)* routine, which sequentially inserts k vertices into the solution (the choice of which ones will be explained shortly) and removes the neighboring vertices as necessary. (We call these *forced insertions*.) It then adds free vertices at random until the solution is maximal. We set $k = 1$ in most iterations, which means a single vertex will be inserted. With small probability $(1/(2 \cdot |S|))$, however, we pick a higher value: k is set to $i + 1$ with probability proportional to $1/2^i$, for $i \geq 1$. We must then choose *which* k vertices to insert. If $k = 1$, we pick a random non-solution vertex. If k

is larger, we start with a random vertex, but pick the j-th vertex (for $j > 1$) among the non-solution vertices within distance exactly two from the first $j - 1$ vertices. (If there is no such vertex, we simply stop inserting.)

We use two techniques for diversification. The first is *soft tabu*. We keep track of the last iteration in which each non-solution vertex was part of the solution. Whenever the *force* routine has a choice of multiple vertices to insert, it looks at κ (an input parameter) candidates uniformly at random (with replacement) and picks the "oldest" one, i.e., the one which has been outside the solution for the longest time. We set $\kappa = 4$ in our experiments. The second diversification technique is employed during local search. If v was the only vertex inserted by the *force* routine, the subsequent local search will only allow v to be removed from the solution after all other possibilities have been tried.

Regarding the third step of the main loop, if the solution S' obtained after the local search is at least as good as S, S' becomes the new current solution. If $|S'| < |S|$, we have observed that always going to S' may cause the algorithm to stray from the best known solution too fast. To avoid this, we use a technique akin to plateau search. If ILS arrives at the current solution S from a solution that was better, it is not allowed to go to a worse solution for at least $|S|$ iterations. If the current solution does not improve in this time, the algorithm is again allowed to go to a worse solution S'. It does so with probability $1/(1 + \delta \cdot \delta^*)$, where $\delta = |S| - |S'|$, $\delta^* = |S^*| - |S'|$, and S^* is the best solution found so far. Intuitively, the farther S' is from S and S^*, the least likely the algorithm is to set $S \leftarrow S'$. If the algorithm does not go to S' (including during plateau search), we undo the insertions and deletions that led to S', then add a small perturbation by performing a 1-swap in S (if possible).

Finally, we consider the stopping criterion. We stop the algorithm when the average number of scans per arc exceeds a predetermined limit (which is the same for every instance within each family we tested). An arc scan is the most basic operation performed by our algorithm: in fact, the total running time is proportional to the number of such scans. By fixing the number of scans per arc (instead of the total number of scans) in each family, we make the algorithm spend more time on larger instances, which is a sensible approach in practice. To minimize the overhead of counting arc scans individually, our code converts the bound on arc scans into a corresponding bound on vertex scans (using the average vertex degree), and only keeps track of vertex scans during execution.

4.2   The GLP Algorithm

We now discuss the algorithm of Grosso, Locatelli, and Pullan [8], which we call GLP. Although it was originally formulated for the maximum clique problem, our description (as well as our implementation) refers to the MIS problem. We implemented "Algorithm 1 with restart rule 2," which seems to give the best results overall among the several variants proposed in [8]. What follows is a rough sketch of the algorithm. See the original paper for details.

The algorithm keeps a *current solution* S (initially empty), and spends most of its time performing plateau search (simple swaps). A simple tabu mechanism ensures that vertices that leave the solution during plateau search do not return during the same iteration, unless they become free and there are no alternative moves. A successful iteration ends when a non-tabu vertex becomes free: we simply insert it into the solution and start a new iteration. An iteration is considered unsuccessful if this does not happen after roughly $|S|$ moves: in this case, the solution is perturbed with the forced insertion of a single non-solution vertex (with at least four solution neighbors, if possible), and a new iteration starts. GLP does not use local search.

Unlike Grosso et al.'s implementation of GLP, ours does not stop as soon as it reaches the best solution reported in the literature. Instead, we use the same stopping criterion as the ILS algorithm, based on the number of arc scans. Although different, both ILS and GLP have scans as their main basic operation. By using the number of arc scans as the stopping criterion, we ensure that both algorithms have similar running times for all instances.

5   Experimental Results

All algorithms were implemented by the authors in C++ and compiled with gcc v. 3.4.4 with the full optimization (-O4) flag. All runs were made on a 3 GHz Pentium IV with 2 GB of RAM running Windows XP Professional. CPU times were measured with the getrusage function, which has precision of 1/60 second. Times do not include reading the graph and building the adjacency lists, since these are common to all algorithms. But they do include the time to allocate, initialize and destroy the data structures specific to each algorithm.

5.1   Instances

The DIMACS family contains maximum clique instances from the 2nd DIMACS Implementation Challenge [10], which have been frequently tested in the literature. It includes a wide variety of instances, with multiple topologies and densities. Since we deal with the MIS problem, we use the complements of the original graphs. For instances with no known optimum, we report the best results available at the time of writing (as listed in [8,15]).

The SAT family contains transformed satisfiability instances from the SAT'04 competition, available at [18] and tested in [8,15]. All optima are known.

The CODE family, made available by N. Sloane [17], consists of challenging graphs arising from coding theory. Each subfamily refers to a different error-correcting code, with vertices representing code words and edges indicating conflicts between them. The best known results for the hardest instances were found by the specialized algorithms of Butenko et al. [3,4].

The last two families, MESH and ROAD, are novel in the context of the independent set problem. MESH is motivated by an application in Computer Graphics recently described by Sander et al. [16]. To process a triangulation efficiently in graphics hardware, their algorithm finds a small subset of triangles that covers all the edges in the mesh. This is the same as finding a small set cover on the corresponding dual graph (adjacent faces in the original mesh become adjacent vertices in the dual). The MESH family contains the duals of well-known triangular meshes. While converting the original primal meshes, we repeatedly eliminated vertices of degree one and zero from the dual, since there is always a maximum independent set that contains them. (Degree-one vertices arise when the original mesh is open, i.e., when it has edges that are adjacent to a single triangle instead of the usual two.) Almost all vertices in the resulting MIS instances (which are available upon request) have degree three.

The ROAD family contains planar graphs representing parts of the road network of the United States, originally made available for the 9th DIMACS Implementation Challenge, on shortest paths [5]. Vertices represent intersections, and arcs represent the road segments connecting them. As in the previous family, these graphs have numerous vertices of degree one. We chose not to eliminate them explicitly, since these instances are already available in full form.

Due to space limitations, we only report results on a few representatives of each family, leaving out easy instances and those that behave similarly to others.

5.2   Local Search

We first evaluate the local search algorithm by itself, in terms of both solution quality and running time. We tested it with three different constructive algorithms. The random algorithm (R) inserts free vertices uniformly at random until the solution is maximal. The greedy algorithm (G) assigns a *cost* to each free vertex equal to the number of free neighbors it has, and in each iteration picks a free vertex with lowest cost. The randomized greedy algorithm (RG) is a variant of G that picks the vertex to insert *uniformly at random* among all minimum-cost free vertices. Both G and RG can be implemented in linear time, but there is some data structure overhead associated with RG. While G keeps the free vertices in buckets (one for each possible cost), RG maintains the vertices sorted by cost, which is more complicated.

For a representative sample of instances, we ran the constructive algorithms by themselves (R, G, and RG) and followed by local search (RL, GL, and RGL). Table 1 shows the average solutions obtained for 999 random seeds, and Table 2 the average running times. Also shown are the number of vertices (n), the average degree (DEG), and the best known solution (BEST) for each graph. Given the somewhat low precision of our timing routine (and how fast the algorithms are in this experiment), we did not measure running times directly. Instead, we ran each subsequence of 111 seeds repeatedly until the total running time was at least 5 seconds, then took the average time per run. Before each timed run, we ran the whole subsequence of 111 once to warm up the cache and minimize fluctuations. (Single runs would be slightly slower, but would have little effect on the relative performance of the algorithms.)

The greedy algorithms (G and RG) find solutions of similar quality, and are usually much better than random (R). Random is consistently faster, however, especially for very dense instances such as p_hat1500-1. While the greedy algorithm must visit every edge in the graph, the random algorithm only traverses

Table 1. Average solutions found by the random, greedy, and randomized greedy constructive algorithms, without (R, G, RG) or with (RL, GL, RGL) local search. The best results among these algorithms are marked in bold. The horizontal lines separate different families, in order: DIMACS, SAT, CODE, MESH, and ROAD.

GRAPH	n	DEG	BEST	R	RL	G	GL	RG	RGL
C2000.9	2000	199.5	80	51.2	59.5	66.0	**68.0**	66.5	67.4
MANN_a81	3321	3.9	1100	1082.1	1082.1	**1096.0**	**1096.0**	1095.5	1095.6
brock400_2	400	100.1	29	16.7	19.4	22.0	**23.0**	22.0	22.4
brock400_4	400	100.2	33	16.7	19.2	**22.0**	**22.0**	21.7	**22.0**
c-fat500-10	500	312.5	126	125.0	125.0	**126.0**	**126.0**	**126.0**	**126.0**
hamming10-2	1024	10.0	512	242.3	412.8	**512.0**	**512.0**	**512.0**	**512.0**
johnson32-2-4	496	60.0	16	**16.0**	**16.0**	**16.0**	**16.0**	**16.0**	**16.0**
keller6	3361	610.9	59	34.4	43.1	48.2	48.9	48.5	**49.6**
p_hat1500-1	1500	1119.1	12	6.8	8.1	10.0	10.0	9.9	**10.4**
p_hat1500-3	1500	369.3	94	42.6	77.7	86.0	**91.0**	85.9	88.3
san1000	1000	498.0	15	7.7	7.7	**10.0**	**10.0**	9.5	9.5
san400_0.7_1	400	119.7	40	19.7	20.6	21.0	21.0	**21.4**	**21.4**
san400_0.9_1	400	39.9	100	44.2	54.0	92.0	**100.0**	81.3	**100.0**
frb59-26-1	1534	165.0	59	39.4	45.8	48.0	48.0	47.6	**48.3**
1et.2048	2048	22.0	316	232.9	268.6	292.2	295.0	292.9	**295.8**
1zc.4096	4096	45.0	379	254.2	293.5	328.5	**329.5**	327.3	328.6
2dc.2048	2048	492.6	24	15.6	18.7	21.0	**22.0**	21.0	21.3
dragon	150000	3.0	66430	56332	61486	64176	64176	64024	**64247**
dragonsub	600000	3.0	282192	227004	256502	277252	**277252**	275611	276451
buddha	1087716	3.0	480664	408215	445100	463914	463914	463303	**464878**
fla	1070376	2.5	549535	476243	523237	545961	545972	545507	**546150**

the adjacency lists of the vertices that end up in the solution. Even after local search, RL is often faster than G or RG, but still finds worse solutions. On sparser instances, RL can be slower than GL or RGL, since the local search has a much worse starting point.

The local search is remarkably fast when applied to the greedy solutions. For large, sparse instances (such as fla and buddha) the local search is much more effective on RG than on G. In fact, G tends to find better solutions than RG, but after local search the opposite is true. We conjecture that the stack-like nature of buckets in G causes it to generate more "packed" solutions than RG. The higher variance of RG helps after local search: over all 999 runs, the best solution found by RGL (not shown in the table) was in most cases at least as good as the best found by any of the other algorithms. (The exceptions were san400_0.7_1, for which RL was superior, and dragonsub, for which G and GL were the best.) This suggests that RGL (or a variant) would be well-suited to multistart-based metaheuristics, such as GRASP [6].

For completeness, we briefly discuss the 3-improvement algorithm (not shown in the tables). Applied to the solutions obtained by RGL, it improved the average solutions of only six instances: 1et.2048 (296.4), 2dc.2048 (21.6), frb59-26-1 (48.6), keller6 (50.2), p_hat1500-1 (10.5) and san1000 (9.7). It also improved RL

Table 2. Constructive algorithms and local search: running times in milliseconds

GRAPH	n	DEG	R	RL	G	GL	RG	RGL
C2000.9	2000	199.5	**0.08**	0.47	8.15	8.57	16.62	16.96
MANN_a81	3321	3.9	**0.18**	0.60	0.53	0.97	0.70	1.13
brock400_2	400	100.1	**0.03**	0.11	0.77	0.85	1.49	1.56
brock400_4	400	100.2	**0.03**	0.11	0.77	0.85	1.48	1.56
c-fat500-10	500	312.5	**0.09**	0.52	2.28	2.71	5.19	5.60
hamming10-2	1024	10.0	**0.07**	0.40	0.29	0.49	0.46	0.66
johnson32-2-4	496	60.0	**0.02**	0.10	0.45	0.52	1.01	1.08
keller6	3361	610.9	**0.11**	0.88	35.32	35.96	71.73	72.43
p_hat1500-1	1500	1119.1	**0.04**	0.31	31.66	31.94	58.66	58.71
p_hat1500-3	1500	369.3	**0.07**	0.61	11.35	11.73	21.26	21.64
san1000	1000	498.0	**0.03**	1.48	8.73	8.87	16.81	17.02
san400_0.7_1	400	119.7	**0.03**	0.18	0.88	1.00	1.69	1.80
san400_0.9_1	400	39.9	**0.03**	0.14	0.35	0.46	0.67	0.79
frb59-26-1	1534	165.0	**0.06**	0.33	4.86	5.10	10.00	10.25
1et.2048	2048	22.0	**0.10**	0.46	1.17	1.46	1.97	2.26
1zc.4096	4096	45.0	**0.17**	0.85	4.06	4.63	7.86	8.44
2dc.2048	2048	492.6	**0.06**	0.45	21.25	21.59	36.15	36.41
dragon	150000	3.0	**17.80**	64.22	33.95	69.28	42.09	77.53
dragonsub	600000	3.0	**123.95**	390.90	193.38	400.14	169.59	377.23
buddha	1087716	3.0	**281.65**	795.98	448.92	854.27	447.78	859.55
fla	1070376	2.5	**299.70**	867.04	521.01	969.91	741.88	1193.48

on these instances, as well as 1et.2048 and brock400_4, but still not enough to make the random algorithm competitive with their greedy counterparts. It only improved the results obtained by GL in one case (1et.2048). On the positive side, the 3-improvement algorithm is reasonably fast. In most cases, it adds less than 20% to the time of RGL, and at most 80% (on johnson-32-2-4). Still, the minor gains and added complexity do not justify using 3-improvement within ILS.

5.3 Metaheuristics

Although local search can improve the results found by constructive heuristics, the local optima it finds are usually far from the best known bounds. For near-optimal solutions, we turn to metaheuristics. We compare our iterated local search (ILS) with our implementation of Grosso et al.'s GLP algorithm. Our version of GLP deals with the maximum independent set problem directly, and its time per operation is comparable to the original implementation.

Tables 3, 4, and 5 present results for DIMACS, CODE, and SAT, respectively. For each instance, we first show its number of vertices, its density, and the best known solution. We then report the minimum, average, and maximum solutions found over nine runs of each algorithm (the numbers in parentheses indicate how many of these runs found the maximum). Finally, we give the average running time in seconds. Both algorithms were run until the average number of scans per arc reached 2^{17}. The best averages are highlighted in bold.

Table 3. DIMACS family. For each algorithm, we show the minimum, average, and maximum solutions found over 9 runs, as well as the average running time in seconds. Both algorithms were run until the average arc was scanned 2^{17} times.

GRAPH				ILS				GLP			
NAME	n DENS BEST			MIN	AVG	MAX	TIME	MIN	AVG	MAX	TIME
C2000.9	2000	0.100	80	77	77.2	78(2)	277	77	**77.9**	79(2)	246
MANN_a45	1035	0.004	345	344	**344.7**	345(6)	8	343	343.6	344(5)	8
MANN_a81	3321	0.001	1100	1100	**1100.0**	1100(9)	20	1097	1097.7	1098(6)	27
brock400_1	400	0.252	27	25	25.0	25(9)	26	25	**25.9**	27(4)	27
brock400_2	400	0.251	29	25	25.4	29(1)	26	25	**26.8**	29(4)	26
brock400_3	400	0.252	31	25	30.3	31(8)	27	31	**31.0**	31(9)	23
brock400_4	400	0.251	33	25	31.2	33(7)	27	33	**33.0**	33(9)	20
hamming10-2	1024	0.010	512	512	**512.0**	512(9)	24	512	**512.0**	512(9)	13
keller6	3361	0.182	59	59	**59.0**	59(9)	1385	59	**59.0**	59(9)	1026
p_hat1500-1	1500	0.747	12	11	11.8	12(7)	345	12	**12.0**	12(9)	1207
san1000	1000	0.498	15	15	**15.0**	15(9)	185	15	**15.0**	15(9)	426

Table 4. Results for the CODE family with 2^{17} scans per arc

GRAPH				ILS				GLP			
NAME	n DENS BEST			MIN	AVG	MAX	TIME	MIN	AVG	MAX	TIME
1dc.1024	1024	0.046	94	93	**93.2**	94(2)	31	93	93.1	94(1)	42
1dc.2048	2048	0.028	172	170	**171.3**	172(6)	76	170	**171.3**	172(6)	95
1et.2048	2048	0.011	316	316	**316.0**	316(9)	38	316	**316.0**	316(9)	57
1tc.2048	2048	0.009	352	352	**352.0**	352(9)	35	352	**352.0**	352(9)	52
1zc.1024	1024	0.064	112	111	111.3	112(3)	23	112	**112.0**	112(9)	40
1zc.2048	2048	0.038	198	196	197.4	198(6)	53	197	**197.7**	198(6)	92
1zc.4096	4096	0.022	379	364	370.7	379(1)	127	367	**373.0**	379(1)	224
2dc.1024	1024	0.323	16	16	**16.0**	16(9)	105	16	**16.0**	16(9)	322
2dc.2048	2048	0.241	24	24	**24.0**	24(9)	388	23	23.8	24(7)	851

Together, the algorithms do rather well on these families. For almost all instances, the best known bound was found at least once. For all four exceptions (C2000.9 and the three largest frb instances) the best solution shown in the tables is within one unit of the best known [8].

The average solutions found by ILS and GLP are usually within 0.1 unit from one another. Among the exceptions, GLP found better solutions on nine (C2000.9, brock*, p_hat1500-1, and 1zc.*) and ILS on four (MANN*, 2dc.2048, and frb53-24-1). The brock instances are dense random graphs with a "hidden" larger clique. C2000.9 is also random, with larger cliques naturally hidden by the large value of n. GLP is clearly better at finding these cliques, probably because of its stronger tabu mechanism. In contrast, GLP does poorly on the MANN instances (sparse graphs with large independent sets), while ILS finds the optimal solution MANN_a81 in only 0.9 seconds on average.

Running times in the tables refer to full executions. When both algorithms found the same solution in every run, it makes sense to compare the average time

Table 5. Results for the SAT family with 2^{17} scans per arc

GRAPH				ILS				GLP		
NAME	n DENS	BEST	MIN	AVG	MAX	TIME	MIN	AVG	MAX	TIME
frb30-15-1	450 0.176	30	30	**30.0**	30(9)	20	30	**30.0**	30(9)	28
frb35-17-1	595 0.158	35	35	**35.0**	35(9)	30	35	**35.0**	35(9)	41
frb40-19-1	760 0.143	40	40	**40.0**	40(9)	42	40	**40.0**	40(9)	57
frb45-21-1	945 0.133	45	44	**44.6**	45(5)	65	44	**44.6**	45(5)	79
frb50-23-1	1150 0.121	50	49	**49.1**	50(1)	86	48	49.0	50(1)	106
frb53-24-1	1272 0.117	53	51	**51.6**	52(5)	98	51	51.1	52(1)	121
frb56-25-1	1400 0.112	56	54	**54.1**	55(1)	118	54	54.0	54(9)	139
frb59-26-1	1534 0.108	59	57	**57.2**	58(2)	137	57	57.1	58(1)	161

to reach it (not shown in the tables). ILS is faster on 2dc.1024 (by a factor of 2), frb40-19-1 (3), and keller6 (13). The algorithms are essentially tied for 1tc.2048. GLP is faster for the remaining instances, usually by a factor of less than four. On san1000 and hamming10_2, GLP was at least 6 times faster.

Although our algorithm does well on these families, GLP is somewhat more robust on DIMACS and CODE. This is not the case for large, sparse graphs, to which we now turn our attention. Table 6 presents results for the MESH family. Because its graphs are much larger, we limit average number of arc scans to 2^{14}.

Table 6. Results for the MESH family with 2^{14} scans per arc

GRAPH		ILS				GLP			
NAME	n	MIN	AVG	MAX	TIME	MIN	AVG	MAX	TIME
dolphin	554	249	**249**	249(9)	1	249	**249**	249(9)	1
mannequin	1309	583	**583**	583(9)	3	583	**583**	583(9)	1
beethoven	4419	2000	**2002**	2004(3)	9	1999	2001	2004(1)	5
cow	5036	2333	2339	2346(2)	11	2335	**2343**	2346(6)	5
venus	5672	2668	2676	2680(2)	11	2680	**2682**	2684(4)	6
fandisk	8634	4057	4068	4072(2)	18	4063	**4069**	4073(1)	11
blob	16068	7232	7236	7239(1)	36	7234	**7239**	7242(1)	21
gargoyle	20000	8843	**8846**	8849(1)	50	8841	8844	8847(1)	32
face	22871	10203	**10206**	10211(1)	51	10203	10205	10207(1)	31
feline	41262	18791	18803	18810(1)	105	18806	**18813**	18822(1)	74
gameguy	42623	20625	20639	20664(1)	104	20635	**20657**	20676(1)	61
bunny	68790	32211	32228	32260(1)	208	32221	**32246**	32263(1)	184
dragon	150000	66399	**66417**	66430(1)	506	66318	66331	66343(1)	507
turtle	267534	122262	**122298**	122354(1)	1001	122133	122195	122294(1)	1185
dragonsub	600000	281942	281972	282002(1)	2006	282100	**282149**	282192(1)	2340
ecat	684496	321881	**321981**	322040(1)	3191	321689	321742	321906(1)	4757
buddha	1087716	480604	**480630**	480664(1)	4773	478691	478722	478757(1)	6795

Even though all instances come from the same application, results are remarkably diverse. The relative performance of the algorithms appears to be correlated with the regularity of the meshes: GLP is better for regular meshes, whereas ILS is superior for more irregular ones. We verified this by visual inspection, but the standard deviation of the vertex degrees in the original (primal) mesh is a rough proxy for irregularity. It is relatively smaller for bunny (0.58) and dragon-sub (0.63), on which GLP is the best algorithm, and bigger for buddha (1.28) and dragon (1.26), on which ILS is superior.[2] Note that dragonsub is a subdivision of dragon: a new vertex is inserted in the middle of each edge, and each triangle is divided in four. Both meshes represent the same model, but because every new vertex has degree exactly six, dragonsub is much more regular.

Although optimal solutions for the MESH family are not known, Sander et al. [16] computed lower bounds on the cover solutions for eleven of their original meshes (the ten largest in Table 6 plus fandisk). These can be easily translated into upper bounds for our (MIS) instances. On average, ILS is within 2.44% of these bounds (and hence of the optimum). The highest gap (3.32%) was observed for face, and the lowest for gameguy (1.22%). The gaps for GLP range from 1.13% (on gameguy) to 3.45% (on buddha), with an average of 2.48%.

Finally, Table 7 presents the results for ROAD, with the average number of scans per arc limited to 2^{12}. Here ILS has clear advantage.

Table 7. Results for the ROAD family with 2^{12} scans per arc

GRAPH			ILS				GLP			
NAME	n	DEG	MIN	AVG	MAX	TIME	MIN	AVG	MAX	TIME
ny	264346	2.8	131421	**131440**	131454(1)	248	131144	131178	131213(1)	293
bay	321270	2.5	166349	**166355**	166360(1)	287	166215	166226	166250(1)	372
col	435666	2.4	225741	**225747**	225754(1)	395	225569	225586	225614(1)	568
fla	1070376	2.5	549508	**549523**	549535(1)	1046	548592	548637	548669(1)	1505

We note that MESH and ROAD are fundamentally different from the previous families. These are large graphs with linear-sized maximum independent sets. Both ILS and GLP start from relatively bad solutions, which are then steadily improved, one vertex at a time. To illustrate this, Figure 1 shows the average solutions found for the two largest instances (buddha and fla) as the algorithms progress. GLP initially finds better solutions, but is soon overtaken by ILS. The third curve in the plots (ILS+plateau) refers to a version of our algorithm that also performs plateau search when the current solution improves (recall that ILS only performs plateau search when the solution worsens). Although faster at first, ILS+plateau is eventually surpassed by ILS. The average solutions it found (after all 2^{12} scans per arc) were 480285 for buddha and 549422 for fla.

For comparison, we also show results for longer runs (2^{20} scans per arc, with nine different seeds) on C2000.9 (from the DIMACS family) and 1zc.4096 (from the CODE family). As before, GLP starts much better. On 1zc.4096, ILS slowly

---

[2] The standard deviation is not always a good measure of regularity. Despite being highly regular, gameguy has a triangulation pattern in which roughly half the vertices have degree 4 and half have degree 8, leading to a standard deviation higher than 2.

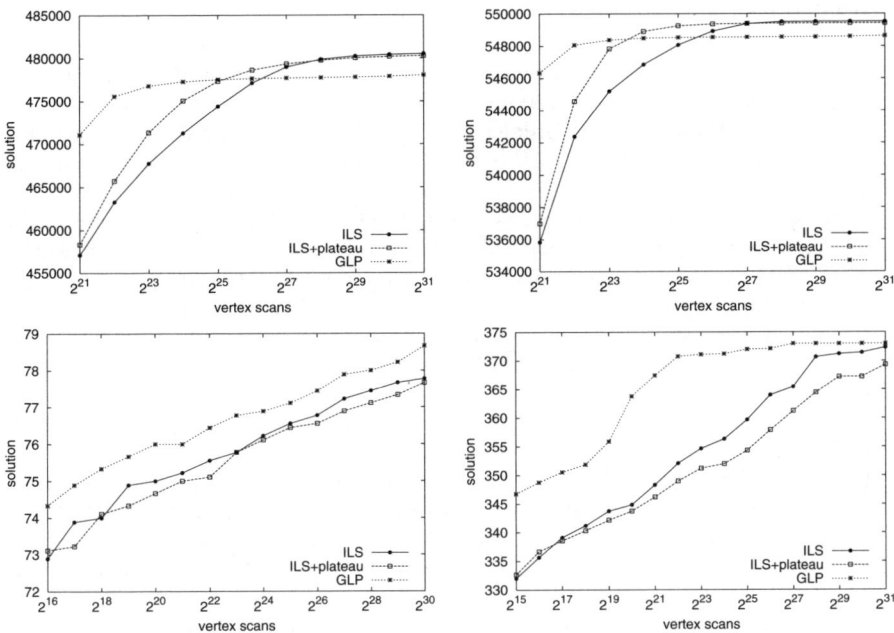

Fig. 1. Average solutions found as the number of scans per vertex increases. Results for buddha (top left), fla (top right), C2000.9 (bottom left), and 1zc.4096 (bottom right).

reduces the gap, but does not quite close it. On C2000.9, GLP is consistently better, even as the number of scans increases.

6   Final Remarks

We have proposed a fast implementation of a natural local search procedure for the independent set problem. Within an iterated local search (a metaheuristic), it provided results competitive with the best methods previously proposed, often matching the best known solutions (including optima) on the DIMACS, CODE, and SAT families. On large, sparse instances (meshes and road networks), its performance is consistently superior to that of GLP, particularly when the graph is irregular. For these large instances, however, we do not know exactly how far our method is from the optimal solution: there may be much room for improvement. It seems reasonable, for example, to deal with these problems more locally. Instead of looking at the entire graph at once, we conjecture that one could do better by focusing at individual regions at a time.

Acknowledgements. We thank D. Nehab and P. Sander for sharing their paper and providing us with the MESH instances, and three anonymous referees for their helpful comments.

References

1. Battiti, R., Protasi, M.: Reactive local search for the maximum clique problem. Algorithmica 29(4), 610–637 (2001)
2. Bomze, I.M., Budinich, M., Pardalos, P.M., Pelillo, M.: The maximum clique problem. In: Du, D.Z., Pardalos, P.M. (eds.) Handbook of Combinatorial Optimization (Sup. Vol. A), pp. 1–74. Kluwer, Dordrecht (1999)
3. Butenko, S., Pardalos, P.M., Sergienko, I., Shylo, V., Stetsyuk, P.: Finding maximum independent sets in graphs arising from coding theory. In: Proceedings of the 2002 ACM Symposium on Applied Computing, pp. 542–546 (2002)
4. Butenko, S., Pardalos, P.M., Sergienko, I., Shylo, V., Stetsyuk, P.: Estimating the size of correcting codes using extremal graph problems. In: Pearce, C. (ed.) Optimization: Structure and Applications, Springer, Heidelberg (2008)
5. Demetrescu, C., Goldberg, A.V., Johnson, D.S.: 9th DIMACS Implementation Challenge: Shortest Paths (2006) (last visited on March 15, 2008), http://www.dis.uniroma1.it/~challenge9
6. Feo, T., Resende, M.G.C., Smith, S.: A greedy randomized adaptive search procedure for maximum independent set. Operations Research 42, 860–878 (1994)
7. Grosso, A., Locatelli, M., Della Croce, F.: Combining swaps and node weights in an adaptive greedy approach for the maximum clique problem. J. Heuristics 10, 135–152 (2004)
8. Grosso, A., Locatelli, M., Pullan, W.: Simple ingredients leading to very efficient heuristics for the maximum clique problem. J. Heuristics (30 October, 2007), doi:10.1007/s10732-007-9055-x
9. Hansen, P., Mladenović, N., Urošević, D.: Variable neighborhood search for the maximum clique. Discrete Applied Mathematics 145(1), 117–125 (2004)
10. Johnson, D.S., Trick, M.: Cliques, Coloring and Satisfiability. DIMACS Series in Discrete Mathematics and Theoretical Computer Science, vol. 26. AMS (1996)
11. Karp, R.: Reducibility among combinatorial problems. In: Miller, R., Thatcher, J. (eds.) Complexity of Computer Computations, pp. 85–103. Plenum Press, New York (1972)
12. Katayama, K., Hamamoto, A., Narihisa, H.: An effective local search for the maximum clique problem. Information Processing Letters 95, 503–511 (2005)
13. Lourenço, H.R., Martin, O., Stützle, T.: Iterated local search. In: Glover, F., Kochenberger, G. (eds.) Handbook of Metaheuristics, pp. 321–353. Kluwer, Dordrecht (2003)
14. Pullan, W.J., Hoos, H.H.: Dynamic local search for the maximum clique problem. J. Artificial Intelligence Research 25, 159–185 (2006)
15. Richter, S., Helmert, M., Gretton, C.: A stochastic local search approach to vertex cover. In: Proceedings of the 30th German Conference on Artificial Intelligence (KI), pp. 412–426 (2007)
16. Sander, P.V., Nehab, D., Chlamtac, E., Hoppe, H.: Efficient traversal of mesh edges (submitted, 2008)
17. Sloane, N.J.A.: Challenge problems: Independent sets in graphs (2000) (last visited on March 15, 2008), http://www.research.att.com/~njas/doc/graphs.html
18. Xu, K.: BHOSLIB: Benchmarks with hidden optimum solutions for graph problems (2004) (last visited on March 15, 2008), http://www.nlsde.buaa.edu.cn/~kexu/benchmarks/graph-benchmarks.htm

Optimal University Course Timetables and the Partial Transversal Polytope

Gerald Lach and Marco E. Lübbecke

Technische Universität Berlin, Institut für Mathematik, MA 5-1,
Straße des 17. Juni 136, 10623 Berlin, Germany
{lach, m.luebbecke}@math.tu-berlin.de

Abstract. University course timetabling is the conflict-free assignment of courses to weekly time slots and rooms subject to various hard and soft constraints. One goal is to meet as closely as possible professors' preferences. Building on an intuitive integer program (IP), we develop an exact decomposition approach which schedules courses first, and matches courses/times to rooms in a second stage. The subset of constraints which ensures a feasible room assignment defines the well-known partial transversal polytope. We describe it as a polymatroid, and thereby obtain a complete characterization of its facets. This enables us to add only strong valid inequalities to the first stage IP. In fact, for all practical purposes the number of facets is small. We present encouraging computational results on real-world and simulated timetabling data. The sizes of our optimally solvable instances (respecting all hard constraints) are the largest reported in the literature by far.

Keywords: integer programming, partial transversal polytope, university course timetabling.

1 Introduction

Timetabling comes in many flavors, in education and sports, in industry and public transport. This diversity and its relevance in practice made timetabling an active research area in operations research; a series of conferences (Practice and Theory of Automated Timetabling, PATAT) is devoted to the topic [1]. In this paper, we aim for optimal solutions to one of the core problems of the field, the NP-complete *university course timetabling problem.*

A university timetable is an assignment of an appropriate number of time slots, or *periods*, and rooms to each weekly occurrence of each course. It is usually valid for one term. Customarily, one distinguishes between hard and soft constraints which have to be respected [2]. Typical hard constraints are: A professor cannot teach two classes at the same time; lectures belonging to the same curriculum must not be scheduled simultaneously; a room cannot be assigned to different courses in the same period; etc. A timetable is infeasible if one of these requirements is violated (which frequently occurs in practice). Soft constraints

C.C. McGeoch (Ed.): WEA 2008, LNCS 5038, pp. 235–248, 2008.

e.g., call for not exceeding a room's capacity; to provide the necessary equipment like beamer/PC; to spread the lectures of one course over the week; etc. A violation of these constraints is tolerated but penalized. Professors express preferences as to when to teach; an optimal timetable minimizes the total deviation from these preferences.

1.1 Our Contribution

This paper makes a contribution to practical problem solving via integer programming, as well as it adds to the theory of combinatorial optimization.

On the practical side, we give a proof-of-concept that optimal timetables can be computed for larger universities in acceptable time. Our focus in on meeting all hard constraints, where we take some of the constraints traditionally considered soft (like room capacity) as hard ones.

In the integer program we propose, instead of simultaneously assigning courses to time slots and rooms, we only schedule rooms, providing for a later feasible room assignment. This is done by interpreting feasible course/room pairs on a bipartite graph, and enforcing the classical Hall's conditions [3], a.k.a. "marriage theorem," on the existence of perfectly matchable sets (or transversals). This allows a simpler formulation and results in much improved solution times.

Hall's conditions directly lead us to an investigation of the partial transversal polytope [4]. We obtain a complete description of its facets by stating it as a polymatroid. Thereby, on the theoretical side, we obtain an interesting strengthening of Hall's Theorem. Finally, we are interested in the number of facets of the partial transversal polytope, and obtain a generating set of facets, of linear size. All facets can be obtained from this set by an intuitive operation.

We tested our approach on instances from the second international timetabling competition [5]. It turns out that we are able to compute optimal solutions within negligible running times. We therefore used simulated data which are almost identical to real data from Technical University of Berlin. To the best of our knowledge, we are the first to obtain optimal solutions to university course timetabling instances of this size.

It is our impression that integer programming has been used for timetabling only because of its modeling power. It was not realized that a deeper understanding of combinatorial properties of the problem may be the key to actually solving large instances to proven optimality. In this sense, we consider our work a significant step forward in this field of research.

1.2 Related Work

University course timetabling problems are well studied, see e.g., the surveys [6, 7]. Much has been written about practical details [8], and the non-negligible human factor of timetabling [9].

Meta heuristics clearly constitute the main solution approach, see [2, 10, 11], and the references therein. Several integer programs were suggested as well [8, 9, 12, 13, 14], however, optimally solvable problem instances are (i) smaller than ours by at least an order of magnitude, or (ii) are much simpler (and thus less realistic) than ours.

Interestingly, complete polyhedral descriptions of problems closely related to finding transversals are well known. We have Edmonds' seminal work on the matching polytope [15]. Also, the perfectly matchable subgraph polytope for bipartite graphs is fully characterized [16]. Yet, we are not aware of any previous attempts to give a strong formulation of the partial transversal polytope.

2 Integer Programs and Decomposition

2.1 An Intuitive Integer Program

We give a generic integer program (IP) for the university course timetabling problem which concentrates on hard constraints (time conflicts and room conflicts). However, it is easy to enhance this IP by soft constraints.

Denote by $\mathcal{C}$ the set of courses, by $\mathcal{R}$ the set of rooms, and by $\mathcal{T}$ the set of time slots. For each course $c \in \mathcal{C}$ we know its eligible time slots $T(c) \subseteq \mathcal{T}$, and eligible rooms $R(c) \subseteq \mathcal{R}$. Further, $R^{-1}(r) \subseteq \mathcal{C}$ is the set of all courses which may take place in room $r \in \mathcal{R}$. Each course $c \in \mathcal{C}$ consists of $\ell(c)$ lectures, that is, we have to provide $\ell(c)$ different time slots for course c. The instructor of course $c \in \mathcal{C}$ assigns a preference $prio(c, t)$ to all eligible time slots $t \in T(c)$; the smaller it is, the better.

Time conflicts are represented via a conflict graph $G_{\mathrm{conf}} = (V_{\mathrm{conf}}, E_{\mathrm{conf}})$: A vertex (c, t) represents an eligible combination of a course c and a timeslot t. Two nodes (c_1, t_1) and (c_2, t_2) are adjacent iff it is forbidden that c_1 is scheduled at t_1 and c_2 at t_2 (typically, $t_1 = t_2$). We see that time conflicts introduce a stable set flavor into our problem.

A binary variable $x_{c,t,r}$ represents whether course c is scheduled at time t in room r, or not. The following IP for the generic university course timetabling problem guarantees a sufficient number of time slots per course (2), avoids room conflicts (3), and time conflicts (4).

$$\min \sum_{c,t,r} prio(c, t) \cdot x_{c,t,r} \tag{1}$$

$$\text{s.t.} \sum_{t \in T(c), r \in R(c)} x_{c,t,r} = \ell(c) \qquad \forall c \in \mathcal{C} \tag{2}$$

$$\sum_{c \in R^{-1}(r)} x_{c,t,r} \leq 1 \qquad \forall t \in \mathcal{T},\, r \in \mathcal{R} \tag{3}$$

$$\sum_{r \in R(c_1)} x_{c_1,t_1,r} + \sum_{r \in R(c_2)} x_{c_2,t_2,r} \leq 1 \qquad \forall((c_1, t_1), (c_2, t_2)) \in E_{\mathrm{conf}} \tag{4}$$

$$x_{c,t,r} \in \{0, 1\} \qquad \forall(c, t) \in V_{\mathrm{conf}},\, r \in \mathcal{R} \tag{5}$$

This integer program will be infeasible for any reasonable practical data since usually some courses cannot be scheduled without conflicts. Thus, one tries to schedule as many courses as possible; a modification to accomplish this is straight forward. However, the computation times and solution qualities (cf. Table 3) do not advise to actually work with this formulation.

2.2 Decomposition into Time and Room Assignment

Instead, we reduce the problem in three dimensions to a problem in two dimensions, implicitly taking care of room conflicts. To this end, we represent eligible combinations of courses and rooms as undirected bipartite graphs $G_t = (\mathcal{C}_t \cup \mathcal{R}_t, E_t)$, one for every time slot $t \in \mathcal{T}$. Courses which may be scheduled at t are given in set $\mathcal{C}_t$; and $\mathcal{R}_t$ denotes the set of all eligible rooms for all courses in $\mathcal{C}_t$. A course c and a room r are adjacent iff r is eligible for c. For ease of exposition let $G = (\mathcal{C} \cup \mathcal{R}, E)$ be the graph consisting of all components G_t, $t \in \mathcal{T}$.

For any subset U of vertices, denote by $\Gamma(U) := \{i \in \mathcal{C} \cup \mathcal{R} \mid j \in U, (i,j) \in E\}$ the neighborhood of U; in particular, $\Gamma(U) \subseteq \mathcal{R}$ for any $U \subseteq \mathcal{C}$. The set of all vertices which are adjacent *only* to vertices in U is denoted by $\Gamma^{-1}(U) := \{i \in \mathcal{U} \cup \mathcal{R} \mid \Gamma(\{i\}) \subseteq U\}$. In particular, $\Gamma^{-1}(U) \subseteq \mathcal{C}$ for any $U \subseteq \mathcal{R}$.

Hall's Theorem [3] states that a bipartite graph $G = (\mathcal{C} \cup \mathcal{R}, E)$ has a matching of all vertices in $\mathcal{C}$ into $\mathcal{R}$ if and only if $|\Gamma(U)| \geq |U|$ for all $U \subset \mathcal{C}$. This enables us to state a simpler integer program which schedules courses in such a way that a later assignment of rooms is possible. It is thus based on binary variables $x_{c,t}$, but obviously has an exponential number of constraints.

$$\min \sum_{(c,t) \in V_{\mathrm{conf}}} prio(c,t) \cdot x_{c,t} \tag{6}$$

$$\text{s.t.} \sum_{t \in T(c)} x_{c,t} = \ell(c) \qquad \forall c \in \mathcal{C} \tag{7}$$

$$\sum_{c \in U} x_{c,t} \leq |\Gamma(U)| \qquad \forall U \subseteq \mathcal{C}, \ t \in \mathcal{T} \tag{8}$$

$$x_{c_1,t_1} + x_{c_2,t_2} \leq 1 \qquad \forall ((c_1,t_1),(c_2,t_2)) \in E_{\mathrm{conf}} \tag{9}$$

$$x_{c,t} \in \{0,1\} \qquad \forall (c,t) \in V_{\mathrm{conf}} \tag{10}$$

Once this IP is solved, the second stage merely consists of solving a sequence of minimum weight bipartite matching problems; clearly, this decomposition approach is exact.

Even though Hall's inequalities (8) can be separated in polynomial time via a maximum flow computation, we would like to work with a strongest possible formulation: We are interested in the facets of the polytope defined by (8) (and non-negativity).

3   The Partial Transversal Polytope

In the context of Hall's Theorem, $\mathcal{C}$ is known as *system of distinct representatives* or *transversal*. A *partial transversal* is a subset of $\mathcal{C}$ which can be perfectly matched (we may assume that all $r \in \mathcal{R}$ will be matched). The *partial transversal polytope* $P(\mathcal{C})$ is the convex hull of all incidence vectors of partial transversals of $\mathcal{C}$. It is full dimensional in $\mathbb{R}^{|\mathcal{C}|}$.

The *deficiency* of a vertex set $U \subseteq \mathcal{C}$ is defined as $\text{def}_G(U) := |U| - |\Gamma(U)|$. The *deficiency of a graph* G is $\text{def}(G) := \max_{U \subseteq \mathcal{C}} \text{def}_G(U)$. We will often consider the deficiency of induced subgraphs $(U \cup \Gamma(U), E)$, and denote it by $\text{def}(U)$, slightly abusing notation. Graph deficiency is known to be supermodular [3], that is, $\text{def}(U \cup V) + \text{def}(U \cap V) \geq \text{def}(U) + \text{def}(V)$ for $U, V \subseteq \mathcal{C}$. Finally, denote by $\nu(G)$ the cardinality of a maximum matching in G.

We consider two equivalent descriptions of the partial transversal polytope $P(\mathcal{C})$. We use the common shorthand notation $x(U) := \sum_{i \in U} x_i$.

Lemma 1 (The Partial Transversal Polytope). *Given a bipartite graph* $G = (\mathcal{C} \cup \mathcal{R}, E)$, *the partial transversal polytope* $P(\mathcal{C}) \subseteq \mathbb{R}^{|\mathcal{C}|}$ *is defined by*

$$
\begin{align}
x(U) &\leq |\Gamma(U)| & \forall U \subseteq \mathcal{C} \tag{11} \\
0 &\leq x \leq 1 \tag{12}
\end{align}
$$

or equivalently by

$$
\begin{align}
x(U) &\leq |U| - \text{def}(U) & \forall U \subseteq \mathcal{C} \tag{13} \\
x &\geq 0 \ . \tag{14}
\end{align}
$$

The advantage of the latter description is that $x \leq 1$ is not explicitly required. This will facilitate characterizing facets.

3.1   Facets

A theorem by Edmonds on the facets of polymatroids [4, Thm. 44.4] allows us to easily give a complete and non-redundant description of the partial transversal polytope. For a consistent presentation we define the set function

$$
f : 2^{\mathcal{C}} \to \mathbb{N}, \ U \mapsto f(U) := |U| - \text{def}(U) \ , \tag{15}
$$

which is submodular by supermodularity of the deficiency. Note also that f is nondecreasing, that is, $f(U) \leq f(T)$ for $U \subseteq T$. Further, $f(\emptyset) = 0$ and $f(\{i\}) > 0$ for $i \in \mathcal{C}$.

A subset $U \subseteq \mathcal{C}$ is called an f-*flat* if $f(U \cup \{i\}) > f(U)$ for all $i \in \mathcal{C} \setminus U$; and U is f-*inseparable* if there are no $U_1, U_2 \neq \emptyset$ with $U_1 \cap U_2 = \emptyset$ and $U_1 \cup U_2 = U$ such that $f(U) = f(U_1) + f(U_2)$.

Edmonds has the following theorem: With the properties of a set function f as given in (15), the facets of $\{x \in \mathbb{R}^{|\mathcal{C}|} \mid x \geq 0, x(U) \leq f(U) \text{ for } U \subseteq \mathcal{C}\}$ are given by (i) $x \geq 0$, and (ii) $x(U) \leq f(U)$ for each nonempty f-inseparable f-flat $U \subseteq \mathcal{C}$.

Definition (Defining $\mathcal{C}$-set). *Given a bipartite graph* $G = (\mathcal{C} \cup \mathcal{R}, E)$, *and* f *as defined in* (15). *A set* $\emptyset \neq U \subseteq \mathcal{C}$ *is called* defining $\mathcal{C}$-set, *iff* U *is an* f-*flat,* *and*

$$\text{def}(U) > \max_{\substack{U_1,U_2 \subseteq U \\ U_1 \cap U_2 = \emptyset}} \{\text{def}(U_1) + \text{def}(U_2)\} \ . \tag{16}$$

This definition reflects the intuition that a $\mathcal{C}$-set is important, if it bears more information than the union of its parts. Inequality (16) will guarantee f-inseparability.

Theorem 2. *Given a bipartite graph $G = (\mathcal{C} \cup \mathcal{R}, E)$, then a set $U \subseteq \mathcal{C}$ is facet inducing for the partial transversal polytope $P(\mathcal{C})$, if and only if U is a defining $\mathcal{C}$-set.*

Proof. To prove necessity, let $V \subseteq \mathcal{C}$ be facet inducing. V is an f-inseparable f-flat by definition. Hence there are no disjoint $V_1, V_2 \neq \emptyset$ with $V = V_1 \cup V_2$ with

$$|V_1| - \text{def}(V_1) + |V_2| - \text{def}(V_2) = |V| - \text{def}(V) \ .$$

Equivalently, for all disjoint $\emptyset \neq V_1, V_2 \subseteq V$:

$$|V| - \text{def}(V) < |V_1| - \text{def}(V_1) + |V_2| - \text{def}(V_2)$$
$$\text{def}(V) > \text{def}(V_1) + \text{def}(V_2) \ ,$$

so V is a defining $\mathcal{C}$-set. For sufficiency, let V be a defining $\mathcal{C}$-set. V is an f-flat by definition. Further it holds that

$$\text{def}(V) > \max_{\substack{U_1,U_2 \subseteq U \\ U_1 \cap U_2 = \emptyset}} \{\text{def}(U_1) + \text{def}(U_2)\} \ .$$

That is, for arbitrary disjoint $\emptyset \neq V_1, V_2 \subseteq V$ with $V_1 \cup V_2 = V$ we have

$$\text{def}(V) > \text{def}(V_1) + \text{def}(V_2)$$
$$|V| - \text{def}(V) < |V_1| - \text{def}(V_1) + |V_2| - \text{def}(V_2)$$
$$f(V) < f(V_1) + f(V_2) \ .$$

So V is facet inducing for the partial transversal polytope. $\qquad\square$

Corollary 3 (Strengthening of Hall's Conditions). *Let $G = (\mathcal{C} \cup \mathcal{R}, E)$ be a bipartite graph, and $D_1, .., D_n \subseteq \mathcal{C}$ the collection of all defining $\mathcal{C}$-sets. There exists a matching covering all elements of $A \subseteq \mathcal{C}$, if and only if for all D_i and for all $X \subseteq A$*

$$|D_i \cap X| \ \leq \ |\Gamma(D_i)| \ . \tag{17}$$

3.2   Generating All Facets, and a Generating Subset

Now that we know how to strengthen constraints (8), we would like to make algorithmic use of this knowledge. We will first see that taking unions of defining $\mathcal{C}$-sets again yields a defining $\mathcal{C}$-set, if we preserve the f-flat property.

Definition (The flat-union $\sqcup$). *Given a bipartite graph $G = (\mathcal{C} \cup \mathcal{R}, E)$ and two sets $U_1, U_2 \subseteq \mathcal{C}$, then the flat-union $\sqcup$ is defined as follows:*

$$U_1 \sqcup U_2 := U_1 \cup U_2 \cup \{c \in \mathcal{C} : \Gamma(\{c\}) \in \Gamma(U_1) \cup \Gamma(U_2)\} .$$

Lemma 4 (The flat-union of defining $\mathcal{C}$-sets). *Given a bipartite graph $G = (\mathcal{C} \cup \mathcal{R}, E)$, a set function f as in (15), and two defining $\mathcal{C}$-sets $U_1, U_2 \subseteq \mathcal{C}$ such that*

$$f(U_1) + f(U_2) > f(U_1 \cup U_2) .$$

Then $U = U_1 \sqcup U_2$ is a defining $\mathcal{C}$-set.

Proof. By definition, U is an f-flat. We assume for contradiction that there are disjoint $V_1, V_2 \neq \emptyset$ with

$$U = V_1 \cup V_2 \tag{18}$$

$$f(U) = f(V_1) + f(V_2) . \tag{19}$$

U_1, U_2 and V_1, V_2 both partition U. Thus, U_1 or U_2 cannot be completly contained in V_1 or V_2, so at least one of U_1, U_2 has to have a non trivial intersection with V_1 and V_2. W.l.o.g., $U_1 \cap V_1 \neq \emptyset$ and $U_1 \cap V_2 \neq \emptyset$. A consequence of (19) is

$$\nu((U_1 \cup \Gamma(U_1))) = \nu(((U_1 \cap V_1) \cup \Gamma(U_1 \cap V_1))) + \nu(((U_1 \cap V_2) \cup \Gamma(U_1 \cap V_2))) \tag{20}$$

which is equivalent to

$$\begin{aligned}
|U_1 \cup \Gamma(U_1)| &- \mathrm{def}((U_1 \cup \Gamma(U_1), E)) \\
&= |U_1 \cap V_1| - \mathrm{def}(((U_1 \cap V_1) \cup \Gamma(U_1 \cap V_1), E)) + \\
&\quad |U_1 \cap V_2| - \mathrm{def}(((U_1 \cap V_2) \cup \Gamma(U_1 \cap V_2), E)) .
\end{aligned}$$

That is, U_1 is not f-inseparable, hence it is not facet inducing, and thus no defining $\mathcal{C}$-set. $\qquad\square$

The number of defining $\mathcal{C}$-sets can be as large as $2^{|\mathcal{R}|} - 1$. Consider $G = (\mathcal{C} \cup \mathcal{R}, E)$ described by the incidence matrix

$$A = \begin{pmatrix}
1 & 1 & 1 & 0 & 0 & 0 & 1 & 0 & 0 & 0 & 0 & 1 & 0 & 0 \\
0 & 0 & 1 & 1 & 1 & 1 & 0 & 0 & 0 & 1 & 0 & 0 & 0 \\
0 & 0 & 0 & 0 & 0 & 1 & 1 & 1 & 1 & 1 & 0 & 0 & 0 & 0 \\
0 & 0 & 0 & 0 & 0 & 0 & 0 & 0 & 1 & 1 & 1 & 1 & 1
\end{pmatrix} .$$

One can see, that for all $R \subseteq \mathcal{R}$, with $|R| \leq 2$, $\Gamma^{-1}(R)$ is a defining $\mathcal{C}$-set. It is a consequence of Lemma 4 that $\Gamma^{-1}(R)$ is a defining $\mathcal{C}$-set for all $R \subseteq \mathcal{R}$.

Even though the number of facets can be large, we will show we can obtain all facets from a (practically small) subset via Lemma 4.

Definition (Atomic defining $\mathcal{C}$-set). *Given a bipartite graph $G = (\mathcal{C} \cup \mathcal{R}, E)$. A defining $\mathcal{C}$-set A is called atomic, if $|A| > 1$, and no two defining $\mathcal{C}$-sets $U_1, U_2 \subseteq A$ exist, such that*

$$\Gamma(U_1) \cup \Gamma(U_2) = \Gamma(A)$$

$$\Gamma(U_1) \cap \Gamma(U_2) \neq \emptyset .$$

All other defining $\mathcal{C}$-sets are called non-atomic.

Theorem 5 (Number of atomic defining $\mathcal{C}$-sets). *Given a bipartite Graph $G = (\mathcal{C} \cup \mathcal{R}, E)$. The number of atomic defining $\mathcal{C}$-sets is at most $\operatorname{def}(\mathcal{C}) = \operatorname{def}(G)$.*

Proof. Proof by induction on $r = |\mathcal{R}|$:
The assertion is easily verified for $r = 1$ and $r = 2$. For the induction step, let $\mathcal{A}$ be the set of all atomic defining $\mathcal{C}$-set and $h : 2^{\mathcal{C}} \to \mathbb{N}$, $U \mapsto |\{C \in \mathcal{A} : C \subseteq U\}|$.

Case I: $\mathcal{C}$ is an atomic defining $\mathcal{C}$-set. Let $A_1, \ldots, A_k \subset \mathcal{C}$ be all inclusion maximal subsets of $\mathcal{C}$, from Lemma 4 we know that

$$\forall A \in \mathcal{A} \; \exists D_i : A \subset D_i \; . \tag{21}$$

Now the assertion is easy to show.

$$\operatorname{def}(\mathcal{C}) \overset{\text{Lemma 12}}{\geq} \sum_{i=1}^{k} \operatorname{def}(D_i) + 1$$

$$\overset{\text{by ind. hyp.}}{\geq} \sum_{i=1}^{k} h(D_i) + 1$$

$$\overset{(21)}{=} |\mathcal{A}|$$

Case II: $\mathcal{C}$ is not a defining $\mathcal{C}$-set or a non-atomic defining $\mathcal{C}$-set. Let $C_i = \mathcal{C} \setminus \{c \in \mathcal{C} : i \in \Gamma(\{c\})\}$. Clearly, we have

$$\mathcal{C} \geq \bigcup_{i=1}^{r+1} C_i \; . \tag{22}$$

Further, one can conclude from Lemma 10 and the f-flat condition of a defining $\mathcal{C}$-set

$$\forall D \in \mathcal{A} \; \exists R \subset \mathcal{R} : D = \Gamma^{-1}(R) \; . \tag{23}$$

By supermodularity of $\operatorname{def}(\cdot)$ we can proof the assertion.

$$\operatorname{def}(\mathcal{C}) \overset{(22)}{\geq} \operatorname{def}\left(\bigcup_{i=1}^{r+1} C_i\right)$$

$$\overset{\text{Lemma 8}}{\geq} \sum_{k=1}^{r+1} (-1)^{k-1} \sum_{1 \leq i_1 < \ldots < i_k \leq r+1} \operatorname{def}(C_{i_1} \cap \ldots \cap C_{i_k})$$

$$\overset{\text{by ind. hyp.}}{\geq} \sum_{k=1}^{r+1} (-1)^{k-1} \sum_{1 \leq i_1 < \ldots < i_k \leq r+1} h(C_{i_1} \cap \ldots \cap C_{i_k})$$

$$\overset{(23)}{=} \sum_{C \in \mathcal{A}} \left(\sum_{k=1}^{r+1-|\Gamma(C)|} (-1)^{k-1} \sum_{1 \leq i_1 < \ldots < i_k \leq r+1} 1 \right)$$

$$\overset{\text{Lemma 9}}{=} \sum_{C \in \mathcal{A}} 1$$

$$= |\mathcal{A}| \qquad \qquad \square$$

3.3 Facet Enumeration

In our present implementation we enumerate all defining $\mathcal{C}$-sets, basically using Lemmas 4 and 11. If the number of defining $\mathcal{C}$-sets is polynomially bounded, the running time of the algorithm is polynomial. As pointed out below, for real-world instances this is a reasonable assumption. Theorem 5 suggests an algorithm which first constructs all atomic $\mathcal{C}$-sets, and repeatedly takes all non disjoint flat-unions. We postpone a detailed description of such an algorithm to the full paper.

4 Consequences

In real-world instances of the *university course timetabling problem* a room can be described by various attributes (or *features*). These may be capacity, location, seating, beamer, blackboard, etc. We distinguish between two types of features, *exclusive* and *inclusive*. Exclusive features cannot be requested at the same time (e.g., different room capacities). It is characteristic to exclusive features that the graph $G_t = (\mathcal{C}_t \cup \mathcal{R}_t, E_t)$ decomposes into independent components. We will show that for each component of G_t the maximum number of defining $\mathcal{C}$-sets only depends on the number of different (inclusive) features.

Lemma 6 (Number of defining $\mathcal{C}$-sets). *If the number of different features in a connected bipartite graph $G = (\mathcal{C} \cup \mathcal{R}, E)$ is ϕ, then the number of defining $\mathcal{C}$-sets in G is at most $2^{\phi} - 1$.*

Proof. Let $\mathcal{F}$ be the set of features and $F \subseteq \mathcal{F}$. We then denote with C_F all the courses, that apply for a room, which has to be provided with all features $f \in F$. Let

$$\mathcal{D} = \{C_F : F \subseteq \mathcal{F}\} \ .$$

We show that if $A \subseteq \mathcal{C}$ is a defining $\mathcal{C}$-set and $|A| > 1$, then $A \in \mathcal{D}$. We assume that $A \subseteq \mathcal{C}$ is a defining $\mathcal{C}$-set and $A \notin \mathcal{D}$. Then there exists a $c \in \mathcal{C} \backslash A$, such that $\Gamma(c) \subseteq \Gamma(A)$.

Case I: $\nu((A \cup \Gamma(A), E)) = |\Gamma(A)|$. Then, $|A \cup c| - \mathrm{def}(A \cup c) = |A| - \mathrm{def}(A)$, so A is not an f-flat and hence no defining $\mathcal{C}$-set.

Case II: $\nu((A \cup \Gamma(A), E)) < |\Gamma(A)|$. Then, A cannot be a defining $\mathcal{C}$-set (Lemma 10). $\qquad\square$

This has important consequences for the applicability of our apporach to real-world instances.

Corollary 7. *For a fixed number of features, the number of defining $\mathcal{C}$-sets is $O(1)$.*

Practical evidence shows that the number of defining $\mathcal{C}$-sets is in fact small. For example, we added a total of about 6400 non-trivial facets to our largest instance, cf. Table 2.

5 Computational Results

All our results were obtained on a 3.2GHz Pentium 4 Linux PC with 1GB memory. Integer programs are solved using CPLEX 10.1. We separately list running times for three steps: (i) facet generation, (ii) solution of the integer program (6)–(10), and (iii) allocating rooms to all assigned periods of all courses via a sequence of perfect matching calculations.

5.1 The Second International Timetabling Competition

Accompanying the PATAT08 conference, the second international timetabling competition, ITC2007, is held. The data of seven problems were published [5]. We present the statistics of our approach for these instances in Table 1. Note that we only report computation times for respecting all given hard constraints. Almost all soft constraints can be easily included in our IP without significantly worsen the running time.

Table 1. Statistics and results for ITC2007 instances

Name	Courses	Course-Slots	Rooms	Violations	Step1	Step2	Step3
comp01	30	160	6	0	< 0.01 sec.	0.05 sec.	0.02 sec.
comp02	82	283	16	0	< 0.01 sec.	0.19 sec.	0.02 sec.
comp03	72	251	16	0	< 0.01 sec.	0.17 sec.	0.02 sec.
comp04	79	286	18	0	< 0.01 sec.	0.24 sec.	0.03 sec.
comp05	54	152	9	0	< 0.01 sec.	0.56 sec.	0.02 sec.
comp06	108	361	18	0	< 0.01 sec.	0.43 sec.	0.04 sec.
comp07	131	434	20	0	< 0.01 sec.	0.57 sec.	0.04 sec.

5.2 Statistics and Results on Simulated Data

As we can see, the ITC2007 instances are no challenge to our approach. To obtain a better idea of its potential performance, we developed a simulation tool which is able to create large problem instances with near real-world character. We present statistics of three representative instances of different sizes, cf. Table 2. The key data (not listed here) of the large instance is almost identical to that of Technical University of Berlin (which is a rather large university). Computation times are acceptable, even though for an interactive timetable design, some tuning is necessary. Almost 80% of instructors teach at their first choice time slots.

Table 2. Statistics and results for simulated instances

Name	Courses	Course-Slots	Rooms	Violations	Step1	Step2	Step3
small	180	420	35	0	45 sec.	9 sec.	3 sec.
medium	950	2100	165	0	307 sec.	52 sec.	6 sec.
large	2100	4640	345	0	1235 sec.	5106 sec.	5 sec.

For comparison, we list in Table 3 the results for the same instances when using the intuitive integer program (1)–(5).

Table 3. Sizes, solution times, and quality for the intuitive integer program (1)–(5)

Name	# Variables	# Constraints	Runtime	Gap
small	13 000	7000	30 sec.	< 2%
medium	100 000	31 000	510 sec.	7 %
large	240 000	80 000	1 day	no solution

6 Discussion

We did not discuss several extensions, which are (or can easily be) incorporated in our practical implementation, most notably practical soft constraints. However, our generic model (in particular using the concept of a conflict graph) is well suited for this purpose. One can model e.g., that two courses have to be scheduled on consecutive time slots, or that no two lectures of the same course are given on the same day, etc. We report experiences with a soft constraint solver in a companion paper [17].

One could think of solving the integer program (6)–(10) via branch-and-cut. However, even for our largest instances, the number of facet inducing Hall inequalities (8) was rather small. This is why we simply added all facet inducing inequalities up-front.

We have access to the courses database at Technical University of Berlin. It comprises 2100 courses (to be scheduled to about 4500 time slots), 345 rooms of about 50 types, and 1550 instructors; there are seven time periods each day. The only reason for using simulated data instead of the real instance, is that the database is severely inconsistent and incomplete [18]. It is planned to manually repair and complete the necessary data in the near future, and to test our implementation for the construction of timetables for the whole university.

References

1. Burke, E., Rudová, H. (eds.): PATAT 2006: Proceedings of the 6th International Conference on the Practice and Theory of Automated Timetabling, Berlin. Lect. Notes Comp. Science. Springer, Heidelberg (2007),
   http://patat06.muni.cz/proceedings.html
2. Burke, E., Jackson, K., Kingston, J.H., Weare, R.: Automated university timetabling: The state of the art. Comput. J. 40(9), 565–571 (1997)
3. Lovász, L., Plummer, M.D.: Matching Theory, p. 544. North-Holland, Amsterdam (1986)
4. Schrijver, A.: Combinatorial Optimization Polyhedra and Efficiency. Springer, Berlin (2003)
5. ITC2007: 2nd International Timetabling Competition,
   http://www.cs.qub.ac.uk/itc2007/

6. Burke, E.K., Petrovic, S.: Recent research directions in automated timetabling. European J. Oper. Res. 140(2), 266–280 (2002)
7. Schaerf, A.: A survey of automated timetabling. Artificial Intelligence Review 13(2), 87–127 (1999)
8. Daskalaki, S., Birbas, T., Housos, E.: An integer programming formulation for a case study in university timetabling. European J. Oper. Res. 153, 117–135 (2004)
9. Schimmelpfeng, K., Helber, S.: Application of a real-world university-course timetabling model solved by integer programming. OR Spectrum 29, 783–803 (2007)
10. Lewis, R.: A survey of metaheuristic-based techniques for university timetabling problems. OR Spectrum (in press, 2007)
11. Meyers, C., Orlin, J.B.: Very large-scale neighborhood search techniques in timetabling problems. In: PATAT 2006: Proceedings of the 6th International Conference on the Practice and Theory of Automated Timetabling, Berlin, pp. 36–52. springer, Heidelberg (2007)
12. Carter, M.: A comprehensive course timetabling and student scheduling system at the university of Waterloo. In: Burke, E., Erben, W. (eds.) PATAT 2000. LNCS, vol. 2079, pp. 64–82. Springer, Heidelberg (2001)
13. Daskalaki, S., Birbas, T.: Efficient solutions for a university timetabling problem through integer programming. European J. Oper. Res. 127(1), 106–120 (2005)
14. Qualizza, A., Serafini, P.: A column generation scheme for faculty timetabling. In: Burke, E.K., Trick, M.A. (eds.) PATAT 2004. LNCS, vol. 3616, pp. 161–173. Springer, Heidelberg (2005)
15. Edmonds, J.: Maximum matching and a polyhedron with 0,1-vertices. J. Res. Nat. Bur. Standards 69B, 125–130 (1965)
16. Balas, E., Pulleyblank, W.: The perfectly matchable subgraph polytope of a bipartite graph. Networks 13, 495–516 (1983)
17. Lach, G., Lübbecke, M.: Curriculum based course timetabling: Optimal solutions to the Udine benchmark instances. Preprint 2008/9, Technische Universität Berlin, Institut für Mathematik (2008)
18. Lach, G.: Modelle und Algorithmen zur Optimierung der Raumvergabe der TU Berlin. Master's thesis, Technische Universität Berlin, Institut für Mathematik (in German, 2007)
19. Grinstead, C., Snell, J.: Introduction to Probability, 2nd edn. American Mathematical Society, Providence, RI (2003),
http://math.dartmouth.edu/~prob/prob/prob.pdf

A   Auxiliary Results

The first two lemmas follow directly from the inclusion-exclusion principle in probability theory [19].

Lemma 8. *Given a set V and a supermodular function $f : 2^V \to \mathbb{N}$, then for arbitrary $A_1, \ldots, A_n \subseteq V$ the following inequality holds:*

$$f(\bigcup_{i=1}^{n} A_i) \geq \sum_{k=1}^{n} (-1)^{k-1} \sum_{1 \leq i_1 < \ldots < i_k \leq n} f(A_{i_1} \cap \ldots \cap A_{i_k}) \qquad (24)$$

Lemma 9. *For each $n \in \mathbb{N}$ the following equation holds:*

$$\sum_{k=1}^{n}(-1)^{k-1}\sum_{1 \leq i_1 < \ldots < i_k \leq n}1 = 1 \tag{25}$$

Lemma 10. *Given a bipartite graph $G = (\mathcal{C} \cup \mathcal{R}, E)$ and a defining $\mathcal{C}$-set $U \subseteq \mathcal{C}$ with cardinality larger one, then we have*

$$\nu((U \cup \Gamma(U), E)) = |\Gamma(U)| \ .$$

Proof. Assume for contradiction that $\nu((U \cup \Gamma(U), E)) < |\Gamma(U)|$. Then there exists at least one unmatched $r \in \mathcal{R}$ for all maximal matchings. So we choose $c \in \Gamma(\{r\})$ and define:

$$U_1 = c$$
$$U_2 = U \backslash \{c\}$$

Clearly, U_1 and U_2 are disjoint. Furthermore,

$$\begin{aligned}
f(U_1) + f(U_2) &= |U_1| - \text{def}(U_1) + |U_2| - \text{def}(U_2) \\
&= |U| - \text{def}(U_1) \\
&= f(U) \ .
\end{aligned}$$

Thus, U does not induce a facet of $P(\mathcal{C})$. This is a contradiction to Lemma 2. $\square$

Lemma 11. *Given a bipartite graph $G = (\mathcal{C} \cup \mathcal{R}, E)$ and an atomic $\mathcal{C}$-set $U \subseteq \mathcal{C}$, then all inclusion maximal defining subsets $D_1, \ldots, D_k \subseteq U$ are disjoint, and furthermore their neighbourhoods $\Gamma(D_1), \ldots, \Gamma(D_k)$ are disjoint.*

Proof. Assume for contradiction that there exist two inclusion maximal defining disjoint subsets D_1, D_2 with $\Gamma(D_1) \cap \Gamma(D_2) \neq \emptyset$.
Case I: $\Gamma(D_1) \cup \Gamma(D_2) = \Gamma(U)$
Then U is not atomic.
Case II: $\Gamma(D_1) \cup \Gamma(D_2) \subsetneq \Gamma(U)$
Then $D_1 \sqcup D_2$ is defining and D_1 is not an inclusion maximal defining subset of U. $\square$

Lemma 12. *Given a bipartite graph $G = (\mathcal{C} \cup \mathcal{R}, E)$, an atomic $\mathcal{C}$-set $U \subseteq \mathcal{C}$ and all inclusion maximal defining subsets $D_1, \ldots, D_k \subsetneq U$ of U, then the following inequality holds:*

$$\text{def}(U) \geq \sum_{i=1}^{k}\text{def}(D_i) + 1 \ .$$

Proof. For all defining $\mathcal{C}$-sets D we have

$$
\begin{aligned}
\operatorname{def}(U) &= |U| - |\Gamma(U)| \\
&= \sum_{i=1}^{k} |D_i| - |\Gamma(D_i)| + |U \backslash \bigcup_{i=1}^{k} D_i| - |\Gamma(U) \backslash \bigcup_{i=1}^{k} \Gamma(D_i)| \\
&\geq \sum_{i=1}^{k} \operatorname{def}(D_i) + 1
\end{aligned}
$$

$\square$

A Basic Toolbox for Constrained Quadratic 0/1 Optimization[*]

Christoph Buchheim[1], Frauke Liers[1], and Marcus Oswald[2]

[1] Universität zu Köln, Institut für Informatik, Pohligstraße 1, 50969 Köln, Germany
[2] Universität Heidelberg, Institut für Informatik, INF 368, 69120 Heidelberg, Germany

Abstract. In many practical applications, the task is to optimize a non-linear function over a well-studied polytope P as, e.g., the matching polytope or the travelling salesman polytope (TSP). In this paper, we focus on quadratic objective functions. Prominent examples are the quadratic assignment and the quadratic knapsack problem; further applications occur in various areas such as production planning or automatic graph drawing. In order to apply branch-and-cut methods for the exact solution of such problems, they have to be linearized. However, the standard linearization usually leads to very weak relaxations. On the other hand, problem-specific polyhedral studies are often time-consuming. Our goal is the design of general separation routines that can replace detailed polyhedral studies of the resulting polytope and that can be used as a black box. As unconstrained binary quadratic optimization is equivalent to the maximum cut problem, knowledge about cut polytopes can be used in our setting. Other separation routines are inspired by the local cuts that have been developed by Applegate, Bixby, Chvátal and Cook for faster solution of large-scale traveling salesman instances. By extensive experiments, we show that both methods can drastically accelerate the solution of constrained quadratic 0/1 problems.

Keywords: quadratic programming, maximum cut problem, local cuts, crossing minimization, similar subgraphs.

1 Introduction

Optimizing a linear objective function over binary variables under additional linear constraints is NP-hard in general. One of the most successful frameworks for solving such problems is branch-and-cut. In order to develop fast branch-and-cut algorithms, it is crucial to determine good outer descriptions of the polytope P consisting of the convex hull of all feasible solutions of the problem at hand. The branch-and-cut approach is well developed, and the facial description of many polytopes corresponding to classical combinatorial optimization problems is well understood. For several problems practically efficient implementations exist.

[*] Financial support from the German Science Foundation is acknowledged under contracts Bu 2313/1-1 and Li 1675/1-1. Partially supported by the Marie Curie RTN Adonet 504438 funded by the EU.

C.C. McGeoch (Ed.): WEA 2008, LNCS 5038, pp. 249–262, 2008.

Instead of a linear objective function, we often desire to optimize a non-linear objective function over P. We consider problems where the non-linearities are locally defined, i.e., where every non-linear term in the objective function depends on few variables. In this paper, we focus on binary quadratic functions, however some of the proposed methods can easily be adapted to general non-linear functions.

The easiest example of a binary quadratic optimization problem is the maximum cut problem, which is equivalent to optimizing a degree-two polynomial over the hyper cube [4]. Many practical applications lead to non-linear objective functions in a natural way. Several crossing minimization problems in automatic graph drawing can be modeled as quadratic optimization problems over linear ordering type polytopes. To give another example, the tool switching problem arising in production planning can be solved by minimizing a polynomial of degree three over a polytope that is closely related to the TSP.

In any integer programming based approach to such non-linear 0/1 problems, the first step is to linearize the problem by introducing artificial variables that model the non-linearities. We thus need to optimize the linearized objective function over a polytope Q defined in a higher-dimensional space instead of optimizing a non-linear objective function over the original polytope P.

It is easy to see that all facets of P yield valid inequalities for Q. A naive branch-and-cut approach for the optimization over Q would use the separation routines known for P, in combination with the constraints modeling the connection between original and new variables, and resort to branching if no violated inequality can be detected any more. According to our experience, the performance of such an approach is very weak. Often, facet-inducing inequalities for P do not induce facets of Q, and the variables modeling non-linear terms change the polyhedral structure significantly. This can even happen if only one product is introduced and linearized.

In view of this, one could decide to undertake a polyhedral investigation of Q and try to develop specialized separation routines. Doing this will – very probably – be time consuming. Instead, much (human and computer) time could be saved by having some effective black-box routines at hand that speed up the solution algorithms but need only very limited knowledge about the problem structure. For quadratic problems, we provide such black-box routines and show that they drastically improve the running time of the solution algorithms.

Assuming that P is well understood, we ask the following question: How can we exploit the knowledge of P for optimizing over Q, without detailed polyhedral studies of Q? Even if the user is willing to invest some specific knowledge of Q, he/she can still combine her own separation strategies with our general methods outlined below. Moreover, the constraints produced by our methods might give some insight into the polyhedral structure of Q and point at important classes of cutting planes, which could be separated right from the start, using tailored separation algorithms.

We address the general separation problem from two complementary directions. First assume that the objective function is quadratic. In case the problem is unconstrained, one can formulate it as a maximum cut problem on an associated graph [4]. Even in the presence of constraints, valid inequalities for the cut polytope remain valid for Q after transformation, and can be separated using the same transformation. In several applications, the transformed constraints of P induce a face of the corresponding cut polytope, which gives some theoretical evidence that the inequalities derived from the cut polytope can be helpful.

On the other hand, we want to exploit the knowledge of the structure of the feasible solutions in P. Our proposed separation routine is inspired by the local cuts that have been developed by Applegate, Bixby, Chvátal and Cook (ABC^2) [1]. With the help of local cuts, they could solve big TSP instances being unsolved before. Recently, we proposed a variant of the local cut generation procedure that has some advantageous features [3]. We call our cutting planes *target cuts*. The main difference to the local cuts lies in a modified LP formulation that makes it possible to avoid the time-consuming tilting steps, as always a facet of the projected polytope is determined that can immediately be lifted to a valid inequality for Q.

For non-linear problems, the local or target cut approach is well-suited, as every non-linear term is determined by the original variables, so that the number of vertices does not change from P to Q. In particular, going from linear to non-linear objective functions does not slow down the cut generation significantly. Another advantage of this approach is that the separation can be implemented as a general framework that applies to all problems in this class. The user only needs to input some information about the structure of the feasible solutions, which is much easier than understanding the structure of the corresponding polytope. This approach can be applied to arbitrary non-linear problems in which the non-linearities are locally defined.

Our main contribution is to show that these approaches are very easy to use and lead to much better performance of general branch-and-cut approaches. By extensive computational experiments we show that not only the number of nodes in the enumeration tree but also the running time decreases dramatically, when compared to an algorithm that only uses the standard separation routines for the well-studied polytope P.

For some classical quadratic 0/1 problems, such as the quadratic knapsack problem or the quadratic assignment problem, special-purpose algorithms and implementations exist that exploit the problem structure and lead to effective algorithms. Clearly, we cannot compete with such problem-specific approaches. In this work, we aim at designing general-purpose methods that help improve the solution algorithms for quadratic problems for which not much is known about their structure. In particular, the reference point for our evaluation is the basic approach using standard linearization and separation for P.

The outline of this paper is as follows. We fix notation in Section 2. In Section 3, we discuss cutting planes derived from the cut polytope. In Section 4, we introduce target cuts and their usage in the context of quadratic problems. In Section 5, we explain the studied applications: the quadratic matching problem in Section 5.1 and the quadratic linear ordering and the linear arrangement problem in Section 5.2 and 5.3. In Section 6 we present experimental results.

2   Definitions

Consider a combinatorial optimization problem on a finite set E with feasible solutions $\mathcal{I} \subseteq 2^E$ and with a linear objective function $c(I) = \sum_{e \in I} c_e$, where $c_e \in \mathbb{R}$ for all $e \in E$. Without loss of generality, we desire to minimize $c(I)$ over all $I \in \mathcal{I}$. Let the polytope $P \subseteq \mathbb{R}^E$ denote the convex hull of all incidence vectors of feasible solutions. The corresponding integer linear program reads

$$
\begin{array}{rl}
\min & \sum_{e \in E} c_e x_e \\
\text{(P)} \quad \text{s.t.} & x \in P \\
& x \in \{0; 1\}^E
\end{array}
$$

In the following, we focus on objective functions that are quadratic in the variables x, i.e., we consider problems of the form

$$
\begin{array}{rl}
\min & \sum_{e \in E} c_e x_e + \sum_{e, f \in E; e \neq f} c_{ef} x_e x_f \\
\text{(QP)} \quad \text{s.t.} & x \in P \\
& x \in \{0; 1\}^E,
\end{array}
$$

For problems defined on a graph $G = (V, E)$ with variables corresponding to edges, and for two edges $e = (i, j)$ and $f = (k, l)$, we will use the notations c_{ef}, $c_{(i,j)(k,l)}$, and c_{ijkl} interchangeably. In order to address (QP) by integer programming techniques, we apply the standard linearization: for each pair $\{e, f\}$ with $c_{ef} \neq 0$, we introduce a binary variable y_{ef} modeling $x_e x_f$, along with the constraints $y_{ef} \leq x_e$, $y_{ef} \leq x_f$, and $y_{ef} \geq x_e + x_f - 1$. The linearized problem then reads

$$
\begin{array}{rll}
\min & \sum_{e \in E} c_e x_e + \sum_{e, f \in E; e \neq f} c_{ef} y_{ef} \\
\text{(LQP)} \quad \text{s.t.} & x \in P \\
& y_{ef} \leq x_e, x_f & \text{for all } \{e, f\} \text{ with } c_{ef} \neq 0 \\
& y_{ef} \geq x_e + x_f - 1 & \text{for all } \{e, f\} \text{ with } c_{ef} \neq 0 \\
& y_{ef} \in \{0; 1\} & \text{for all } \{e, f\} \text{ with } c_{ef} \neq 0 \\
& x \in \{0; 1\}^E .
\end{array}
$$

We are interested in the polytope Q spanned by all feasible solutions of (LQP).

Note that other methods for linearizing (QP) have been proposed in the literature. Nevertheless, we focus on the standard linearization, as it is the most natural and popular way to linearize (QP) and as it can easily be implemented.

3   Cutting Planes from Maxcut

Consider a graph $G = (V, E)$ with edge weights w_e. For $W \subseteq V$, the cut $\delta(W)$ is defined as

$$
\delta(W) = \{(u, v) \in E \mid u \in W, v \notin W\} .
$$

Its weight is $\sum_{e \in \delta(W)} w_e$. The maximum cut problem asks for a cut of maximum weight and is NP-hard for general graphs. The corresponding cut polytope, i.e., the convex hull of incidence vectors of cuts, is well studied [2,5], and practically effective branch-and-cut implementations exist for its solution [6,7].

It is a well-known result that the problem of optimizing a binary quadratic function without further constraints is equivalent to determining a maximum cut in an auxiliary graph $G_{\text{lin}} = (V_{\text{lin}}, E_{\text{lin}})$ [4]. The latter contains a node for each variable x_e. For each quadratic term $x_e x_f$ occuring in the objective function with $c_{ef} \neq 0$, the edge set E_{lin} contains an edge between the nodes corresponding to x_e and x_f. Furthermore, an additional root node and edges from this node to all nodes in V_{lin} are introduced. Now there exists a simple linear transformation between the edge variables in the maximum cut setting and the linear variables or products in the unconstrained quadratic optimization setting under which P is isomorphic to the cut polytope of G_{lin} [4].

If P is the unit hypercube, solving (LQP) thus amounts to determining a maximum cut in G_{lin}, i.e., to optimizing over a cut polytope defined in the E_{lin}-dimensional space. If P is a strict subset of the unit hypercube, i.e., if additional constraints are present, these constraints can be transformed as well and we derive that P is isomorphic to a cut polytope with further linear constraints. In particular, all inequalities valid for the cut polytope still yield inequalities valid for (LQP) and can be used in a cutting plane approach.

Clearly, intersecting the cut polytope with arbitrary hyperplanes in general yields a non-integer polytope. The structure of the resulting polytope can be very different from a cut polytope. In this case it is not clear whether the knowledge about the cut polytope can help solving the constrained optimization problem. However, several relevant applications exist in which the intersection of the cut polytope with a set of hyperplanes cuts out a face of a cut polytope, at least if certain product variables are present, e.g., for quadratic assignment and quadratic matching. The proof for the quadratic matching is a slight modification of the proof for the quadratic assignment polytope.

In any case, we obtain a correct separation algorithm for (LQP) based on cut separation. Within a branch-and-cut framework, we can always work in the original model and apply other separation algorithms as desired. When it comes to the cut separation, we build the graph $G_{\text{lin}} = (V_{\text{lin}}, E_{\text{lin}})$, transform the fractional point, and separate the inequalities known for the cut polytope. Found cutting planes are transformed back to yield cutting planes for (LQP).

4   Target Cuts for Quadratic 0/1 Problems

Usually, separation routines aim at generating faces or facets of some polytope in question that share similar structure. They are said to follow the *template paradigm*. Recently, ABC² proposed some general separation routine yielding so-called *local cuts* that are inequalities outside the template paradigm for which the structure is not known [1]. The size of the problem is first reduced by projecting the incidence vectors of feasible solutions onto a small-dimensional space (ABC^2 do this by shrinking nodes into supernodes).

For $r \leq m$, let π denote a projection $\mathbb{R}^m \to \mathbb{R}^r$ and let $\overline{Q} = \pi(Q) \subseteq \mathbb{R}^r$ denote the convex hull of the projected feasible solutions. Let $x^* \in \mathbb{R}^m$ be the point to be separated and $\overline{x}^* = \pi(x^*)$ be its projection to $\mathbb{R}^r$. A face-inducing inequality that separates some projected fractional point from $\overline{Q}$ can be obtained by solving an appropriately chosen linear program. Its size is basically determined by the number of its vertices. Thus, if the dimension of $\overline{Q}$ is not too big, this is fast in practice. Furthermore, the size of the linear program can be reduced by several considerations, and by delayed column generation only necessary feasible solutions are enumerated. A found local cut is then sequentially lifted and tilted until it becomes a facet for $\overline{Q}$ and then lifted to become feasible for the original TSP polytope.

Recently, we proposed a variant of the local cuts that we call *target cuts* [3]. The local cut framework can easily be adapted to target cuts, however the time-consuming tilting steps can be omitted. The reason for this is that we propose a different cut-generating linear program that generates a facet of $\overline{Q}$ right away. Furthermore, the volume of the generated facet is expected to be big. In the following, we briefly explain the target cuts separation. Details can be found in [3]. Subsequently, we will show that their use is favorable in the context of quadratic problems.

Assuming for now that $\overline{Q}$ is full-dimensional, we choose a point $\overline{q}$ in the interior of $\overline{Q}$. In case the projected non-feasible point $\overline{x}^*$ is not contained in $\overline{Q}$, we want to return a cutting plane that separates $\overline{x}^*$ from $\overline{Q}$. We argue in [3] that a facet from $\overline{Q}$ can be obtained by solving the following linear program:

$$\begin{aligned}
\max \quad & a^\top(\overline{x}^* - \overline{q}) \\
\text{s.t.} \quad & a^\top(\overline{x}_i - \overline{q}) \leq 1 \quad \text{for all } i = 1, \dots, s \\
& a \in \mathbb{R}^r
\end{aligned} \tag{1}$$

Here, $x_1, \dots, x_s$ are the vertices of $\overline{Q}$. A facet for $\overline{Q}$ violated by $\overline{x}^*$ can be read off the optimum solution of (1) as follows. If the optimum value of (1) is greater than 1, the corresponding inequality $a^\top(x - \overline{q}) \leq 1$ is violated by $\overline{x}^*$. Otherwise, $\overline{x}^*$ is contained in $\overline{Q}$.

In case the dimension of $\overline{Q}$ is smaller than r, the linear program (1) can be unbounded. In this case, $a^\top(x - \overline{q}) = 0$ is a valid equation for $\overline{Q}$ violated by $\overline{x}^*$, if a is an unbounded ray in (1).

In order to reduce the size of LP (1), we adapted the delayed column generation procedure proposed for local cuts to the target cut case. The procedure requires an oracle for maximizing any linear function over $\overline{Q}$. Having this at hand, one starts with a small, possibly empty, set of vertices $\overline{x}_1, \dots, \overline{x}_h$. Then a target cut $a^\top(x - \overline{q}) \leq 1$ is produced for the polytope $\overline{Q}_h = \text{conv}\{\overline{x}_1, \dots, \overline{x}_h\}$, by solving the corresponding linear program. Then, the oracle is called to maximize the left-hand side of the inequality. In case the maximum is bigger than 1, we add the maximal solution as a new $\overline{x}_{h+1}$ to (1). Otherwise we stop the procedure, having found a valid target cut. This process is iterated until the generated inequality is found to be valid. The number of columns added in this procedure is usually much smaller than the number of vertices of $\overline{Q}$.

In order to use target cuts for quadratic problems, we need to specify which projection to choose. In general, there is no easy answer to this question, and the user might have to test the performance of different projections in order to find which one gives best results. The projection needs to allow fast recognition or determination of the points in $\mathbb{R}^r$ that can be extended to feasible solutions in the original space. For several applications this is possible with the trivial, i.e., orthogonal, projection onto some sub-graph or sub-space, or the projection through shrinking of nodes into supernodes. For a given (linear) projection, lifting of a found inequality is trivial.

For some problems, certain projections seem to be favorable to others. For example, in a problem in which the global structure is important, as is the case for the TSP, a projection through shrinking should be prefered in case it allows to characterize the points in $\mathbb{R}^r$ having a preimage in $\mathbb{R}^m$ under π. On the other hand, there are problems in which the local structure seems to be characteristic of the problem, as, e.g., for the matching problem. In the latter, trivial projections can be used.

The usage of target cuts allows the implementation of a general framework in which only the projection and the oracle need to be specified for the particular application; everything else is problem-independent. Moreover, target cuts are well-suited for quadratic 0/1 problems: the size of the cut generating program (1) remains moderate, as there is a 1-1 correspondence between the vertices of the polytope P and the polytope Q. Therefore, the projected linearized polytope $\overline{Q}$ has the same number of vertices as $\pi(P)$, so that the number of rows of (1) does not grow with the introduction of product variables. In other words, the additional product variables do not affect the performance considerably, which allows to deal with non-trivial chunk sizes.

5   Applications

Applications of constraint quadratic binary optimization problems abound. One of the more traditional examples is the quadratic assignment problem; more recently also the quadratic knapsack problem has attracted some interest. In the following, we consider two other problems: the quadratic matching and the quadratic linear ordering problem. More precisely, we consider applications that are naturally modeled as such problems. In Section 5.1, we look at the problem of finding highly similar subgraphs, which can be modeled as a quadratic (bipartite) matching problem. In Section 5.2 and 5.3, we discuss two applications of quadratic linear ordering: the bipartite crossing minimization problem and the linear arrangement problem.

5.1   Finding Highly Similar Subgraphs – Quadratic Matching

Assume we are given two graphs $G_1 = (V_1, E_1)$ and $G_2 = (V_2, E_2)$, and we want to get insight into how similar the two graphs are. This problem occurs in several practical applications, e.g., in automatic graph drawing and computational biology. The task is to determine a matching of a subset or all nodes of G_1 to those of G_2 such that as many edges as possible in the two graphs are mapped onto each other. Obviously, this problem is a generalization of the graph isomorphism problem in which we decide whether there exists a matching of the nodes in V_1 to those in V_2 such that all edges in E_1 are mapped onto edges in E_2, and vice versa.

In the generalization we are concerned with, we also allow but penalize the case in which $u_1 \in V_1$ is matched on $u_2 \in V_2$ and $v_1 \in V_1$ on $v_2 \in V_2$, but exactly one of the edges (u_1, v_1) or (u_2, v_2) exists. A straight-forward model for this problem is the following quadratic matching formulation

$$
\begin{array}{rl}
\max & \sum_{i \in V_1, j \in V_2} x_{ij} + \sum_{i,k \in V_1, j,l \in V_2} c_{ijkl} x_{ij} x_{kl} \\
\text{(QMP)} \quad \text{s.t.} & \sum_{i \in V_1} x_{ij} \leq 1 \qquad \forall j \in V_2 \\
& \sum_{j \in V_2} x_{ij} \leq 1 \qquad \forall i \in V_1 \\
& x_{ij} \in \{0; 1\} \ \forall i \in V_1, j \in V_2
\end{array}
$$

with costs $c_{ijkl} < 0$ if either $(i, k) \in E_1$ or $(j, l) \in E_2$, but not both. Otherwise $c_{ijkl} \geq 0$. In this model, $x_{ij} = 1$ means that node $i \in V_1$ is matched with node $j \in V_2$.

5.2   Bipartite Crossing Minimization – Quadratic Linear Ordering I

Consider a bipartite graph $G = (V_1 \cup V_2, E)$. We want to draw G in the plane so that the nodes of V_1 and V_2 are placed on two parallel horizontal lines. The task is to minimize the number of crossings between edges, assuming that all edges are drawn as straight lines. Several applications exist in the area of automatic graph drawing. Clearly, the number of crossings only depends on the orders of vertices on the two lines.

First, we assume that the nodes V_1 on the upper level are layouted in some fixed order, whereas the nodes on the lower level are allowed to permute within the layer. The permutation of the nodes in V_2 has to be chosen such that the number of edge crossings is minimal. Let $i, j \in V_1$, $k, l \in V_2$ and edges $(i, k), (j, l)$ be present. Assume i is before j in the fixed order. No crossing exists in case k is before l on the second level, otherwise there is a crossing.

Hence the bipartite crossing minimization problem with one fixed layer can easily be formulated as a linear ordering problem. Now let us formulate the problem with two free layers as a quadratic optimization problem over the linear ordering polytope. For i, j, k, l chosen as above, there is no crossing in case i is before j and k is before l, or j is before i and l is before k. Let us introduce variables x_{uv} that take value 1 if u is drawn before v, and 0 otherwise. Then we have to solve the problem

$$
\begin{array}{rl}
\max & \sum_{(i,k),(j,l) \in E} x_{ij} x_{kl} \\
\text{(QLO}_1\text{)} \quad \text{s.t.} & x \in P_{LO} \\
& x_{ij} \in \{0; 1\} \ \forall i, j \in V_1 \text{ or } i, j \in V_2 ,
\end{array}
$$

where P_{LO} is the linear ordering polytope.

5.3   Linear Arrangement – Quadratic Linear Ordering II

The linear arrangement problems is given as follows. We are looking for a permutation of n objects in such a way that a linear function c on the differences of positions of the objects is minimized. More precisely, we desire to determine a permutation π of $\{1, \ldots, n\}$ minimizing

$$\sum_{1 \le i,j \le n} c_{ij} |\pi(i) - \pi(j)| \ .$$

To this end we use the fact that the distances of the positions of two elements i and j with respect to a permutation π can be expressed in terms of betweenness variables. This distance equals 1 plus the number of elements lying between i and j, i.e., $|\pi(i) - \pi(j)| = 1 + \sum_k x_{ik} x_{kj}$ where x_{ij} is the usual linear odering variable modeling whether $\pi(i) < \pi(j)$ or not. Therefore, up to a constant, the linear arrangement problem can be rewritten as

$$\begin{array}{lll} & \max & \sum_{i \ne j \ne k \ne i} c_{ij} x_{ik} x_{kj} \\ (\text{QLO}_2) & \text{s.t.} & x \in P_{LO} \\ & & x_{ij} \in \{0;1\} \ \forall i,j \in \{1 \ldots n\}, i \ne j. \end{array}$$

Note that for this application only products of linear ordering variables are required that are of the type $x_{ik} x_{kj}$, which are only $O(n^3)$ many.

6   Experiments

We implemented the two separation approaches discussed in Section 3 and 4 within the branch-and-cut framework **ABACUS**, using CPLEX 11. All test runs were performed on Xeon machines with 2.66 GHz.

For each application we addressed, we start a branch-and-cut algorithm with the linear programming relaxation of the linearized problem (LQP). Separation routines for the polytopes P are assumed to be readily available. We compare the performance of this basic approach with the same approach extended by appropriately used maximum cut separation as described in Section 3 and the target cut separation as introduced in Section 4. For the tested applications, we used trivial projections onto subsets of variables, called *chunks*.

The chunks were chosen randomly in the sense that we first generate a subgraph randomly and then project onto all those linear and product variables that are completely determined by the subgraph. For the maximum cut separation, we separate the cycle inequalities [2]. We aimed at developing one relatively abstract implementation that can easily be used for all quadratic problems of type (QP) without having to incorporate many changes. Only the target-cut oracle and the test whether some vector represents a feasible solution are specific to the problem and have to be implemented separately for each application. We tested our approaches on randomly generated instances.

6.1   The Quadratic Matching Problem

For the quadratic matching problem, we studied instances defined on complete graphs. Note that a product $x_{ij}x_{kl}$ is necessarily zero if i, j, k, l are not pairwise distinct. We create random instances where for given pairwise distinct i, j, k, l the weight c_{ijkl} is non-zero with a given probability p. In this case, the weight is randomly chosen from $\{-1000, \ldots, 1000\}$. All linear weights c_{ij} are also chosen randomly from $\{-1000, \ldots, 1000\}$. An instance is thus defined by the number of nodes n, the percentage p of products with non-zero coefficient, and a random seed r for the weights.

Our implementations either determine a maximum quadratic matching or a minimum perfect quadratic matching. In the basic branch-and-bound approach, we separate the blossom inequalities that are known to be the only non-trivial facets of the matching polytope. We compare this basic approach with a branch-and-cut algorithm that uses separation of cutting planes derived from the cut polytope and of target cuts on varying chunk sizes, as explained in Section 3 and 4. We also test an implementation with both separation routines.

It turns out that better performance can be achieved if the maximum-cut separation procedure is only called in the root node of the branch-and-bound tree, and not after branching has been done. For the target cuts, the extendable solutions under a trivial projection are the incidence vectors of (not necessarily perfect) quadratic matchings. For their generation, two oracles are implemented: First, a heuristic greedy oracle tries to identify fast necessary incidence vectors of quadratic matchings. In case it is successful, the delayed column generation procedure continues. In case it is not successful, we test whether a violating vector exists by calling an exact oracle. In the latter, the integer programming formulation for the quadratic matching problem on the small chunk is solved exactly. The column generation procedure is iterated until no more violating vector is found by the exact oracle.

We show some running times in Table 1. We report the cpu time in seconds and the number of subproblems needed to solve the instance to optimality. IP refers to the basic algorithm, $MxTy$ means that we apply cut generation if $x = 1$, and target cut separation with chunk size y.

As can be expected, in practice the number of found blossom inequalities is very small, often none of them is violated, and so the basic implementation solves the problem basically via branching. Only very small instances can be solved to optimality. It is obvious that the separation of inequalities from the cut polytope considerably improves the running times. Also the target cut separation strongly reduces the number of subproblems, the number of linear problems and the running time. The best improvement is achieved when both separation routines are included.

The optimal size of the chunks depends on the size of the instances. Clearly, using too large chunks can increase the total runtime, since the effect of having to solve less subproblems is foiled by the long running time needed to compute the target cuts; the latter increases exponentially in the size of the chunk. For the larger instances we considered, the best results where obtained with chunks of 5 to 7 nodes.

Table 1. Results for the quadratic matching problem, perfect matchings (top) and general matchings (bottom)

|V|	p	IP	MC0 T5	MC0 T6	MC0 T7	MC1 T0	MC1 T5	MC1 T6	MC1 T7
14	0.4	1:06 505	0:48 267	0:47 49	7:42 13	1:23 297	0:55 91	2:01 33	9:23 11
16	0.4	29:25 5145	13:42 2021	6:29 377	23:30 69	21:21 2187	9:06 703	7:12 221	21:57 65
18	0.1	0:27 331	0:27 281	0:27 195	0:17 73	0:22 159	0:26 123	0:22 83	0:20 59
20	0.1	2:47 1021	2:30 807	1:49 525	1:04 203	1:42 425	1:42 211	1:04 153	1:01 81
16	0.2	0:22 255	0:16 119	0:23 69	0:20 1	0:11 33	0:46 5	0:22 39	0:51 1
16	0.2	1:33 877	1:15 591	0:48 257	0:47 99	1:50 65	1:16 363	0:57 203	2:36 47
16	0.2	0:41 473	0:37 379	0:19 71	0:28 43	1:25 23	0:44 391	0:35 83	1:38 21
16	0.2	1:08 751	1:06 667	0:37 231	0:42 87	1:33 45	0:54 335	0:45 197	2:18 39
18	0.1	0:21 267	0:27 281	0:27 195	0:22 83	0:26 79	0:22 159	0:25 123	0:45 49
18	0.1	0:20 261	0:19 225	0:15 125	0:18 103	0:22 91	0:16 149	0:20 147	0:33 65
18	0.1	0:17 211	0:18 185	0:13 107	0:14 45	0:17 51	0:12 69	0:16 63	0:25 35
18	0.1	0:20 225	0:19 197	0:18 155	0:14 45	0:31 111	0:20 169	0:23 131	0:33 73
18	0.2	16:16 3501	12:42 2463	4:06 527	3:53 327	5:11 259	10:29 1583	12:42 2623	7:08 205
18	0.2	12:39 3043	10:28 2387	2:09 245	2:27 175	3:35 205	7:58 1337	4:24 373	4:24 123
18	0.2	10:10 2359	8:29 1749	3:33 513	3:03 147	5:13 277	8:41 1251	5:01 415	5:19 123
18	0.2	20:13 4497	17:23 3343	6:56 1193	5:15 485	8:14 583	16:28 2663	8:57 1257	8:18 339
20	0.1	2:38 1063	2:29 807	1:48 525	1:04 153	1:19 199	1:42 425	1:41 211	1:16 71
20	0.1	3:49 1129	4:05 1117	2:21 615	1:32 141	2:16 507	2:44 543	2:45 521	1:44 171
20	0.1	3:54 1259	4:04 1089	2:47 767	1:40 235	2:11 453	2:55 679	2:37 541	1:40 215
20	0.1	2:40 915	2:39 807	1:54 419	1:28 121	1:28 255	2:09 477	2:02 269	1:35 81

Table 2. Results for the quadratic linear ordering problem, random instances (top) and linear arrangement instances (bottom). Instances marked with −− could not be solved within 2 GB of RAM.

$|V|$	$|E|$	IP		MC0 T5		MC0 T6		MC1 T0		MC1 T5		MC1 T6	
10	45	0:13:45	10995	0:00:43	1	0:13:55	1	0:23:18	341	0:00:34	1	0:10:40	1
10	45	0:15:50	12349	0:00:58	1	0:21:15	1	0:16:06	301	0:00:39	21	0:12:55	1
10	45	0:14:39	13319	0:00:41	1	0:14:01	1	0:23:52	527	0:00:29	1	0:09:14	1
10	45	0:11:18	8871	0:00:55	1	0:19:27	1	0:31:47	491	0:00:40	1	0:13:34	1
11	55	>3:00:00	−−	0:02:31	1	0:37:15	1	2:50:34	965	0:01:51	1	0:21:59	1
11	55	>0:30:00	−−	0:02:55	1	0:44:25	1	6:06:02	1955	0:02:05	1	0:24:57	1
11	55	>0:32:00	−−	0:02:00	1	0:25:27	1	3:15:18	1003	0:01:34	1	0:18:25	1
11	55	>0:29:00	−−	0:02:15	1	0:30:15	1	2:23:34	867	0:01:37	1	0:20:51	1
14	29	0:00:16	1049	0:00:07	189	0:00:23	25	0:00:02	15	0:00:02	5	0:00:10	1
14	38	>0:06:23	−−	0:02:03	2347	0:02:27	3	0:00:55	15	0:00:30	3	0:01:38	1
14	47	>0:09:08	−−	0:00:13	3	0:01:54	1	0:00:34	17	0:00:10	1	0:00:58	1
14	56	>0:11:40	−−	0:00:22	3	0:02:30	1	0:01:49	29	0:00:18	1	0:02:12	1
14	65	>0:14:30	−−	0:01:54	27	0:15:04	11	0:03:19	21	0:02:06	3	0:13:48	3
14	74	>0:17:27	−−	0:00:09	3	0:02:08	1	0:01:38	41	0:00:15	1	0:02:10	1
14	83	>0:20:19	−−	0:00:19	3	0:06:56	3	0:14:48	329	0:00:29	3	0:05:43	1
16	37	>0:09:33	−−	0:03:11	2447	0:07:44	381	0:01:04	33	0:00:23	1	0:01:07	1
16	46	>0:13:11	−−	0:00:38	55	0:01:58	3	0:00:39	1	0:00:14	1	0:01:03	1
16	55	>0:16:48	−−	0:01:10	3	0:04:29	1	0:03:49	27	0:00:53	3	0:02:39	1
16	64	>0:20:36	−−	0:01:04	3	0:05:07	3	0:04:51	9	0:00:56	1	0:03:44	1
16	73	>0:24:19	−−	0:02:24	3	0:14:58	1	0:07:06	3	0:02:48	3	0:12:04	1
16	82	>0:28:52	−−	0:02:09	9	0:15:21	1	0:11:27	11	0:02:17	3	0:14:43	3
16	91	>0:33:43	−−	0:02:23	3	0:28:02	5	0:43:05	71	0:03:03	3	0:14:59	1
16	100	>0:40:05	−−	0:00:33	1	0:07:40	1	0:18:40	17	0:00:57	1	0:09:05	1
16	103	>0:42:10	−−	0:01:43	1	0:23:00	3	0:36:19	61	0:02:23	3	0:18:32	1
20	38	0:03:29	3049	0:02:12	1597	0:02:58	1019	0:00:15	1	0:00:13	1	0:00:26	1
20	41	>0:22:24	−−	>0:26:10	−−	0:30:13	6917	0:02:04	15	0:00:57	3	0:01:16	3
20	44	>0:24:49	−−	>0:31:19	−−	1:00:08	8961	0:04:36	7	0:03:10	1	0:03:09	1

6.2   The Quadratic Linear Ordering Problem

According to our experience, separating inequalities known to be valid for the polytope P does not speed up the optimization over Q considerably, and so our basic branch-and-cut algorithm for the solution of (QLO_1) and (QLO_2) only separates the 3-dicycle inequalities. The latter are known to be facets for the linear ordering polytope. In contrast to the quadratic matching case, max-cut separation turns out to be very effective for the quadratic linear ordering problem, and so it is called in every node of the branch-and-bound tree. The target cut separation is again performed on randomly chosen chunks that are generated via trivial projection. The vectors that are extendable under the trivial projection are again linear orderings on the chunks.

We studied instances defined on complete graphs. Again, weights of linear and product variables are chosen randomly in $\{-1000, \dots, 1000\}$. All products are generated. An instance is defined by the number of nodes n of the complete graph and a random seed r for the randomly chosen weights. Moreover, we created linear arrangement instances defined by random graphs, see Section 5.3.

In Table 2 we show some running times for both types of instances. As above, we report the cpu time in seconds and the number of subproblems needed to solve the instance to optimality.

The results are similar to the quadratic matching case: the basic implementation solves the problem essentially via branching, only very small instances can be solved. Again it is obvious that the separation of inequalities from the cut polytope considerably improves the running time. Also the target cut separation strongly reduces the number of subproblems, the number of linear problems and the running time. The best chunk sizes are 5 to 6.

In summary, our results show that both presented separation methods improve the performance of the basic branch-and-cut approach significantly.

7   Conclusion

We present and evaluate two methods for improving the performance of branch-and-cut approaches to general quadratic 0/1 optimization problems, addressing the problem from two different directions. The first method addresses the quadratic structure, exploiting separation routines for cut polytopes, while the second implicitly takes into account the specific structure of the underlying polytope, applying a technique similar to local cut generation. Our results show that the total running time can be decreased significantly by both techniques.

References

1. Applegate, A., Bixby, R., Chvátal, V., Cook, W.: TSP cuts which do not conform to the template paradigm. In: Jünger, M., Naddef, D. (eds.) Computational Combinatorial Optimization. LNCS, vol. 2241, pp. 261–304. Springer, Heidelberg (2001)
2. Barahona, F., Mahjoub, A.R.: On the cut polytope. Mathematical Programming 36, 157–173 (1986)

3. Buchheim, C., Liers, F., Oswald, M.: Local cuts revisited. Operations Research Letters (2008), doi:10.1016/j.orl.2008.01.004
4. De Simone, C.: The cut polytope and the Boolean quadric polytope. Discrete Mathematics 79, 71–75 (1990)
5. Deza, M., Laurent, M.: Geometry of Cuts and Metrics. Algorithms and Combinatorics, vol. 15. Springer, Heidelberg (1997)
6. Liers, F., Jünger, M., Reinelt, G., Rinaldi, G.: Computing Exact Ground States of Hard Ising Spin Glass Problems by Branch-and-Cut. New Optimization Algorithms in Physics, pp. 47–68. Wiley-VCH, Chichester (2004)
7. Rendl, F., Rinaldi, G., Wiegele, A.: A branch and bound algorithm for Max-Cut based on combining semidefinite and polyhedral relaxations. In: Fischetti, M., Williamson, D.P. (eds.) IPCO 2007. LNCS, vol. 4513, pp. 295–309. Springer, Heidelberg (2007)

Empirical Investigation of Simplified Step-Size Control in Metaheuristics with a View to Theory

Jens Jägersküpper* and Mike Preuss

Technische Universität Dortmund, Fakultät für Informatik,
44221 Dortmund, Germany

Abstract. Randomized direct-search methods for the optimization of a function $f\colon \mathbb{R}^n \to \mathbb{R}$ given by a black box for f-evaluations are investigated. We consider the cumulative step-size adaptation (CSA) for the variance of multivariate zero-mean normal distributions. Those are commonly used to sample new candidate solutions within metaheuristics, in particular within the CMA Evolution Strategy (CMA-ES), a state-of-the-art direct-search method. Though the CMA-ES is very successful in practical optimization, its theoretical foundations are very limited because of the complex stochastic process it induces. To forward the theory on this successful method, we propose two simplifications of the CSA used within CMA-ES for step-size control. We show by experimental and statistical evaluation that they perform sufficiently similarly to the original CSA (in the considered scenario), so that a further theoretical analysis is in fact reasonable. Furthermore, we outline in detail a probabilistic/theoretical runtime analysis for one of the two CSA-derivatives.

1 Introduction

The driving force of this paper is the desire for a theoretical runtime analysis of a sophisticated stochastic optimization algorithm, namely the *covariance matrix adaptation evolution strategy (CMA-ES)*. As this algorithm is very hard to analyze theoretically because of the complex stochastic process it induces, we follow an unorthodox approach: We decompose it into its main algorithmic components: the *covariance matrix adaptation (CMA)*, and the *cumulative step-size adaptation (CSA)*. While the CMA is, in some sense, supposed to handle ill-conditionedness in the optimization problems, it is the duty of the CSA to cope with a challenge that every algorithm for real-valued black-box optimization faces: step-size control, i.e. the adaptation of step-sizes when approaching an optimum. The idea behind this decomposition is to substitute one of the components by an alternative mechanism that is more amenable to a theoretical analysis. While doing so, we rely on experimentation to assess how far we depart from the original algorithm. Here simpler mechanisms for step-size control are substituted for CSA. The desired outcome of this process is a modified

---

* Supported by the German Research Foundation (DFG) through the collaborative research center "Computational Intelligence" (SFB 531).

C.C. McGeoch (Ed.): WEA 2008, LNCS 5038, pp. 263–274, 2008.

algorithm that is both: tractable by theoretical analysis techniques, and, tested empirically, still working reasonably similar to the original one. At the same time we aim at better understanding the core-mechanisms of the original algorithm. (A simplified algorithm may also be interesting for practitioners, as simple methods often spread much faster than their complicated counterparts, even if the latter have slight performance advantages.) The full theoretical analysis itself, however, is not part of this work, but is still pending. Rather we discuss why a theoretical analysis seems feasible and outline in detail how it could be accomplished. Note that proving (global) convergence is *not* the subject of the theoretical investigations. The subject is a probabilistic analysis of the random variable corresponding to the number of steps necessary to obtain a predefined reduction of the approximation error. For such an analysis to make sense, the class of objective functions covered by the analysis must necessarily be rather restricted and simple. This paper provides evidence by experimental and statistical evaluation that a theoretical analysis for the simplified algorithm does make sense to yield more insight into CSA, the step-size control within the CMA-ES.

In a sense, our approach can be seen as algorithm re-engineering, a viewpoint which is to our knowledge uncommon in the field of metaheuristics. Therefore, we also strive for making a methodological contribution that will hopefully inspires other researchers to follow a similar path. To this end, it is not necessary to be familiar with the latest scientific work on metaheuristics to understand the concepts applied herein. It is one of the intrinsic properties of stochastic algorithms in general that even simple methods may display surprising behaviors that are not at all easy to understand and analyze. We explicitly focus on practically relevant dimensions. As it may be debatable what relevant is, we give a 3-fold categorization used in the following: Problems with up to 7 dimensions are seen as small, whereas medium sized ones from 8 to 63 dimensions are to our knowledge of the highest practical importance. Problems with 64 dimensions and beyond are termed large.

In the following section the original CSA is described in detail. The two new simplified CSA-derivatives are presented in Section 3, and their technical details and differences as well as some related work are discussed in Section 4. Furthermore, a detailed outline of a theoretical runtime analysis is given. The experimental comparison (including statistical evaluation) of the three CSA-variants is presented in Section 5. Finally, we conclude in Section 6.

2   Cumulative Step-Size Adaptation (CSA)

The CMA-ES of Hansen and Ostermeier [1] is regarded as one of the most efficient modern stochastic direct-search methods for numerical black-box optimization, cf. the list of over 100 references to applications of CMA-ES compiled by Hansen [2]. Originally, CSA was designed for so-called *(1,λ) Evolution Strategies,* in which *1* candidate solution is iteratively evolved. In each iteration, λ new search points are sampled, each independently in the same way, namely by adding a zero-mean multivariate normal distribution to the current candidate solution. When CMA is *not* used, like here, each of the λ samples is generated by independently adding a zero-mean normal distribution with variance σ^2 to each of its n components. The best of the λ samples becomes the next candidate solution—irrespective of whether this best of λ amounts to an improvement or not (so-called *comma selection*; CSA does not work well for so-called *elitist selection* where the best sample becomes the next candidate solution only if it is at least as good as the current one). The idea behind CSA is as follows: Consecutive steps of an iterative direct-search method should be orthogonal. Therefore, one may recall that steps of steepest descent (a gradient method) with perfect line search (i.e., the truly best point on the line in gradient direction is chosen) are indeed orthogonal when a positive definite quadratic form is minimized. Within CSA the observation of mainly positively [negatively] correlated steps is taken as an indicator that the step-size is too small [resp. too large]. As a consequence, the standard deviation σ of the multivariate normal distribution is increased [resp. decreased]. Since the steps' directions are random, considering just the last two steps does not enable a smooth σ-adaptation because of the large variation of the random angle between two steps. Thus, in each iteration the so-called *evolution path* is considered, namely its length is compared with the length that would be expected when the steps were orthogonal. Essentially, the evolution path is a recent part of the trajectory of candidate solutions. Considering the complete trajectory is not the most appropriate choice, though. Rather, a certain amount of the recent history of the search is considered. In the original CSA as proposed by Hansen and Ostermeier, σ is adapted continuously, i.e. after each iteration. The deterministic update of the evolution path $p \in \mathbb{R}^n$ after the ith iteration works as follows:

$$p^{[i+1]} := (1 - c_\sigma) \cdot p^{[i]} + \sqrt{c_\sigma \cdot (2 - c_\sigma)} \cdot m^{[i]}/\sigma^{[i]} \tag{1}$$

where $m^{[i]} \in \mathbb{R}^n$ denotes the displacement (vector) of the ith step, and the fixed parameter $c_\sigma \in (0,1)$ determines the weighting between the recent history of the optimization and its past within the evolution path. We use $c_\sigma := 1/\sqrt{n}$ as suggested by Hansen and Ostermeier [1]. Note that $m^{[i]}$ is one of λ vectors each of which was independently chosen according to a zero-mean multivariate normal distribution with standard deviation $\sigma^{[i]}$. The length of such a vector follows a scaled (by $\sigma^{[i]}$) χ-distribution. Initially, $p^{[0]}$ is chosen as the all-zero vector. The σ-update is done deterministically as follows:

$$\sigma^{[i+1]} := \sigma^{[i]} \cdot \exp\left(\frac{c_\sigma}{d_\sigma} \cdot \left(\frac{|p^{[i+1]}|}{\bar{\chi}} - 1\right)\right) \tag{2}$$

where the fixed parameter d_σ is called *damping factor*, and $\bar{\chi}$ denotes the expectation of the χ-distribution. Note that σ is kept unchanged if the length of the evolution path equals $\bar{\chi}$. We used $d_\sigma := 0.5$ because this leads to a better performance than $d_\sigma \in \{0.25, 1\}$ for the considered function scenario. Naturally, there is interdependence between d_σ and c_σ, and moreover, an optimal choice depends (among others) on the function to be optimized, of course.

3   Two Simplified CSA-Derivatives

In this section we introduce two simplifications of the original CSA, created by subsequently departing further from the defining Equations (1) and (2). The first simplification (common to both CSA-derivatives) will be to partition the course of the optimization into phases of a fixed length in which σ is not changed. Such a partitioning of the process has turned out useful in former theoretical analyses, cf. [3,4] for instance. Thus, both variants use phases of a fixed length, after each of which σ is adapted—solely depending on what happened during that phase, respectively. Therefore, recall that $c_\sigma = 1/\sqrt{n}$ in the path update, cf. Eqn. (1). Since $(1 - 1/\sqrt{n})^i = 0.5$ for $i \asymp \sqrt{n} \cdot \ln 2$ (as n grows), the half-live of a step within the evolution path is roughly $0.5\sqrt{n}$ iterations for small dimensions and roughly $0.7\sqrt{n}$ for large n. For this reason we choose the *phase length* k as $\lceil \sqrt{n} \rceil$ a priori for the simplified versions to be described in the following. The second simplification to be introduced is as follows: Rather than comparing the length of the displacement of a phase with the expected length that would be observed if the steps in the phase were completely orthogonal, the actual correlations of the steps of a phase in terms of orthogonality are considered directly and aggregated into a criterion that we call *correlation balance*.

pCSA. The "p" stands for *phased*. The run of the ES is partitioned into phases lasting $k := \lceil \sqrt{n} \rceil$ steps, respectively. After each phase, the vector corresponding to the total movement (in the search space) in this phase is considered. The length of this displacement is compared to $\ell := \sqrt{k} \cdot \sigma \cdot \bar{\chi}$, where $\bar{\chi}$ is the expectation of the χ-distribution with n degrees of freedom. Note that ℓ equals the length of the diagonal of a k-dimensional cube with edges of length $\sigma \cdot \bar{\chi}$, and that $\sigma \cdot \bar{\chi}$ equals the expected step-length in the phase. Thus, if all k steps of a phase had the expected length, and if they were completely orthogonal, then the length of the displacement vector in such a phase would just equal ℓ. Depending on whether the displacement's actual length is larger [or smaller] than ℓ, σ is considered as too small (because of positive correlation) [resp. as too large (because of negative correlation)]. Then σ is scaled up [resp. down] by a predefined scaling factor larger than one [resp. by the reciprocal of this factor].

CBA. "CBA" stands for *correlation-balance adaptation*. The optimization is again partitioned into phases each of which lasts $k := \lceil \sqrt{n} \rceil$ steps. After each phase, the k vectors that correspond to the k movements in the phase are considered. For each pair of these k vectors the correlation is calculated, so that we obtain $\binom{k}{2} = k(k - 1)/2$ correlation values. If the majority of these values are positive [negative], then the σ used in the respective phase is considered as too small [resp. as too large]. Hence, σ is scaled up [resp. down] after the phase by some predefined factor larger than one [resp. by the reciprocal of this factor].

4   Related Work, Discussion, and a View to Theory

Evolutionary algorithms for numerical optimization usually try to *learn* good step-sizes, which may be implemented by self-adaptation or success-based rules, the most prominent of which may be the 1/5-rule to increase [decrease] step sizes if more [less] than 1/5 of the samples result in an improvement. This simple deterministic adaptation mechanism, which is due to Rechenberg [5] and Schwefel [6], has already been the subject of a probabilistic analysis of the (random) number of steps necessary to reduce the approximation error in the search space. The first results from this viewpoint of analyzing ES like "usual" randomized algorithms were obtained in [7] for the simplest quadratic function, namely $x \mapsto \sum_{i=1}^{n} x_i^2$, which is commonly called SPHERE. This analysis has been extended in [8] to quadratic forms with bounded condition number as well as to a certain class of ill-conditioned quadratic forms (parameterized in the dimensionality of the search space) for which the condition number grows as the dimensionality of the search space increases. The main result of the latter work is that the runtime (to halve the approximation error) increases proportionally with the condition number. This drawback has already been noticed before in practice, of course. As a remedy, the CMA was proposed which, in some sense, learns and continuously adjusts a preconditioner by adapting the covariance matrix of the multivariate normal distribution used to sample new candidate solutions. As noted above, within the CMA-ES the CSA is applied for step-size control, which is neither a self-adaptive mechanism nor based on a success rule. In the present work, we exclusively deal with this CSA mechanism. Therefore, we consider a spherically symmetric problem, so that the CMA mechanism is dispensable. We demand that the simplified CSA-derivatives perform sufficiently well on this elementary type of problem at least. If they did not, they would be unable to ensure local convergence at a reasonable speed, so that efforts on a theoretical analysis would seem questionable.

CSA, pCSA, and CBA are basically $(1, \lambda)$ ES, and results obtained following a dynamical-system approach indicate that the expected spatial gain towards the optimum in an iteration is bounded above by $O(\ln(\lambda) \cdot d/n)$, where d denotes the distance from the optimum, cf. [9, Sec. 3.2.3, Eqn. (3.115)]. As this result was obtained using simplifying equations to describe the dynamical system induced by the ES, however, it cannot be used as a basis for proving theorems on the runtime of $(1, \lambda)$ ES. Nevertheless, this is a strong indicator that a $(1, \lambda)$ ES converges, if at all, at best linearly at an expected rate $1 - O(\ln(\lambda)/n)$. From this, we can conjecture that the expected number of iterations necessary to halve the approximation error is bounded below by $\Omega(n/\ln \lambda)$. This lower bound has in fact been rigorously proved in [10, Thm. 1] for a framework of iterative methods that covers pCSA as well as CBA. (Unfortunately, CSA is not covered since the factor for the σ-update in Eqn. (2) depends on the course of the optimization, namely on the length of the evolution path.) Actually, it is proved that *less* than $0.64n/\ln(1+3\lambda)$ iterations suffice to halve the approximation error only with an exponentially small probability $\mathrm{e}^{-\Omega(n)}$ (implying the $\Omega(n/\ln \lambda)$-bound on the expected number of steps). Moreover, [10, Thm. 4] provides a rigorous proof that a $(1, \lambda)$ ES using a simple σ-adaptation mechanism (based on the relative frequency of improving steps) to minimize SPHERE needs with very high probability at most $O(n/\sqrt{\ln \lambda})$ iterations to halve the approximation error. This upper bound is larger than the lower bound by a factor of order $\sqrt{\ln \lambda}$, which is actually due to the σ-adaptation: It adapts σ such that it is of order $\Theta(d/n)$, whereas $\sigma = \Theta(\sqrt{\ln \lambda} \cdot d/n)$ seems necessary. For the original CSA minimizing SPHERE, a result obtained in [11, Eqn. (20)] (using the dynamical-system approach again) indicates that the expected spatial gain in an iteration tends to $(\sqrt{2} - 1) \cdot (c_{1,\lambda})^2 \cdot d/n = \Theta(\ln(\lambda) \cdot d/n)$ as the dimensionality n goes to infinity, where $c_{1,\lambda} = \Theta(\sqrt{\ln \lambda})$ is the so-called $(1,\lambda)$-progress coefficient (obtained for optimal σ, cf. [9, Eqn. (3.114)]); σ is reported to tend to $\sqrt{2} \cdot c_{1,\lambda} \cdot d/n$ as the dimensionality n goes to infinity, which is indeed $\Theta(\sqrt{\ln \lambda} \cdot d/n)$ [11, Eqn. (19)].

So, the long-term objective is a probabilistic analysis which rigorously proves that the $(1,\lambda)$ ES using CSA needs only $O(n/\ln \lambda)$ iterations to halve the approximation error (for SPHERE) with very high probability, which is optimal w.r.t. the asymptotic in n and λ. As this seems intractable at present because of the involved stochastic dependencies due to the evolution path, the intermediate objective may be to prove the $O(n/\ln \lambda)$-bound for CBA. This makes (the most) sense if CBA behaves sufficiently similarly to the original CSA, of course. And this is what we investigate in the present paper. (If CBA performed much worse than CSA, however, we would have to reconsider whether it makes sense to try to prove the $O(n/\ln \lambda)$-bound for CBA as it might just not adapt σ such that it is $\Theta(\sqrt{\ln \lambda} \cdot d/n)$.) The partition of the optimization process into phases in each of which σ is not changed, and after each of which σ is deterministically updated solely depending on what happened in that phase, enables the following line of reasoning in a formal proof:

If σ is too small at the beginning of a phase, i.e., $\sigma < c_1 \cdot \sqrt{\ln \lambda} \cdot d/n$ for an appropriately chosen constant $c_1 > 0$, then it is up-scaled after the phase with very high probability (w.v.h.p.). If, however, σ is too large at the beginning of a phase, i.e., $\sigma > c_2 \cdot \sqrt{\ln \lambda} \cdot d/n$ for another appropriately chosen constant $c_2 > c_1$, then it is down-scaled after the phase w.v.h.p. With these two lemmas, we can obtain that $\sigma = \Theta(\sqrt{\ln \lambda} \cdot d/n)$ for any polynomial number of steps w.v.h.p. once σ is of that order. Subsequently, we show that, if $\sigma = \Theta(\sqrt{\ln \lambda} \cdot d/n)$ in a step, then the actual spatial gain towards the optimum in this step is $\Omega(\ln \lambda \cdot d/n)$ with probability $\Omega(1)$; this can be proved analogously to [10, Sec. 5]. Thus, given that at the beginning of a phase (with $\sqrt{n}$ steps) $\sigma = \Theta(\sqrt{\ln \lambda} \cdot d/n)$, the expected number of steps in the phase each of which actually reduces the approximation error by at least an $\Omega(\ln \lambda/n)$-fraction is $\Omega(\sqrt{n})$. Using Chernoff's bound, in a phase there are $\Omega(\sqrt{n})$ such steps w.v.h.p., so that a phase reduces the approximation error at least by an $\Omega(\ln \lambda/\sqrt{n})$-fraction w.v.h.p. Finally, this implies that $O(\sqrt{n}/\ln \lambda)$ phases, i.e. $O(n/\ln \lambda)$ steps, suffice to reduce the approximation error by a constant fraction w.v.h.p. This directly implies the $O(n/\ln \lambda)$-bound on the number of iterations to halve the approximation error.

The proof of that CBA ensures $\sigma = \Theta(\sqrt{\ln \lambda} \cdot d/n)$ (w.v.h.p. for any polynomial number of steps) remains. Therefore, note that the probability that two steps are exactly orthogonal is zero (because of the random directions). Thus, in a phase of k steps the number of positively correlated pairs of steps equals (with probability one) $\binom{k}{2}$ minus the number of negatively correlated pairs. For a theoretical analysis of CBA, for each phase $\binom{k}{2}$ 0-1-variables can be defined. Each of these indicator variables tells us whether the respective pair of steps is positively correlated ("1") or not ("0"). (Recall that in CBA the decision whether to up- or down-scale σ is based on whether the sum of these indicator variables is larger than $\binom{k}{2}/2$ or smaller.) There are strong bounds on the deviation of the actual sum of 0-1-variables from the expected sum, in particular when the variables are independent—which is not the case for CBA. This can be overcome by stochastic dominance arguments, so that we deal with independent Bernoulli trials, rather than with dependent Poisson trials. We merely need to know how the success probability of a Bernoulli trial depends on σ.

All in all, CBA is a candidate for a theoretical runtime analysis of an ES using this simplified CSA-variant. It was an open question, whether CBA works at all and, if so, how well it performs compared to the original CSA (and pCSA). Thus, we decided for an experimental comparison with statistical evaluation.

5   Experimental Investigation of the CSA-Variants

Since the underlying assumption in theory on black-box optimization is that the evaluation of the function f to be optimized is by far the most expensive operation, the number of f-evaluations (λ times the iterations) is the sole performance measure in the following comparison of the three CSA-variants. To find out the potentials of the σ-adaptation mechanisms described above, we focus on the simplest unimodal function scenario, namely the minimization of the distance from a fixed point. This is equivalent (here) to the minimization of a perfectly (pre)conditioned positive definite quadratic form. One of these functions, namely $x \mapsto \sum_{i=1}^{n} x_i^2$, is the commonly investigated SPHERE (level sets of such functions form hyper-spheres). We decided for a (1,5) Evolution Strategy, i.e. $\lambda := 5$. The reason for this choice is that, according to Beyer [9, p. 73], five samples are most "effective" for comma selection, i.e. allow maximum progress per function evaluation. Thus, differences in the adaptations' abilities (to choose σ as close to optimal as possible) should be most noticeable for this choice.

Experiment: Do the CSA-derivatives perform similar to the original CSA?

Pre-experimental planning. In addition to the adaptation rule and the phase length, the scaling factor by which σ is increased or decreased after a phase had to be fixed for pCSA and CBA. For both CSA-derivatives, the σ-scaling factor was determined by parameter scans, cf. Figure 1, whereas the phase length k was chosen a priori as $\lceil \sqrt{n} \rceil$ (for the reason given above). Interestingly, the parameter scans show that the σ-scaling factor should be chosen identically as $1 + 1/n^{1/4}$ for pCSA as well as for CBA.

Task. The hypothesis is that the three CSA-variants perform equally well in terms of number of iterations. As the data can not be supposed to be normally distributed, we compare two variants, namely their runtimes (i.e. number of iterations), by the Wilcoxon rank-sum test (as implemented by "wilcox.test" in "R"), where a p-value of 0.05 or less indicates a *significant* difference.

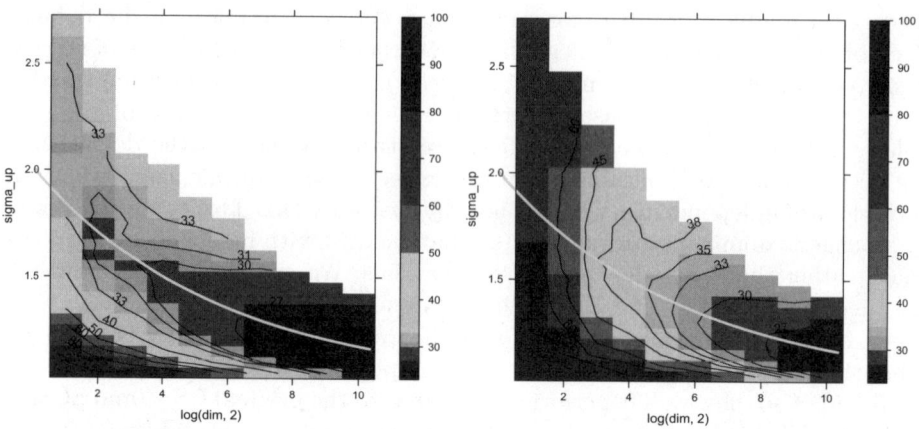

Fig. 1. Parameter scan of σ-scaling factors for pCSA (left) and CBA (right) over 2 to 1024 dimensions. The shading corresponds to the median number of steps divided by dimensionality; 101 runs in the same setting as the final experiment. The solid line represents $1 + 1/n^{1/4}$, whereas the thin lines show the contour.

Setup. The initial distance from the optimum is 2^{20} and the stopping criterion is a distance of less than 1 from the optimum, i.e., we measure the number of iterations to halve the approximation error in the search space 20 times. The initial σ is set to $2^{20} \cdot 1.225/n$ in each case. We investigate ten search-space dimensions, namely $n = 2^i$ for $i \in \{1, 2, \ldots, 10\}$. Each of the three algorithms is run 1001 times.

Results. Figure 2 shows median runtimes and hinges for the three algorithms. Additionally, the σ-adaptation within typical runs (i.e. median runtime) is shown in Figure 3. Note that σ is considered well-chosen (after normalization w.r.t. dimension and distance from the optimum) when it lies between the two horizontal lines, which correspond to a normalized σ of 1.0 (blue) resp. 2.0 (red).

Observations. Independently of the search-space dimension, i.e. for *all* ten dimensions investigated, namely dimension $n = 2^i$ for $i \in \{1, \ldots, 10\}$, we find:

1. The original CSA performs significantly better than pCSA and CBA.
2. Moreover, CBA performs significantly worse than pCSA.

As can be seen in Figure 2, transferring the continuous CSA-mechanism to a phased one, namely to pCSA, leads to an increase in the runtimes by clearly less than 50% in all ten dimensions. Concerning the runtimes at least, pCSA is closer to the original CSA than CBA.

Discussion. Despite the reported findings, we attest that CBA does not fail, it ensures a reliable σ-adaptation—it is merely worse. It particular, we are interested in how much worse CBA is compared to the original CSA. Actually, in 2-dimensional search space, the ratio between the medians of the runtimes of

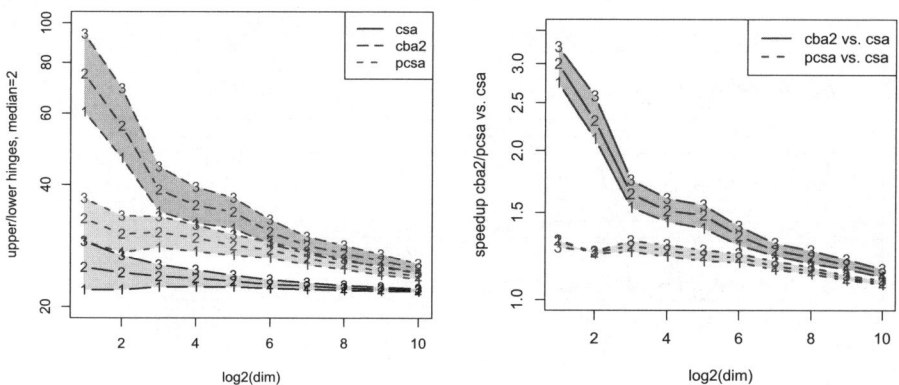

Fig. 2. Left: Number of steps divided by dimensionality for CBA, pCSA, CSA, where (1) lower hinge, (2) median, (3) upper hinge of 1001 runs, respectively. Right: Ratio between (1) lower hinge, (2) median, (3) upper hinge of the 1001 runtimes of CBA vs. CSA and pCSA vs. CSA.

CBA to CSA equals 3. And in fact, when we multiply each of the 1001 runtimes of CSA by 3, then the Wilcoxon test does *not* tell us a significant difference between CSA and CBA (p-value 0.85). Thus, we may conclude that CBA is slower than CSA by roughly a factor of 3. However, "roughly" may be considered too vague. Rather, a confidence interval should be given: For the factor 2.91 as well as for the factor 3.1, the p-value drops below 2.5%, respectively. Thus, the 95%-confidence interval for the factor by which CBA is slower than CSA for the 2-dimensional problem is (at most) [2.91, 3.1], which has a spread of only $3.1/2.91 < 1.066$. The respective confidence intervals for all ten dimensions investigated are given in Table 1. As one might expect, they are pretty much concentrated around the ratio of the median runtimes, respectively.

Concerning the performance differences observed, we think that CBA—as the only variant for which a theoretical analysis seems feasible at present—still works sufficiently similar to CSA to legitimate such analysis, at least for the practically relevant and large dimensions: For eight and more dimensions the number of function evaluations (five times the iterations) that CBA spends is larger than for CSA by (at most) 67%, decreasing for larger dimensions down to (at most) 13.4% in 1024 dimensions.

Clearly, the differences in the runtimes are due to differences in the ability to adapt σ. Therefore, let $\widehat{\sigma} := \sigma \cdot n/d$ denote the normalized σ (w.r.t. dimensionality and distance from the optimum point). Note that, for the simple function

Table 1. Confidence intervals for the ratios between the runtimes of CBA and CSA

CBA vs. CSA	2D	4D	8D	16D	32D	64D	128D	256D	512D	1024D
conf. interval runtime ratio	2.91–3.10	2.25–2.37	1.61–1.67	1.49–1.54	1.46–1.50	1.33–1.36	1.25–1.27	1.219–1.226	1.171–1.182	1.127–1.134
spread	<6.6%	<5.4%	<3.8%	<3.4%	<2.8%	<2.2%	1.6%	<0.6%	<1%	<0.7%

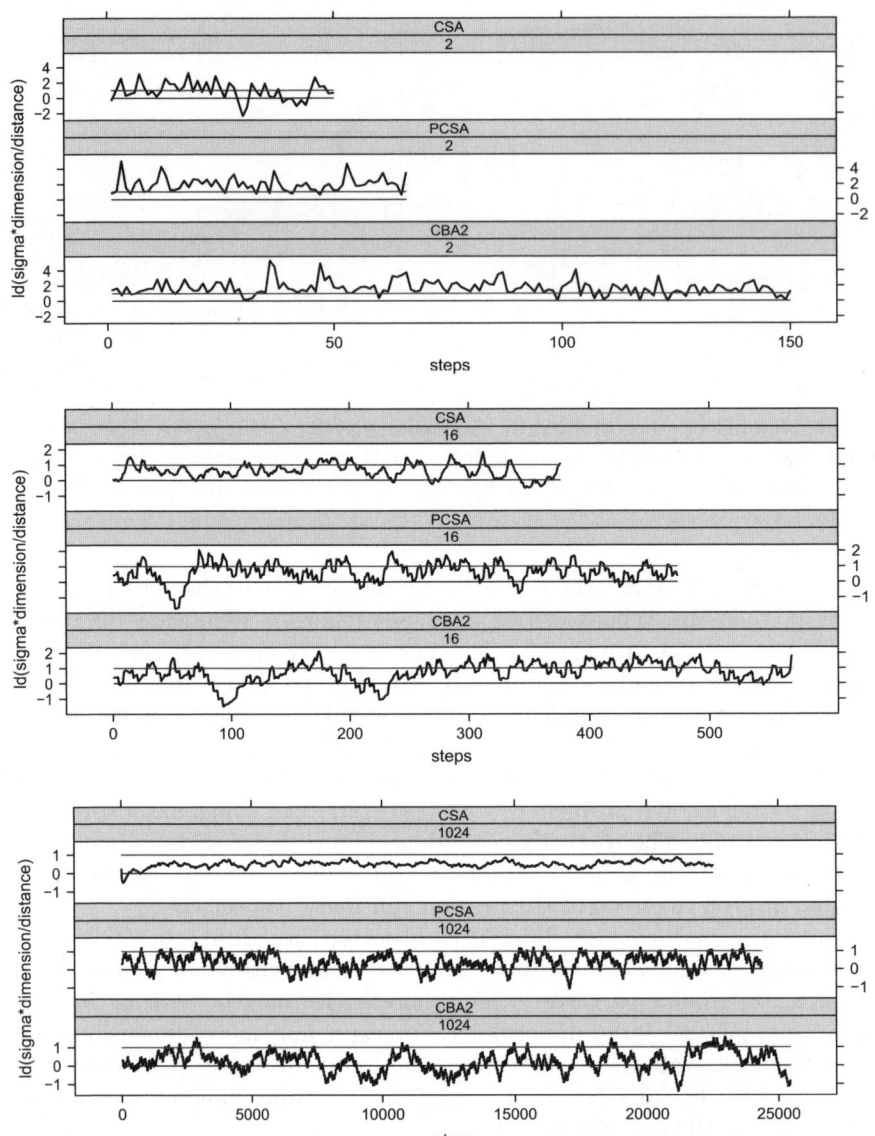

Fig. 3. Normalized σ (w.r.t. dimensionality and distance from optimum) in a typical run (i.e. median runtime) of each variant on 2-/16-/1024-dimensional Sphere

scenario considered, for each dimension there is a unique value of $\hat{\sigma}$ resulting in maximum expected reduction of the approximation error per iteration (which is not known exactly, unfortunately). Figure 3 shows the course of $\hat{\sigma}$ in $\log_2$-scale for each of the three CSA-variants for typical runs in 2-/16-/1024-dimensional space, respectively, where "typical" means median runtime. In Table 2 the mean

Table 2. Means and standard deviations of $\log_2(\hat{\sigma})$ in the typical runs shown in Figure 3. Additionally, the **log-mean** $\hat{\sigma}$, namely $2^{\wedge}\text{mean}(\log_2 \hat{\sigma})$, is given.

mean±std of $\log_2(\hat{\sigma})$	2D	16D	1024D
CSA	0.914 ± 1.183, **1.884**	0.615 ± 0.454, **1.370**	0.492 ± 0.171, **1.406**
pCSA	2.012 ± 0.947, **4.033**	0.642 ± 0.617, **1.560**	0.408 ± 0.417, **1.327**
CBA	1.699 ± 0.923, **3.247**	0.734 ± 0.649, **1.663**	0.236 ± 0.549, **1.178**

and the standard deviation of $\log_2(\sigma^*)$ for these runs are given, as well as the log-mean of $\hat{\sigma}$, which we refer to as the average $\hat{\sigma}$.

For the 16- and 1024-dimensional problems, the original CSA adapts σ much more smoothly than its two derivatives, which is apparently due to its continuous σ-adaptation (most obvious in 1024 dimensions). For the 2-dimensional problem, however, CSA shows a larger standard deviation from the average $\hat{\sigma}$. Taking CSA as a reference, pCSA as well as CBA adapt σ such that it is too large on average in 16 dimensions, whereas in 1024 dimensions they adapt σ such that it is too small on average—besides the larger fluctuations. For 16 dimensions, which we consider practically relevant, as well as for 1024 dimensions, the average $\hat{\sigma}$ of pCSA lies right between the ones of CSA and CBA, respectively, which fits well with the differences in the runtimes. For the 2-dimensional problem, however, correlations between the average $\hat{\sigma}$ and the runtimes can hardly be found, which becomes especially clear for pCSA. In two dimensions the runs are actually quite short and, in addition, the step-lengths can deviate strongly from the expectation $\sigma \cdot \bar{\chi}$, so that the data might just be too noisy. Alternatively, there might be a completely different reason for the good performance of pCSA in 2-dimensional space, which would be very interesting to reveal.

6   Conclusions and Outlook

The main aim of this work has been to develop an algorithm for step-size control in black-box optimization which closely resembles the original CSA mechanism, but which is sufficiently simple to enable a theoretical/probabilistic analysis. From the experimental results obtained and from the construction details of the two proposed CSA-variants we conclude that CBA does fulfill both criteria.

Additionally, some interesting facts have been unveiled. One of these affects the change from continuous σ-update to phases in which σ is kept constant. Contrary to what we expected, this modification does not explain the large runtime differences in small dimensions. Those may rather be due to the large fluctuations observed in the step-sizes adjusted by the two CSA-derivatives; CSA step-size curves are obviously much smoother in higher dimensions. Furthermore, it has been found that an appropriate σ-scaling factor for both CSA-derivatives seems to follow a double square root (namely $1 + 1/n^{1/4}$) rather than a single square root function (like $1 + 1/\sqrt{n}$) as one might expect.

Currently, we work on the details of a theoretical/probabilistic analysis of CBA following the approach outlined in Section 4. Furthermore, we are going to experimentally review the performance of the two CSA-derivatives on more complex—actually, less trivial—functions in comparison to the original CSA (also and in a particular in combination with CMA), but also to classical direct-search methods like the ones reviewed in [12].

Acknowledgment. We thank Thomas Stützle for helping us improve the paper.

References

1. Hansen, N., Ostermeier, A.: Adapting arbitrary normal mutation distributions in evolution strategies: The covariance matrix adaptation. In: Proc. IEEE Int'l Conference on Evolutionary Computation (ICEC), pp. 312–317 (1996)
2. Hansen, N.: List of references to various applications of CMA-ES (2008), http://www.bionik.tu-berlin.de/user/niko/cmaapplications.pdf
3. Droste, S., Jansen, T., Wegener, I.: On the analysis of the (1+1) Evolutionary Algorithm. Theoretical Computer Science 276, 51–82 (2002)
4. Jägersküpper, J.: Algorithmic analysis of a basic evolutionary algorithm for continuous optimization. Theoretical Computer Science 379, 329–347 (2007)
5. Rechenberg, I.: Cybernetic solution path of an experimental problem. Royal Aircraft Establishment (1965)
6. Schwefel, H.-P.: Numerical Optimization of Computer Models. Wiley, New York (1981)
7. Jägersküpper, J.: Analysis of a simple evolutionary algorithm for minimization in Euclidean spaces. In: Baeten, J.C.M., Lenstra, J.K., Parrow, J., Woeginger, G.J. (eds.) ICALP 2003. LNCS, vol. 2719, pp. 1068–1079. Springer, Heidelberg (2003)
8. Jägersküpper, J.: How the (1+1) ES using isotropic mutations minimizes positive definite quadratic forms. Theoretical Computer Science 361, 38–56 (2005)
9. Beyer, H.-G.: The Theory of Evolution Strategies. Springer, Heidelberg (2001)
10. Jägersküpper, J.: Probabilistic runtime analysis of $(1\overset{+}{,}\lambda)$ ES using isotropic mutations. In: Proc. 2006 Genetic and Evolutionary Computation Conference (GECCo), pp. 461–468. ACM Press, New York (2006)
11. Beyer, H.-G., Arnold, D.V.: Performance analysis of evolutionary optimization with cumulative step-length adaptation. IEEE Transactions on Automatic Control, 617–622 (2004)
12. Kolda, T.G., Lewis, R.M., Torczon, V.: Optimization by direct search: New perspectives on some classical and modern methods. SIAM Review 45, 385–482 (2004)

Reconstructing Phylogenetic Networks with One Recombination

Ernst Althaus and Rouven Naujoks

Max-Planck-Institut für Informatik,
Stuhlsatzenhausweg 85, 66123 Saarbrücken, Germany
{althaus, naujoks}@mpi-inf.mpg.de

Abstract. In this paper we propose a new method for reconstructing phylogenetic networks under the assumption that recombination events have occurred rarely. For a fixed number of recombinations, we give a generalization of the maximum parsimony criterion. Furthermore, we describe an exact algorithm for one recombination event and show that in this case our method is not only able to identify the recombined sequence but also to reliably reconstruct the complete evolutionary history.

1 Introduction

A fundamental class of problems in Computational Biology is the reconstruction of phylogenetic trees which reads as follows: Given a set of species one wants to determine their ancestral relationship along a tree. In order to build such a tree we compare specific features of the species under the natural assumption that species with similar features are closely related. In modern phylogeny these features are defined by DNA or protein sequences.

In the presence of recombination events, trees are not sufficient to describe the ancestral relationship. In our setting, recombinations are evolutionary events that cause horizontal transfer of genetic data. There has been active research to automatically compute phylogenetic networks from sequence data, see e.g. [8,7,13,5,4,6,18,9]. In Section 3, we discuss in detail existing approaches and relate our work to them. For more information on the reconstruction of phylogenetic networks, we refer to [10,15]. For a list of available tools see [19] or [20].

We assume that we are given a set N of aligned sequences, i.e. a set of sequences with equal number of characters in each of them. Our aim is to construct an optimal ancestral relationship for a given number of recombination events under a generalization of the maximum parsimony criterion based on Ockham's razor principle. It slightly differs from those in [8,7,13,5,4]. We denote by *phylogenetic network with k recombinations* an ancestral relationship with k recombination events. Following Ockham's razor principle a phylogenetic network for small k with minimal cost should be a correct reconstruction of the original phylogenetic network. We propose to use our method when recombination events are assumed to happen only occasionally.

In Section 2, we describe an exact algorithm that computes the optimal phylogenetic network with one recombination and show in Section 5 that the

C.C. McGeoch (Ed.): WEA 2008, LNCS 5038, pp. 275–288, 2008.

algorithm computes the correct phylogenetic network even for sequences with (still reasonable) long evolutionary distances. Our experiments show that we can compute the correct phylogenetic network for almost the same data sets where the maximum parsimony criterion is able to reconstruct the correct phylogenetic tree when it is given the subsequences that correspond to a specific tree in the evolution.

2   Model

A phylogenetic tree for a given set of species N is a binary tree whose leaves are the set N. In the literature rooted and unrooted trees are considered. Because we consider in the following rooted trees we have a single node with no incoming edges and all other nodes having exactly one incoming edge. Every inner node, i.e. a node that is not contained in N, has exactly two outgoing edges.

In a phylogenetic network, we allow nodes to have indegree two. A node with indegree two has two ancestors and we refer to these nodes as *recombination nodes*. In order to have a valid ancestral relationship, we require the network to be acyclic. Note that this definition is equivalent to those in [8,5,4]. We denote by k-*recombination topology for N* a phylogenetic network for N with exactly k recombination nodes.

We propose to compute all optimal k-recombination topologies for N and to predict the phylogenetic network by (either manually or automatically) inspecting these topologies. In order to be able to speak of an optimal topology, we introduce a method for evaluating a given topology.

In the case of phylogenetic trees, the two probably most favorite and prominent versions are the *maximum likelihood* and the *maximum parsimony* methods. In the maximum parsimony problem we find sequences for the inner nodes of the tree such that the number of changes along the edges of the tree are minimized. These sequences can be computed with Fitch's algorithm [3] which we will discuss in more detail later. We extend the parsimony criterion by defining a cost for recombination events, referred to as the *parsimony score*.

Again, we have to find sequences for the inner nodes that minimize the cost of the k-recombination topology. Assume, the sequences for the inner nodes are fixed. Our cost function is easiest described by charging the cost to the nodes. The cost of the root is 0. The cost of a non-recombination node u is the number of changes between the label of u and the label of the ancestor of u. This is the same cost as it was assigned to the edge from the ancestor of u to u in the maximum parsimony model and is also referred to as the *Hamming distance* between the two sequences. The cost of a recombination node u is defined as follows. Roughly speaking, the recombinant sequence can have the genetic code of either of its ancestors, paying a certain cost for changing the ancestor. This cost reflects the fact that recombination does not happen by randomly choosing genetic code from the sequences, but it is assumed that there are only a few *jumps*, where the source for the recombinant sequence changes. Formally the cost of a node u is defined as follows: Let s be the sequence assigned to u, and let t^1 and t^2 be the sequences of the two ancestors of u and let n be number of characters in the sequences. The cost of u is then

$$\min_{d \in \{0,1\}^n} \left(\sum_{i=1}^{n} (d_i \| s_i, t_i^1 \| + (1 - d_i) \| s_i, t_i^2 \|) + \alpha |\{1 \leq i < n \mid d_i \neq d_{i+1}\}| \right)$$

where $\| \cdot \|$ denotes the Hamming distance. The same definition of the cost for a recombination was used by Maydt and Lengauer [11], where the authors try to explain a given sequence by recombinations of other input sequences. Note that the cost of a non-recombination node is exactly the cost of the in-going edge in the maximum parsimony model. Thus we will sometimes refer to the cost of an edge if it is clear that the corresponding node is a non-recombination node. The cost of a k-recombination network is then just the sum of the costs of its nodes.

Although maximum likelihood methods are sometimes assumed to perform better than parsimony methods, the parsimony method is often used in practice as the optimal phylogenetic tree can be computed much more efficiently. Our generalization of the method to phylogenetic networks is a straight forward generalization to the case of phylogenetic networks. The reconstructed network is exactly the network that explains the data with the smallest possible number of mutations and jumps, where the cost of a jump is scaled by α. To the best of our knowledge, our algorithm is the first that optimizes over all phylogenetic networks to make its prediction.

In Section 4, we show that the cost of a 1-recombination topology as well as sequences for the inner nodes can be computed efficiently. As computing the tree with maximum parsimony is already NP-hard, computing the optimal k-recombination topology is NP-hard for all k, as setting α to infinity will lead to the fact that the cost of the recombination node will become the Hamming distance to one of its ancestors. Nevertheless, our algorithm optimizes over all 1-recombination topologies and all sequences of the inner nodes to find our reconstruction (see Section 4).

3   Related Work

There are many tools for detecting recombination events in sequence data. Posada et al. [15] divide these tools into 5 classes: similarity methods, distance methods, phylogenetic methods, compatibility methods, substitution distribution methods. Most methods only try to detect the existence of a recombination event. Moreover it is frequently required that the recombinant sequence is contained in the data set. The only exception are phylogenetic methods. Most of these methods detect a recombination by inferring different phylogenetic trees in different parts of the data, i.e. they analyze the data by a sliding window approach obtaining different trees and by trying to combine these trees to a prediction. The main drawback is that it is not possible to restrict the different trees such that the trees differ only by a small number of recombination events. A list of available tools can be found at [19].

Posada et al. [15] conclude that most methods have trouble detecting rare recombinational events, especially when sequence divergence is low, i.e. exactly the case where our algorithm is aimed for.

Huson and Kloepper [6], Wang et al. [18], Gusfield et al. [4] and Kececioglu and Gusfield [9] use an extension of the maximum parsimony criterion originally proposed by Hein [5]. The essential difference is that they allow only one jump when constructing a sequence from its two ancestors. If there are several splits, one has to assign several recombination events to a node. Our algorithm can easily be adopted to find the optimal phylogenetic network in this model. Their algorithm does not compute an optimal phylogenetic network in our sense, but it explains the data under a minimal number of recombinations under the infinite state assumption, i.e. there is no back-exchange of a genetic code. As back-exchange is assumed to happen in real evolution, their algorithm typically overestimates the number of recombination events.

The extension of the maximum parsimony criterion proposed Jin et al. [8,7,13] is basically our model with $\alpha = 0$. As shown in our experiments in Section 5, the prediction of the topology is much less accurate for this model. Furthermore, their algorithm does not compute an optimal phylogenetic network, but it starts with a tree, only adding edges to a tree. Hence this approach can only work if the starting tree is contained in the phylogenetic network. As observed by Ruths and Nakhleh [17], phylogenetic tree algorithms fail to reconstruct data when recombination has occurred. Thus it is unclear how to obtain a phylogenetic tree to start with.

Note that the maximum parsimony problem is equivalent to the Steiner minimum tree problem in Hamming metric. The most successful algorithm for solving the Steiner minimum tree (SMT) problem in graphs is based (among other techniques) on successfully pruning parts of the graph. All known techniques are described in the PhD thesis of Polzin [14]. We will adopt some of these pruning techniques for our algorithm.

4   The Algorithm

As noted before the k-Recombination Phylogeny Network (k-RPN) problem is NP-hard. Thus, unless $P = NP$, there is no polynomial time algorithm for solving it exactly. As in the case of the SMT-problem the k-RPN problem can be solved for small input instances by enumerating the set of possible solutions and by determining among them the one which minimizes the total network length. In our algorithm we focus on the case where $k = 1$. We call the phylogeny network spanning all sequences that are offsprings from the sequence that emerged from this recombination event the *recombinant*. In the following we write N to denote the set of terminals, i.e. the set of input sequences.

4.1   Preliminaries

Before we discuss the algorithm it is necessary to recall several things. For $k = 0$ the k-RPN problem becomes the so called parsimony problem in which one tries to seek the shortest (shortest in the sense of the Hamming metric) tree connecting the input sequences N by allowing for an augmentation with additional sequences. That is the 0-RPN problem is exactly the so called Steiner minimum tree problem in Hamming metric where the additional sequences are called Steiner points. Without loss of generality one can assume that an optimal SMT always corresponds to a binary tree T where the leaves are exactly the terminals of the input set N and where the inner nodes are exactly the Steiner points. We denote by $span(T)$ the set of terminals spanned by a topology T. When we talk about the size of a topology we mean $|span(T)|$. In most cases T is considered to be rooted by inserting an additional Steiner point r along some edge in T and by choosing r to be the root of T. When we are not interested in the labels of the inner nodes of a binary tree but only in its connectivity structure we talk about a topology $\mathcal{T}$. A very important ingredient in this context is Fitch's algorithm (see [3]). It computes for a given topology $\mathcal{T}$ labels for all inner nodes of $\mathcal{T}$, such that the resulting tree has minimal length. When we talk about the cost or length $|\mathcal{T}|$ of a topology $\mathcal{T}$ we mean the length of the resulting tree after calling Fitch's algorithm for $\mathcal{T}$. The algorithm consists of two phases. To discuss these phases we have to introduce the notion of a range sequence. A range sequence is a sequence not consisting only of single characters but of sets of characters from the input alphabet. For sake of simplicity we also call the input sequences range sequences by considering a single character to be a set of cardinality 1. Given a node n in a tree we write s_n for the range sequence associated with n. With "choosing a sequence from a range sequence" we mean to extract a sequence such that each character is taken from a corresponding set of the range sequence. In the first phase of Fitch's algorithm ranges are assigned in a bottom up manner to the inner nodes i of $\mathcal{T}$. In the second phase it chooses ranges from the sequences in the nodes of $\mathcal{T}$ in a top down fashion starting from the root node $r(\mathcal{T})$ such that the length of the tree with this labeling is minimal. The choice for the root range in the second phase does not change the length of the resulting tree. Furthermore the range sequence of $r(\mathcal{T})$ is maximal that is there is no other range that would minimize the tree length than the ones extracted from this range sequence. Note that each inner node of $\mathcal{T}$ is the root node of its subtree and thus the second phase could also be applied to any inner node of $\mathcal{T}$ to determine the optimal length of its subtree. Fitch's algorithm takes $O(|span(\mathcal{T})| \cdot d)$ set operations where d is the length of an input sequence (all of them are considered to be equally long). Since the alphabet size of the input is usually a constant (for example in the case of DNA data this is 4) these set operations take only $O(1)$ time.

4.2   Evaluation of a Recombination Network

First we describe how we can compute the parsimony score of a recombination network T. Recall that the cost c_e of an edge $e = (u, v)$ in the parsimony model is the Hamming distance of the two sequences s_u, s_v, i.e. $c_e = \sum_{i=1}^{dim} \|s_u^i, s_v^i\|$ where $\|s_u^i, s_v^i\|$ is the cost considering only the i-th character in s_u and s_v. That is, the cost function has the nice property that the cost of the characters in all occurring sequences in the network can be considered independently.

Thus any recombination network T can be seen as an overlay of two trees T' and T'' that differ only in the Steiner points c', c'' over which the recombinant T_R is connected to rest of these trees (see Figure 1 for an example). Because T' and T'' can be considered independently we can determine for c' and c'' range sequences with maximal Fitch-ranges i.e. we consider the trees T' and T'' as rooted at c' respectively c''. Connecting the recombinant T_R with c' and c'' optimally can be done in $O(d)$ time by the following simple dynamic program. We call this operation the ◇-operator and denote by $|\diamond(s_u, s_v, s_r)|$ the cost of the recombination event in which sequence s_r was recombined out of the sequences s_u and s_v.

Here s_r is the range sequence of $r(R)$ after applying the first phase of Fitch's algorithm and let α be the cost of switching from one sequence to other. Altogether the cost of the recombination network T and is then given by $|\mathrm{SMT}(\overline{R})| + |\mathrm{SMT}(R)| + |\diamond(s_{c'}, s_{c''}, s_r)|$.

Algorithm 1. ◇-operator: $\diamond(s_{c'}, s_{c''}, s_r)$

$OPT' := 0; \quad OPT'' := 0$
for $d = 1 \ldots dimension$ **do**
  $tent' := 0; \quad tent'' := 0$
  if $s_{c'}^d \cap s_r^d \neq \emptyset$ **then**
    $tent' = \min(OPT', OPT'' + \alpha)$
  else
    $tent' = \min(OPT', OPT'' + \alpha) + 1$
  end if
  if $s_{c''}^d \cap s_r^d \neq \emptyset$ **then**
    $tent'' = \min(OPT' + \alpha, OPT'')$
  else
    $tent'' = \min(OPT' + \alpha, OPT'') + 1$
  end if
  $OPT' = tent'; \quad OPT'' = tent''$
end for
return $\min(OPT', OPT'')$

4.3 Enumeration Process

Before we discuss the enumeration process itself we observe that any recombination network RT is composed of two binary trees: T_R corresponding to the recombinant and T spanning all the leaves that are not spanned by T_R. RT can be seen as constructed from T and T_R by connecting T_R via the ◇-operator with two splitted edges in T. Hence one can reduce the enumeration of recombination networks to the enumeration of binary trees.

Enumerating all binary trees can be accomplished in several ways. For the practical performance of the algorithm it is important to do this in such a way

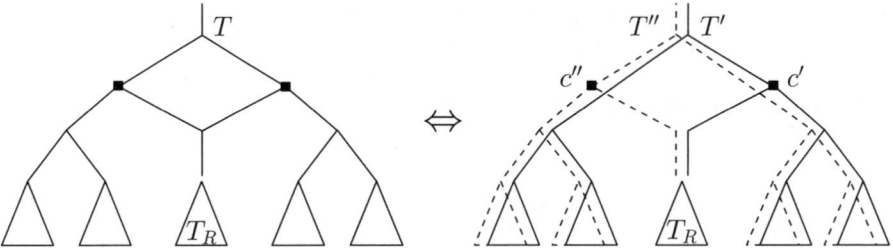

Fig. 1. A recombination network as an overlay of two binary trees

that the search space can strongly be cut. In Algorithm 2 we start with the set X consisting only of the topologies of size 1 namely the terminals itself. We then inductively construct all topologies of size k by combining the ones out of X of size less than k. We do this until all topologies of size $|N| - 1$ are built. In section 4.5 we will see why it is sufficient to enumerate only topologies up to a size of $|N| - 1$.

Cutting down the search space works now as follows: In each step when a topology T was built we run several pruning tests to decide whether an optimal solution for the RPN-problem can contain T as a sub-topology without violating optimality. In the following subsection we will discuss briefly some of these tests. In the following $T_i \cdot T_j$ describes the topology that emerges when we connect the trees T_i and T_j over their root nodes over a newly inserted root node. In pseudo-code the enumeration part now reads as follows:

Algorithm 2. enumeration process

$X := N$
for $i = 2 \ldots |N| - 1$ **do**
   for all T_k, $T_j \in X$ **do**
      if $span(T_k) \cap span(T_l) = \emptyset$ and
         $|span(T_k)| + |span(T_l)| = i$ **then**
         if $\neg$ prunable$(T_k \cdot T_l)$ **then**
            $X = X \cup \{T_k \cdot T_l\}$
         end if
      end if
   end for
end for

4.4 Pruning the Search Space

Before we describe the pruning tests we have to recall several things. We denote by $bnsd(S)$ the so called *bottleneck Steiner distance* of a set $S \subseteq N$ which is defined to be the length of the longest edge in a minimum spanning tree of S (short: MST(S)). It is a well known fact that no edge in SMT(S) can be longer than $bnsd(S)$. We call $lb_{\mathrm{SMT}}(S)$ a lower bound on the length of SMT(S)

and $lb_{\mathrm{RPN}}(S)$ a lower bound on the length of RPN(S). Equivalently we write $ub_{\mathrm{SMT}}(S)$ and $ub_{\mathrm{RPN}}(S)$ for upper bounds on these problems.

Due to space limitations we can only discuss a one of the pruning tests mentioned in the previous section that we use in our implementation of the algorithm:

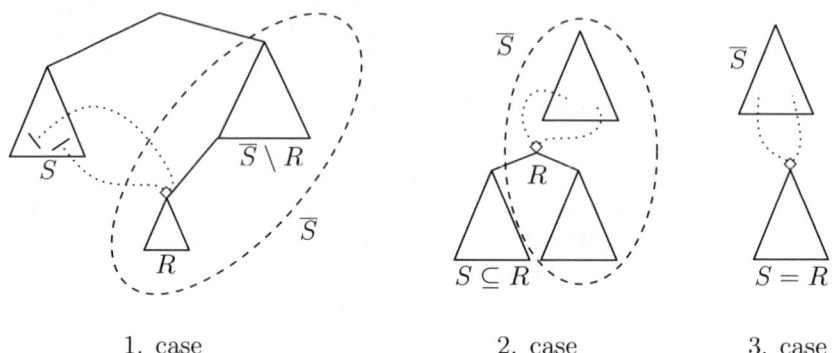

1. case   2. case   3. case

Fig. 2. The three different cases occurring in the lower bound pruning step. The dotted edges represent the incoming edges of the ⬦-operator.

lower bound test: You are given a topology T_S spanning a terminal set $S \neq N$. Consider now an optimal solution for the RPN-problem with recombinant T_R. We have to discuss several cases (see Figure 2):

1. $S \cap R = \emptyset$: We can prune T_S if $|T_S| + lb_{\mathrm{SMT}}(\overline{S}) - bnsd(\overline{S}) > ub_{\mathrm{RPN}}(N)$.
   Proof: any solution with T_S as a subtree where $S \cap R = \emptyset$ clearly has a cost of at least $|T_S| + lb_{\mathrm{SMT}}(R) + lb_{\mathrm{SMT}}(\overline{S} \setminus R)$. Since we do not know what R will be in an optimal solution we have to lower bound the term $lb_{\mathrm{SMT}}(R) + lb_{\mathrm{SMT}}(\overline{S} \setminus R)$. It is easy to see that $lb_{\mathrm{SMT}}(R) + lb_{\mathrm{SMT}}(\overline{S} \setminus R) \geq lb_{\mathrm{SMT}}(\overline{S}) - bnsd(\overline{S})$ following from the property of the bottleneck Steiner distance.
2. $S \subsetneq R$: T_S can be pruned if $|T_S| + lb_{\mathrm{SMT}}(\overline{S}) - bnsd(\overline{S}) > ub_{\mathrm{RPN}}(N)$.
3. $S = R$: T can be pruned if $|T_S| + lb_{\mathrm{SMT}}(\overline{S}) > ub_{\mathrm{RPN}}(N)$.

Note that the last cases follow the same arguments as the first case does. Since we do not know in advance which case occurs in an optimal solution we have to take the weakest condition $|T_S| + lb_{\mathrm{SMT}}(\overline{S}) - bnsd(\overline{S}) > ub_{\mathrm{RPN}}(N)$. The key idea why pruning topologies in such a way leads to a fast enumeration algorithm is that once a topology is pruned it cannot be part of any of the topologies that are constructed in one of the following steps. Thus pruning a topology in an early step cuts off an exponential big part of the search space. We have shown in [1] in detail how tight lower bounds for the SMT-problem can be computed efficiently.

Experiments conducted on real life data sets from [2] have shown that the search space can be cut down very efficiently by factors of about 10^{-6} and more.

4.5   Recombination Phase

The last part of our algorithm is the recombination phase in which we construct the recombination networks by combining the enumerated topologies:

Algorithm 3. recombination process

for all $T \in X$ with: $|span(T)| \in [3, .., |N| - 1]$ **do**
    for all $T_R \in X$ with:
    $|span(T) \uplus span(T_R)| = N$ **do**
        for all edge pairs e_1, $e_2 \in T$ **do**
            combine T and T_R along e_1 and e_2 using the $\diamond$-operator
        end for
    end for
end for

For each edge e in T we have to compute the maximal Fitch-ranges as noted already in the network evaluation section. That is we have to insert a new Steiner point c in e, to re-root the tree T such that c becomes the new root and to apply Fitch's algorithm. Since we have to do this for all edges in T we end up with a total running time of $O(d \cdot span(T)^2)$. This turns out to become the bottleneck of this procedure in practice. Instead of computing these root-ranges one after the other we propose an algorithm which computes for a given topology T the root-ranges for all edges in time proportional to one single call of Fitch's algorithm:

Let us consider T to be unrooted. We associate with each node n in T three range sequences rs_i associated with the three adjacent nodes v_i for $i \in \{0, .., 2\}$. For all $j \in \{0, .., 2\}$ do the following: consider T to be a rooted version such that v_j lies on the path going from n to the root (let $j = 0$ w.l.o.g.). Now perform the first phase of Fitch's algorithm computing range sequences for the subtrees rooted at v_1 and v_2 and store these sequences in v_1 and v_2 in the variables corresponding to node n. Repeat this procedure now for all nodes n in T. The amortized cost of this procedure will be $O(d \cdot |span(T)|)$ since each variable is computed only once and and the computational cost for computing a variable is $O(d)$. To compute the now the range sequence of a root we consider the edge $e = (u, v)$ in which the root would be inserted and compute out of the variables associated with the endpoints u and v the sequence for the root at an additional cost of $O(d)$.

As it turns out that in practice the most time consuming part in the algorithm is this recombination phase we have derived a second pruning step to speed up this part significantly. The goal is to reduce the number of $\diamond$-operator calls since they are the most costly operations in this phase - recall that usually the dimension is quite large and the running time of this function grows linearly in the dimension. Because of the space limitations we again discuss only two examples of such a pruning:

Lower bound test: Reconsider the lower bound tests for the previous subsection. In the last case, i.e. $S = R$. We have seen that in such a case the tree T_S can be pruned if $|T_S| + lb_{\mathrm{SMT}}(\overline{S}) > ub_{\mathrm{RPN}}(N)$. Note that this condition can be much stronger than the more general condition which involves the subtraction of the $bnsd(\cdot)$ term which can be as big as the dimension of the problem. If a topology T_R satisfies now this condition we know that T_R cannot be the recombinant part of an optimal solution, i.e. we can skip this call of the for-loop without calling the $\diamond$-operator.

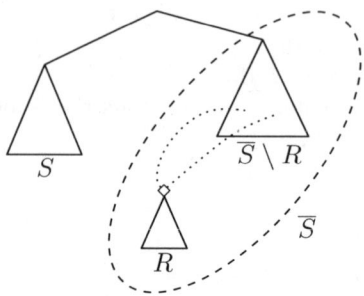

Fig. 3. Pruning by minors

Minor test: Assume you know that the $\diamond$-operator will not connect T_R to any edge in T_S in an optimal solution. Then for any topology T in the recombination process (see Algorithm 3) which contains T_S as a sub-topology the edges in this sub-topology can be neglected (see Figure 3). But when does T_S has such a property? Clearly T_S cannot be part of an optimal solution if $|T_S| + lb_{\mathrm{RPN}}(\overline{T_S}) > ub_{\mathrm{RPN}}(N)$. To derive lower bounds for $lb_{\mathrm{RPN}}(\overline{T_S})$ we compute in a preprocessing step $\mathrm{RPN}(X)$ for small instances X - which can be done by brute force or by calling our algorithm recursively - and store this information in a set M. Then we check for $\overline{T_S}$ if there is a $X \in M$ such that $span(X) \subseteq span(\overline{T_S})$. It is easy to see that $\mathrm{RPN}(X)$ is a lower bound for $lb_{\mathrm{RPN}}(\overline{T_S})$.

5   Experiments

Performing experiments in the context of reconstructing phylogenetic networks with recombination events is a difficult task since there is always some uncertainty about whether recombination events actually took place, see [11] for references. On the other hand synthetically created data sets have the advantage that we know exactly in advance what we expect from our method to return as a phylogenetic network. Furthermore it is easy to create instances of arbitrary size and complexity.

In order to obtain data with a known topology, we use NETGEN [12] to simulate an evolution with exactly one recombination event. NETGEN is an event-driven simulator that creates phylogenetic networks by extending the birth-death model to include diploid hybridizations.

Recall that a recombination network with exactly one recombination event can be interpreted as two phylogenetic trees in different parts of the sequence. We use Seq-Gen [16] to simulate the evolution of gene sequences along the trees of the recombination networks. Seq-Gen is a program that simulates the evolution of nucleotide or protein sequences along a phylogeny, using common models of the substitution process. A range of models of molecular evolution are implemented, including the general reversible model.

The data for the two trees is then merged by randomly choosing a jump point in the data for either tree and using the data alternating from the two trees. Note that one jump point for each of the trees will lead to three jump points in our data.

5.1   Fixed Recombination Scenarios

To validate our model we generate in each of the discussed cases 40 phylogenetic networks with exactly one recombination. For each network N we simulate with Seq-Gen gene sequences of length 1000 along N partitioning the root sequence of the recombinant randomly such that the expected size of a consecutive block of characters from one tree sequence is 250. As the substitution model we choose the "general reversible process" model (Yang, 1994) for the nucleotide sequences since it is the most general model that is still consistent with the requirement of being reversible.

To show that our method is capable of handling strongly correlated as well as highly uncorrelated input data with high accuracy we scale the branch lengths of the tree obtained by the NETGEN program using the "-s" parameter of Seq-Gen before the sequences are computed. We show that our model can handle a wide range for this scaling parameter.

Note that we make no limiting assumptions. All sequences involved in the recombination event can be still part of our input but they do not have to be there. Furthermore the size of the recombinant is not limited as well. This is in stark contrast to recombination detection methods like [11] where all fixed recombination scenarios contain recent recombinations.

To evaluate our approach we reconstruct the phylogenetic network for the input sequences and compare the resulting network with the original one using the program Treedist which is part of the phylip program package (see [21]). Treedist implements the symmetric difference method (Robinson and Foulds, 1981) to measure tree distances. Notice that as discussed before we can compare the networks by just comparing the two trees that build the network.

Note that this is a much stronger condition than just identifying the recombinant since it involves the proper reconstruction of the non-recombinant part of the network. As far as we know there are no tools available that would handle the general case. Therefor we compare the reliability of our method with the reliability of the parsimony method for trees. The goal is to show that on data sets on which our method fails, the parsimony method for trees would also return false results. To do so we take the data constructed by Seq-Gen and split it along the two trees T_1 and T_2 out of which we have constructed the network.

Then we compute for these sets of sequences the maximum parsimony trees and test via treedist if they represent the topologies T_1 and T_2.

For instances that still contain the recombinant in and for which the size of the recombinant is exactly one, we compare our method with the state-of-the-art recombination detection method of Maydt and Lengauer [11] Recco that is specialized for these scenarios. One has to point out that this test is not perfectly fair for any of the two methods, since both programs are not really optimized to output the correct recombinant. It is possible that the correct network is constructed internally but the wrong recombinant is reported. Since we cannot access the internal data structures of Recco we just compare the answers to the question: is the right recombinant sequence reported by the program? Since both implementations suffer from this fact, the experiments are not biased in the direction of any of these approaches.

5.2   Results

Even if it is only a special case in our model our experiments clearly show the superiority of our approach to the method of Maydt and Lengauer when the sequence divergence is low (see table 1). For small branch lengths we report the correct recombinant more often than the algorithm of Maydt and Lengauer does. So our method seems to work very good in the cases that Posada et al. reported to be hard to detect [15], i.e. in cases where the divergence is low. Note that for these experiments we have chosen a rather dull parameter value of 2 for α in the ◇-operator. We are sure that the results can be better when α is chosen more carefully.

Table 1. comparison of Recco and Recomb: the table gives the number of correctly reported recombinants depending on the branch length scaling factor s. Note that for each choice of s 40 tests have been conducted.

s=	0.005	0.01	0.05	0.1	0.5	1.0	1.5
Recomb	31	24	30	31	33	20	15
Recco	18	17	25	26	30	32	21

Table 2. Reliability of Recomb and maximum parsimony in test data: the small tables in the big one represent the relation between correct reconstruction of the parsimony trees and the correct reconstruction of the recombination network depending of branch length scaling factor s. Here R∧P means the correct reconstruction of both structures, ¬ (R∧P) denotes that both reconstructions failed and ¬R respectively ¬P means that only the reconstruction of the recombination network respectively the reconstruction of the parsimony trees failed. Note that for each choice of s 40 tests have been conducted.

s=			0.1		0.25		0.5		0.75		1.0		1.25	
R∧P	¬P		40	0	37	0	18	4	12	1	5	1	1	0
¬R	¬ (R∧P)		0	0	1	2	11	7	11	16	9	25	5	34

Table 3. Reliability of Recomb and maximum parsimony in test data: the table show how often the given implication holds. Note that for each choice of s 40 tests have been conducted.

s=	0.1	0.25	0.5	0.75	1.0	1.25
$\neg R \Rightarrow \neg P$	40	39	29	29	31	35
$\neg P \Rightarrow \neg R$	40	40	36	39	39	40

If we consider table 2 we can see that the reliability of our method is strongly connected to the reliability of parsimony reconstructions. On one hand if you consider the resulting implications of non-reliability you can see in table 3 that almost independently of the branch length scaling factor the unreliability of our method is induced by the unreliability of the parsimony method. On the other hand if the parsimony method fails our method almost always fails as well. Thus one can conclude that the reliability of our reconstruction method almost corresponds to the reliability of the parsimony method for reconstructing evolutionary trees.

6   Conclusion and Future Work

We have presented a new model to compute phylogenetic networks for data, where only a small number of recombination events are assumed to have taken place. Furthermore, we have given an exact algorithm for the case of non or exactly one recombination event. We have shown in our experimental study the high accuracy of our model.

Currently our tool can only deal with a single recombination event. In future work we want to extend our algorithm to more than one recombination event and we want to improve its speed so that we can handle larger instances (for example by implementing a heuristic instead of an exact algorithm). Additionally we want to extend it from the unit cost model for a mutation to an arbitrary cost matrix and to work on more sophisticated methods for the choice of our cost parameter α to make it data dependent. Finally, we plan to develop automated methods for detecting the correct number of recombination events. Preliminary experiments have shown that the decrease in the cost of an optimal 0-recombination network (i.e. the phylogenetic tree) to the cost of an optimal 1-recombination network is more significant if there was indeed a recombination event. This observation could be a good starting point for deriving such a method. Overall we think that this can lead to an extremely helpful tool for the computation of phylogenetic networks with a small number of recombination events.

References

1. Althaus, E., Naujoks, R.: Computing steiner minimum trees in hamming metric. In: Proceedings of the Seventeenth Annual ACM-SIAM Symposium on Discrete Algorithms, pp. 172–181 (2006)
2. Chare, E., Gould, E., Holmes, E.: Phylogenetic analysis reveals a low rate of homologous recombination in negative-sense rna viruses. J. Gen. Virol. 84 (2003)

3. Fitch, W.M.: Toward defining the course of evolution: minimum change for a specified tree topology. Systematic Zoology 20 (1971)
4. Gusfield, D., Eddhu, S., Langley, C.: Efficient reconstruction of phylogenetic networks with constrained recombination. In: Proceedings of the IEEE Computer Society Conference on Bioinformatics (2003)
5. Hein, J.: A heuristic method to reconstruct the history of sequences subject to recombination. J. Mol. Evol. 36 (1993)
6. Huson, D.H., Kloepper, T.H.: Computing recombination networks from binary sequences. Bioinformatics 21(2), 159–165 (2005)
7. Jin, G., Nakhleh, L., Snir, S., Tuller, T.: Inferring phylogenetic networks by the maximum parsimony criterion: A case study. Molecular Biology and Evolution 24(1) (2005)
8. Jin, G., Nakhleh, L., Snir, S., Tuller, T.: Efficient parsimony-based methods for phylogenetic network reconstruction. Bioinformatics 23(2) (2007)
9. Kececioglu, J., Gusfield, D.: Reconstructing a history of recombinations from a set of sequences. Discrete Appl. Math. 88(1-3), 239–260 (1998)
10. Makarenkov, V., Kevorkov, D., Legendre, P.: Phylogenetic network reconstruction approaches. Bioinformatics 6 (2006)
11. Maydt, J., Lengauer, T.: Recco: recombination analysis using cost optimization. Bioinformatics 22(9), 1064–1071 (2006)
12. Morin, M.M., Moret, B.M.E.: Netgen: generating phylogenetic networks with diploid hybrids. Bioinformatics 22(15) (2006)
13. Nakhleh, L., Jin, G., Zhao, F., Mellor-Crummey, J.: Reconstructing phylogenetic networks using maximum parsimony. In: Proceedings of the 2005 IEEE Computational Systems Bioinformatics Conference, Stanford (2005)
14. Polzin, T.: Algorithms for the Steiner Problem in Networks. PhD thesis, Universität des Saarlandes (2003)
15. Posada, D., Crandall, K.A., Holmes, E.C.: Recombination in evolutionary genomics. Annu. Rev. Genet. 36 (2002)
16. Rambaut, A., Grassly, N.C.: Seq-gen: an application for the monte carlo simulation of dna sequence evolution along phylogenetic trees. Computer Applications in the Biosciences 13(3), 235–238 (1997)
17. Ruths, D., Nakhleh, L.: Recombination and phylogeny: Effects and detection. International Journal on Bioinformatics Research and Applications 1(2) (2005)
18. Wang, L., Zhang, K., Zhang, L.: Perfect phylogenetic networks with recombination. Journal of Computational Biology 8 (2001)
19. http://www.bioinf.manchester.ac.uk/recombination/programs.shtml
20. http://evolution.genetics.washington.edu/phylip/
    software.html#Recombinant
21. http://evolution.genetics.washington.edu/phylip/software.html

Exact Algorithms for Cluster Editing: Evaluation and Experiments

Sebastian Böcker[1,2], Sebastian Briesemeister[3], and Gunnar W. Klau[4,5]

[1] Institut für Informatik, Friedrich-Schiller-Universität Jena, Germany
boecker@minet.uni-jena.de
[2] Jena Centre for Bioinformatics, Jena, Germany
[3] Div. for Simulation of Biological Systems, ZBIT/WSI, Eberhard Karls Universität
Tübingen, Germany
briese@informatik.uni-tuebingen.de
[4] Department of Mathematics and Computer Science, Freie Universität Berlin,
Germany
gunnar@math.fu-berlin.de
[5] DFG Research Center MATHEON, Berlin, Germany

Abstract. We present empirical results for the CLUSTER EDITING problem using exact methods from fixed-parameter algorithmics and linear programming. We investigate parameter-independent data reduction methods and find that effective preprocessing is possible if the number of edge modifications k is smaller than some multiple of $|V|$. In particular, combining parameter-dependent data reduction with lower and upper bounds we can effectively reduce graphs satisfying $k \leq 25 |V|$.

In addition to the fastest known fixed-parameter branching strategy for the problem, we investigate an integer linear program (ILP) formulation of the problem using a cutting plane approach. Our results indicate that both approaches are capable of solving large graphs with 1000 vertices and several thousand edge modifications. For the first time, complex and very large graphs such as biological instances allow for an exact solution, using a combination of the above techniques.

1 Introduction

The CLUSTER EDITING problem is defined as follows: Let $G = (V, E)$ be an undirected, loop-less graph. Our task is to find a set of edge modifications (insertions and deletions) of minimum cardinality, such that the modified graph consists of disjoint cliques.

The CLUSTER EDITING problem has been considered frequently in the literature since the 1980's. In 1986, Křivánek and Morávek [11] showed that the problem is NP-hard. The problem was rediscovered in the context of computational biology [14]. Clustering algorithms for microarray data such as CAST [1] and CLICK [15] rely on graph-theoretical intuition but solve the problem only heuristically. Studies in computational biology indicate that exact solutions of CLUSTER EDITING instances can be highly application-relevant,

C.C. McGeoch (Ed.): WEA 2008, LNCS 5038, pp. 289–302, 2008.

see for instance [18]. This is even more the case for the weighted version of the problem, WEIGHTED CLUSTER EDITING: Given an undirected graph with modification costs for every vertex tuple, we ask for a set of edge modifications with minimum total cost such that the modified graph consists of disjoint cliques.

The CLUSTER EDITING problem is APX-hard [4] and has a constant-factor approximation of 2.5 [17]. In this article, we empirically investigate the power of methods that solve the problem to *provable optimality*. In 1989, Grötschel and Wakabayashi [8] presented a formulation of the CLUSTER EDITING problem as an Integer Linear Program (ILP) and pointed out a cutting plane approach for its solution. Recently, the parameterized complexity of unweighted and weighted CLUSTER EDITING, using the number (or total cost) of edge modifications as parameter k, has gained much attention in the literature [2,6,7]. Dehne et al. [5] present an empirical evaluation of parameterized algorithms from [7]. The fastest fixed-parameter algorithm for unweighted CLUSTER EDITING actually transforms the problem into its weighted counterpart [3]. Guo [9] presents parameter-independent data reduction rules for unweighted instances that reduce an instance to a "hard" problem kernel of size $4k_{opt}$. A reduction from unweighted to weighted instances of size at most $4k_{opt}$ is presented in [3]. These reductions allow us to shrink an instance even before any parameter k has been considered.

Our contributions. In the first part of our paper, we evaluate the performance of two parameter-independent data reduction strategies for unweighted CLUSTER EDITING. We find that the efficiency of reduction is governed mostly by the ratio $k/|V|$. The unweighted kernel from [9] efficiently reduces nearly transitive graphs, but fails to reduce graphs with $k \geq \frac{1}{2}|V|$. We then present and evaluate parameter-independent reduction rules data for weighted graphs and find it to be even more effective in application. We combine the latter reduction with parameter-dependent reduction rules plus upper and lower bounds. This downsizes input graphs even more and fails to reduce graphs only when $k > 25|V|$ for large graphs.

To solve reduced instances, we implemented a branch-and-cut algorithm for WEIGHTED CLUSTER EDITING based on the ILP formulation proposed by Grötschel and Wakabayashi [8]. The ILP formulation of the problem has frequently been reported in the literature as being too slow for application, see for instance [10]. In contrast, we find that the cutting plane approach in [8] is capable of optimally solving large instances reasonably fast. We compare the performance of the fastest branching strategy in [3] and the cutting plane algorithm. We apply these methods to weighted instances resulting from unweighted graphs that have been fully reduced in advance using our data reduction. The FPT algorithm solves instances with $k = 5n$ in about an hour, where n, k are size and parameter of the reduced instance. The ILP approach solves instances with $n = 1\,000$ in about an hour, almost independently of k. These approaches are particularly important for weighted input data, because we find data reduction to be less effective here.

Summarized, our experiments show that one can solve CLUSTER EDITING instances on large graphs with several thousands of edge modifications in reasonable running time to provable optimality. In particular, feasible parameters k are orders of magnitude higher than what worst-case running times of the FPT approach suggest.

2   Preliminaries

Throughout this paper, let $n := |V|$. We write uv as shorthand for an unordered pair $\{u, v\} \in \binom{V}{2}$. For weighted instances, let $s : \binom{V}{2} \to \mathbb{R}$ encode the input graph: For $s(uv) > 0$ an edge uv is present in the graph and has deletion cost $s(uv)$, while for $s(uv) \leq 0$ the edge uv is absent from the graph and has insertion cost $-s(uv)$. We call edges with $s(uv) = \infty$ "permanent" and with $s(uv) = -\infty$ "forbidden". A graph G is a disjoint union of cliques if and only if there exist no conflict triples in G: a *conflict triple* consists of three vertices vuw such that uv and uw are edges of G but vw is not. Such graphs are also called *transitive*.

As a quality measure for data reduction we use the *reduction ratio* $\frac{n - n_{\mathrm{red}}}{n}$ where n_{red} denotes the number of vertices after reduction. A reduction ratio of close to 1 corresponds to a strong reduction whereas a reduction ratio of 0 corresponds to no reduction at all.

3   Data Reduction and Branching Algorithm

We now present methods for the parameter-independent data reduction of (unweighted and weighted) CLUSTER EDITING instances. We describe various polynomial-time reduction rules and apply these rules over and over again until no further rule will apply. Since the presented data reduction is parameter-independent, we can apply it during preprocessing without considering any particular parameter k. Afterwards, we can solve the reduced graph with *any* algorithm for WEIGHTED CLUSTER EDITING.

Parameter-independent data reduction. A *critical clique* C in an unweighted graph is an induced clique such that any two vertices $u, v \in C$ share the same neighborhood, $N(u) \cup \{u\} = N(v) \cup \{v\}$, and C is maximal. For unweighted CLUSTER EDITING one can easily see that all vertices of a critical clique of the input graph end up in the same cluster of an optimal clustering [9]. Furthermore, there are at most $4k_{\mathrm{opt}}$ critical cliques in a graph, where k_{opt} is the cost of an optimal solution. Guo [9] uses critical cliques to construct a kernel for unweighted CLUSTER EDITING of size $4k_{\mathrm{opt}}$. For brevity, we omit the details of this reduction, and only note that it is based on inspecting the neighborhood (and second neighborhood) of large critical cliques. In the following, we call this the *unweighted kernel*.

We can encode an unweighted CLUSTER EDITING instance using a weighted graph with edge weights ± 1. In a weighted graph we can *merge* vertices u, v into a new vertex u' when edge uv is set to "permanent": For each vertex $w \in V \setminus \{u, v\}$ we join uw, vw such that $s(u'w) \leftarrow s(uw) + s(vw)$. Moreover, in case w is a non-common neighbor of u, v we can reduce k by $\min\{|s(uw)|, |s(vw)|\}$ [2].

For unweighted instances, all vertices of a critical clique C must end up in the same cluster: This implies that we can merge all vertices in C for the corresponding weighted instance [3]. Doing so, we have reduced an unweighted instance to a weighted one of size at most $4k_{\mathrm{opt}}$. In addition, we may use the following reduction rules for *any weighted* instance:

Rule 1 (heavy non-edge rule). If an edge uv with $s(uv) < 0$ satisfies $|s(uv)| \geq \sum_{w \in N(u)} s(uw)$ then set uv to forbidden.

Rule 2 (heavy edge rule, single end). If an edge uv satisfies $s(uv) \geq \sum_{w \in V \setminus \{u,v\}} |s(uw)|$ then merge vertices u, v.

Rule 3 (heavy edge rule, both ends). If an edge uv satisfies $s(uv) \geq \sum_{w \in N(u) \setminus \{v\}} s(uw) + \sum_{w \in N(v) \setminus \{u\}} s(vw)$, then merge u, v.

Rule 4 (almost clique rule). For $C \subseteq V$ let k_C denote the min-cut value of the subgraph of G induced by vertex set C. If

$$k_C \geq \sum_{u,v \in C, s(uv) \leq 0} |s(uv)| + \sum_{u \in C, v \in V \setminus C, s(uv) > 0} s(uv)$$

then merge C.

Rule 5 (similar neighborhood). For an edge uv we define $N_u := N(u) \setminus (N(v) \cup \{v\})$, $N_v := N(v) \setminus (N(u) \cup \{u\})$ as the exclusive neighborhoods, and set $W := V - (N_u \cup N_v \cup \{u, v\})$. For $U \subseteq V$ set $s(v, U) := \sum_{u \in U} s(v, u)$. Let $\Delta_u := s(u, N_u) - s(u, N_v)$ and $\Delta_v := s(v, N_v) - s(v, N_u)$. If uv satisfies

$$s(uv) \geq \max_{C_u, C_v} \min\{s(v, C_v) - s(v, C_u) + \Delta_v, s(u, C_u) - s(u, C_v) + \Delta_u\} \quad (1)$$

where the maximum runs over all subsets $C_u, C_v \subseteq W$ with $C_u \cap C_v = \emptyset$, then merge uv.

Rule 4 cannot be applied to all subsets $C \subseteq V$ so we greedily choose reasonable subsets: We start with a vertex $C := \{u\}$ maximizing $\sum_{v \in V \setminus \{u\}} |s(uv)|$, and successively add vertices such that in every step, vertex $w \in V \setminus C$ with maximal connectivity $\sum_{v \in C} s(vw)$ is added. In case the connectivity of the best vertex is twice as large as that of the runner-up, we try to apply Rule 4 to the current set C. We cancel this iteration if the newly added vertex u is connected to more vertices in $V \setminus C$ than to vertices in C.

Proving the correctness of Rule 5 is rather involved, we defer the details to the full paper. This rule turns out to be highly efficient but its computation is expensive: For integer-weighted graphs, we can find the maximum (1) using dynamic programming in time $O(|W| Z)$ where $Z := \sum_{w \in W}(s(uw) + s(vw))$. For real-valued edge weights we can only approximate the calculation by multiplying with a blowup factor and rounding. In practice, we use Rule 5 only in case no other rules can be applied.

Using parameter-dependent data reduction. We use the parameter-dependent data reduction for WEIGHTED CLUSTER EDITING from [2]: We define induced costs $icf(uv)$ and $icp(uv)$ for setting uv to "forbidden" or "permanent" by

$$icf(uv) = \sum_{w \in N(u) \cap N(v)} \min\{s(uw), s(vw)\}, \quad icp(uv) = \sum_{w \in N(u) \triangle N(v)} \min\{|s(uw)|, |s(vw)|\},$$

where $A \triangle B$ denotes the symmetric set difference of A and B. If $icp(uv)$ or $icf(uv)$ exceed k, we can set uv to "forbidden" or "permanent", respectively. In the latter case, we merge u,v and reduce k by $icp(uv)$ accordingly. We can also remove isolated cliques.

As an algorithm-engineering technique, we now describe fast methods to compute a lower bounds on the cost of a weighted instance. Clearly, such bounds can be used to stop search tree recursion more efficiently. Assume that there exist t conflict triples in our instance G, k. For every tuple uv let $t(uv)$ denote the number of conflict triples in G that contain uv, and let $r(uv) := |s(uv)|/t(uv)$. To resolve t conflicts in our graph we have to pay at least $t \cdot \min_{uv}\{r(uv)\}$. A more careful analysis shows that we can sort tuples uv according to the ratio $r(uv)$, then go through this sorted list from smallest to largest ratio. This leads to a tighter lower bound but requires more computation time.

Our third lower bound proved to be most successful in applications: Let CT be a set of edge-disjoint conflict triples. Then, $\sum_{vuw \in CT} \min\{s(uv), s(uw), -s(vw)\}$ is a lower bound for solving all conflict triples. Since finding the set CT maximizing this value is computationally expensive, we greedily construct a set of edge-disjoint conflict triples CT and use the above sum as a lower bound.

We can use such lower bounds to make induced costs $icf(uv)$ and $icp(uv)$ tighter: let $b(G, uv)$ be a lower bound that *ignores* all edges uw and vw for $w \in V \setminus \{u, v\}$ in its computation. Then, we can set an edge to "forbidden" or "permanent" if $icp(uv) > k - b(G, uv)$ or $icf(uv) > k - b(G, uv)$ holds, resp.

To use this powerful reduction during (parameter-independent) preprocessing, we generate a problem instance (G, k) from G by using an *upper bound* for the modification costs of G as our parameter k. There exist a multitude of possibilities to compute such upper bounds, because we can use any heuristic for the problem and compute the cost of its solution, see for instance [18]. For this study, we calculate an upper bound using a greedy approach that iteratively searches for edges where reduction rules almost apply. We find this reduction to be extremely effective in applications.

Branching strategy. After parameter-independent data reduction, the remaining instance can be solved using a branching tree strategy. In these algorithms, we identify a conflict triple and then branch into sub-cases to repair this conflict. In practice, branching strategies that do merge vertices clearly outperform branching strategies that do not [2]. The fastest known branching strategy for CLUSTER EDITING, both in theory and in practice, is surprisingly simple [3]: Let uv be an edge of a conflict triple vuw. Then, (a) set uv to forbidden, or (b) merge uv. If we always choose the edge uv with minimal branching number,[1]

---

[1] The branching number is the root of the characteristic polynomial and governs the asymptotic size of the search tree, see e.g. [12] for details.

then the resulting search tree has size $O(2^k)$. To find an optimal solution we call the algorithm repeatedly, increasing k in an interval defined by lower and upper bound for this problem instance. While traversing the search tree, we apply reduction rules in every node of the search tree. The simple Rules 1–3 and parameter-dependent rules are applied in every node of the search tree, whereas the two more involved Rules 4 and 5 are applied only every sixth step. To find an edge with minimal branching number, we approximate log branching numbers using two rational functions.

4   Integer Linear Programming and Branch-and-Cut

In this section we describe an algorithm for WEIGHTED CLUSTER EDITING, which is based on mathematical optimization. It relies on the following integer linear programming (ILP) formulation due to Grötschel and Wakabayashi [8].

Let x be a binary decision vector with $x_e = 1$ if edge e is part of the solution and $x_e = 0$ otherwise, for all $e \in E$. Then, an optimal solution to WEIGHTED CLUSTER EDITING can be found by solving

$$\text{minimize} \quad \sum_{e \in E} s(e) - \sum_{1 \leq i < j \leq n} s(ij)x_{ij} \tag{2}$$

$$\text{subject to} \quad +x_{ij} + x_{jk} - x_{ik} \leq 1 \qquad \text{for all } 1 \leq i < j < k \leq n \tag{3}$$

$$+x_{ij} - x_{jk} + x_{ik} \leq 1 \qquad \text{for all } 1 \leq i < j < k \leq n \tag{4}$$

$$-x_{ij} + x_{jk} + x_{ik} \leq 1 \qquad \text{for all } 1 \leq i < j < k \leq n \tag{5}$$

$$x_{ij} \in \{0, 1\} \qquad \text{for all } 1 \leq i < j \leq n \ . \tag{6}$$

The $3\binom{n}{3}$ *triangle inequalities* (3)–(5) of the ILP ensure that no conflict triple occurs in the solution. The above ILP formulation can already be used to solve instances of WEIGHTED CLUSTER EDITING to provable optimality.

A faster algorithm can be obtained by a mathematical analysis of the corresponding problem polyhedron. Using methods from polyhedral combinatorics, Grötschel and Wakabayashi have studied the facial structure of the corresponding *clique partitioning polytope*. They have identified a number of classes of facet-defining inequalities. As proposed by the authors, we concentrate on the *2-partition inequalities*

$$\sum_{i \in S, j \in T} x_{ij} - \sum_{i \in S, j \in S} x_{ij} - \sum_{i \in T, j \in T} x_{ij} \leq \min\{|S|, |T|\} \ ,$$

where S and T are disjoint and nonempty subsets of V.

There is an exponential number of 2-partition inequalities. We therefore do not generate them at once but follow a *cutting plane* approach, adding 2-partition inequalities only if they are violated by a current fractional solution. We have implemented a variant of the iterative cutting plane method proposed by Grötschel and Wakabayashi. We start optimizing the LP relaxation (2) with an empty constraint set. Let x^* denote the vector corresponding to an intermediate solution of the linear programming relaxation. We first check whether x^* violates any triangle inequalities. If this is the case, we add the violated inequalities, resolve, and iterate. Otherwise, we check whether x^* is integral. If so, we stop, and x^* is an optimal solution. If x^* has fractional entries, we heuristically try to find violated 2-separation inequalities in the following manner:

For every node $i \in V$ we look at the nodes in $W := \{j \in V \setminus \{i\} \mid x_{ij}^* > 0\}$. Then, we pick a node $w \in W$ and iteratively construct a subset T of W, setting initially $T = \{w\}$ and adding nodes $k \in W$ to T if $x_{ik}^* - \sum_{j \in T} x_{jk}^* > 0$. Finally, we check whether

$$\sum_{j \in T} x_{ij}^* - \sum_{j \in T} \sum_{k \in T, k \neq j} x_{jk}^* > 1 \ .$$

If this is the case, we add the violated 2-partition inequality

$$\sum_{j \in T} x_{ij} - \sum_{j \in T} \sum_{k \in T, k \neq j} x_{jk} \leq 1 \ .$$

If we find cutting planes in the separation procedure we iterate, otherwise we branch.

5   Datasets

In the absence of publicly available datasets that meet our requirements (note that the datasets used in [5] are far too small for our evaluations) we concentrate on the following two datasets:

Random unweighted graphs. Given a number of nodes n and parameter k, we uniformly select an integer $i \in [1, n]$ and define i nodes to be a cluster. We proceed in this way with the remaining $n \leftarrow n - i$ nodes until $n \leq 5$ holds: In this case, we assign all remaining n nodes to the last cluster. Starting from this transitive graph $G = (V, E)$ we choose k' distinct vertex tuples $uv \in \binom{V}{2}$ and delete or insert the edge uv in G. Let k denote the minimum number of modifications to make G transitive, then $k \leq k'$. For instances where we cannot compute exact modification costs k, we estimate k using upper, lower bounds, and general observations.

Protein similarity data. We also apply our algorithms to *weighted* instances that stem from biological data. Rahmann et al. [13] present a set of graphs derived from protein similarity data: The vertices of our graph are more than 192 000 protein sequences from the COG database [16]. The similarity $S(u, v)$ of two proteins u, v is calculated from $\log_{10}$ E-values of bidirectional BLAST hits. We use an E-value of 10^{-10} as our threshold indicating that two proteins are "sufficiently similar", so $s(uv) := S(u, v) - 10$. See [13] for more details.

The graph encoded by s contains 50 600 connected components: 42 563 components are of size 1 or 2, and 4 073 components are cliques of size ≥ 3. The remaining 3 964 components serve as our evaluation instances. Only 11 instances have more than 600 vertices. As a side comment, we mention that Wittkop et al. [18] evaluate several clustering methods for this application, and find that WEIGHTED CLUSTER EDITING methods show the best clustering quality.

Evaluation platform. All algorithms were implemented in C++, the branch-and-cut algorithm (ILP) uses the Concert interface to the commercial CPLEX solver 9.03. Running times were measured on an AMD Opteron-275 2.2 GHz with 6+ GB of memory.

6   Data Reduction Results

We now compare the performance of the unweighted kernel [9] and the weighted data reduction from Sec. 3 on the dataset of random *unweighted* graphs. To allow for a fair comparison with the weighted data reduction, we merge all permanent edges of the unweighted kernel, resulting in an integer-weighted graph with even fewer vertices. This seems reasonable since both ILP and edge branching can handle integer-weighted input graphs. For the weighted data reduction, we first merge all critical cliques in the input graph. Next, we use weighted reduction rules plus the parameter-dependent reduction rules as described in Sec. 3. Despite the additional reduction steps, the reduced graph can have $4k_{opt}$ vertices for both approaches: A disjoint union of k paths of length 3 is not reduced by any reduction rule.

For our first evaluation, we concentrate on the weighted reduction strategy. For fixed $k = 2\,000$ and varying $n = 100, \ldots, 5\,000$ we study reduction ratio and absolute size of the resulting graph for 11\,000 random instances. Results for n up to 1\,000 are shown in Fig. 1. Similar results were obtained for larger n and other choices of k, data not shown. As one can see, the larger the graphs get, the better the reduction ratio on average. Most graphs are either reduced down to a

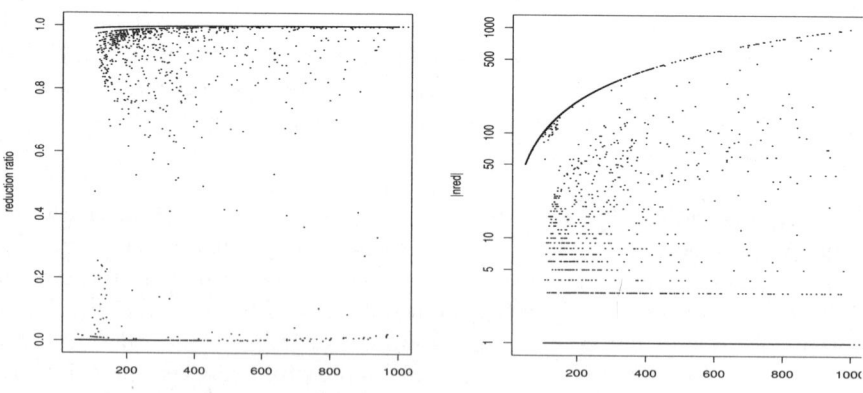

Fig. 1. Data reduction for fixed $k = 2\,000$ and variable graph size n: Left plot shows reduction ratio vs. n, right plot shows reduced graph size n_{red} vs. n. Both plots show 11000 instances.

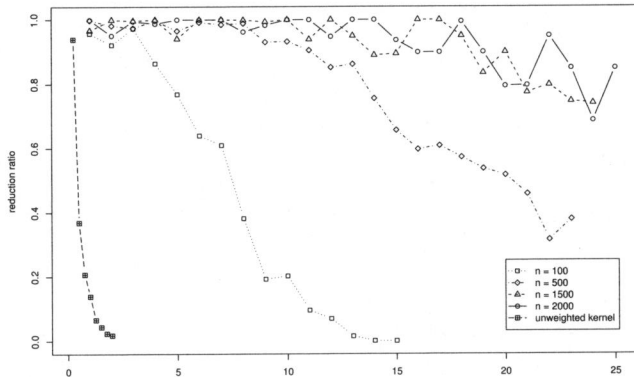

Fig. 2. Average reduction ratio vs. ratio k/n for $n = 100, 500, 1\,500, 2\,000$. Note that the unweighted kernel is practically independent from graph size n.

few vertices or stay unreduced. Only a few reduced graphs end up in a "twilight zone" between these extremes. This effective reduction is *not* due to the upper bound $n \le 4k = 8\,000$: In fact, the absolute size of reduced graphs gets smaller when input graphs get larger. This might seem counterintuitive at first glance, but larger graphs show smaller relative defects, which allows weighted reduction rules to more "aggressively" merge or delete edges.

The above evaluation indicates that reduction results do not only depend on k and n directly, but even more so on the ratio k/n. In our second evaluation, we choose $n \in \{100, 500, 1\,500, 2\,000\}$ and set $k := c \cdot n$, for varying factors $c \in \{0.25, 0.5, \ldots, 2.0\}$. For every combination of n and k we create 10 input graphs and apply the unweighted kernel. See Fig. 2 for resulting reduction ratios. We find reduction ratios of the unweighted kernel to be mostly independent of the actual graph size. The unweighted kernel is very effective for graphs with $k \le \frac{1}{2}n$, and graphs are downsized to half of their original size on average. For $k \ge 2n$ no reduction is observed. To evaluate the weighted data reduction we again set $k := c \cdot n$, for factors $c \in \{1, 2, \ldots, 25\}$. For every combination of k and graphs size with $n < 1\,000$ ($n \ge 1\,000$) we create 50 (20) input graphs. See again Fig. 2 for reduction ratios. We observe that the weighted data reduction is much more effective than the unweighted kernel. Here, the reduction ratio depends strongly on the ratio k/n and, less pronounced, also on the graph size n. We observed that large graphs of size $n = 2\,000$ are reduced by 80% for $k = 25$ and by more than 90% for $k = 18n$.

Figure 3 shows the ratio of input graphs being reduced by more than 90%. For the weighted data reduction, we vary the number of vertices n and set $k := cn$ for $c = 5, 10, 15, 20$.[2] For the unweighted kernel we observe significantly reduced graphs only for $c = 0.25$. Most interestingly, for the weighted data reduction, the ratio of significantly reduced graphs increases for larger graphs.

---

[2] We also performed experiments for all $c = 0.25, 0.5, 0.75, 1, 2, 3, \ldots, 25$ but find that results follow the same trend, data not shown.

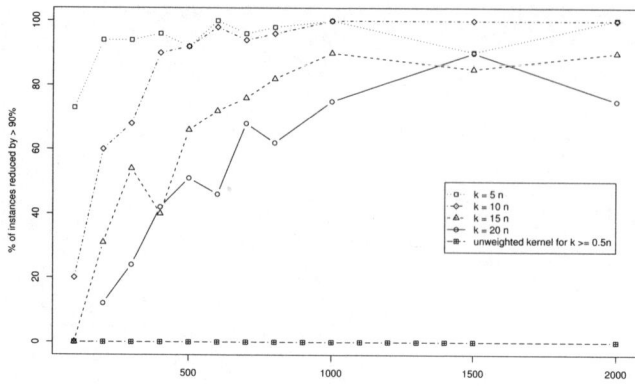

Fig. 3. Percentage of instances which are reduced more than 90% for varying graph size n and $k = cn$ for $c = 5, 10, 15, 20$

In case we only use parameter-independent reduction rules from Sec. 3, the weighted reduction is only slightly better than the unweighted kernel, data not shown. We find the combination of parameter-dependent data reduction and lower/upper bounds to be the reason for the effective reduction. To this end, we estimate the accuracy of our lower and upper bound. We find that our lower bound has a relative error of 1.7 % on average, and the upper bound had a relative error of 17.9 % on average. Calculating tighter upper bounds by, say, a heuristic such as FORCE [18] will further improve the performance of our weighted data reduction.

Running times of data reduction. Using the unweighted kernel, most of the instance were reduced in less than a minute, instances of size 2 000 in about one hour computation time. Graphs with k around n need more computation time than graphs with lower or greater k since reduction rules are checked very often but rarely applied. Running times of the weighted data reduction are equally high for k around $5n$, whereas for k around $20n$ running times are slightly higher. Making the data reduction run fast has not been the focus of our research, because we assumed running times of data reduction to be negligible to the following (exponential-time) step of the analysis. We do not report details and just note that reducing graphs of size 500 took 51.23 seconds on average but at most 9.09 minutes, whereas reducing graphs of size 2 000 took 1.61 hours on average and at most 23.69 hours. Our experiments show that many graphs are reduced to trivial or very small instances, so the exponential-time step of the algorithm has very small running times. We believe that by optimizing our data reduction algorithm we can achieve significantly reduced running times in the future.

Data reduction results for weighted instances. We also apply our weighted data reduction strategy to the protein similarity data. In this case, however, parameter k does not reflect the complexity of the instance: here, edges might have modification costs ≤ 1 and, hence, the total modification costs may equal 1 even if thousands of edge modifications are necessary. Instead, we use the number of edge modifications as a complexity measure of an instance. Table 1 shows results of the weighted data reduction. We find that the data reduction reduces weighted instances not as much as unweighted instances. This is mainly caused by the fact that our lower and upper bounds are not as tight as for the unweighted case. In detail, our lower bound has a relative error of 3.6 % on average, and the upper bound had a relative error of 54.7 % on average. In contrast to our findings for unweighted instances, we observe that larger graphs are reduced less effectively than smaller graphs. This can be attributed to the fact that the number of edge modifications is growing faster than linear. Furthermore, parameter-independent reduction rules are less efficient on large weighted graphs, since it gets less likely that an edge weight is greater than a sum over $O(n)$ other edge weights.

Table 1. Protein similarity data: Average reduction ratio for different graph size n

graph size n	3 - 49	50 - 99	100 - 149	150 - 199	200 - 249	250-299	300+
No. of instances	3453	341	78	22	24	20	25
av. reduction ratio	0.84	0.89	0.73	0.68	0.66	0.58	0.35

7 Integer Linear Programming and Search Tree Results

We want to compare the performance of the FPT branching algorithm approach and the ILP-based branch-and-cut method. For this evaluation, we use random unweighted graphs and reduce them by the weighted data reduction. Reduced graphs are sorted into bins for sizes $n \approx 100, 200, \ldots, 900$ and costs $k \approx 1n, 2n, \ldots, 10n$, and every bin contains 28 graphs on average. As described in Sec. 6, most graphs are either reduced completely or not at all, so building these reduced graphs is computationally expensive. For each reduced instance we apply the FPT branching algorithm and ILP with an upper limit of 6 hours of running time. For average running times, we count unfinished instances as 6 hours. Figure 4 shows the resulting running times.

Running times of the fixed-parameter algorithm most strongly depend on the ratio k/n and, to a smaller extent, on the actual parameter k. Instances with modification cost $k \approx 5n$ need about one hour of computation to be solved. Note that running times for FPT branching are much better than worst-case running time analysis suggests, and dependence on the actual parameter k is much less pronounced than expected. We believe that this is mainly due to the good lower bound estimation for the parameter-dependent data reduction used in interleaving, and also the vertex merging operation.

The limiting factor for the ILP algorithm is the size of the input graph whereas dependence on modification costs k is much less pronounced. Small instances with only 100 vertices are solved within seconds, and medium graphs of size 500 are solved within minutes. We find that ILP is well-suited for medium-size

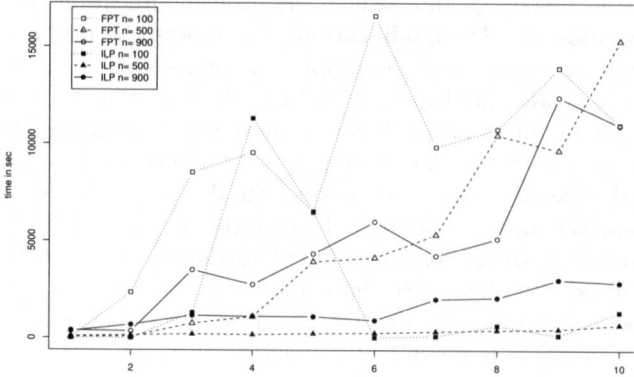

Fig. 4. Running times of FPT branching and ILP branch-and-cut in seconds, for varying ratio k/n and $n = 100, 500, 900$

Table 2. Running times on reduced protein similarity data for FPT branching and ILP. Instances that did not finish after 24 hours of computation were ignored for average running time computation.

Size red. instance	3–49	50–99	100–149	150–199	200–249	250–299	300–1400
No. red. instances	297	52	16	10	9	2	19
Unfinished FPT	0	0	1	1	2	2	15
time FPT	125 ms	23.9 s	44.1 min	4.52 min	47.3 min	n/a	8.98 min
Unfinished ILP	0	0	0	0	1	1	10
time ILP	17 ms	6.97 s	5.30 min	18.20 min	76.2 min	6.85 min	1.67 h

CLUSTER EDITING instances and clearly outperforms the fastest fixed-parameter algorithm for these instances. We stress that ILP requires preprocessing by parameter-independent data reduction since its performance is solely dependent on the input graph size. Only for large graphs with very low modification costs $k \leq 2n$, the FPT algorithm may outperform the cutting plane algorithm. High running times of the cutting plane approach for large instances are, however, mostly *not* due to their structural complexity but to the large number of triangle inequalities that have to be checked in the current implementation. Once a better separation strategy has been found, we expect the branch-and-cut algorithm to perform well even on larger instances.

Results for weighted instances. We now compare the performance of FPT branching and ILP using protein similarity data. We reduced all instances in the protein dataset using our weighted data reduction strategy, resulting in 365 non-trivial instances. In Tab. 2 we report running times of the two methods. The FPT branching algorithm is usually fast enough for graphs with up to 200 vertices, but for most larger graphs, no solution can be computed within 24 hours. In contrast, the ILP algorithm was able to solve most instances with less than 500 vertices in only some minutes.

8   Conclusion

Our results demonstrate that computing exact solutions of CLUSTER EDITING instances is no longer limited to small or almost transitive graphs, thus invalidating what has often been reported in previous work. Using data reduction for WEIGHTED CLUSTER EDITING in combination with parameter-dependent rules and lower/upper bounds strongly improves the ability to shrink down input instances in polynomial running time. Even complex input graphs that are far from transitive and that have modification costs much larger than the number of vertices, can often be reduced very effectively.

We also compared the fastest known FPT branching algorithm for CLUSTER EDITING against a branch-and-cut approach for this problem, based on the ILP formulation by Grötschel and Wakabayashi. Both algorithms perform well, and reduced graphs with hundreds of vertices and thousands of edge modifications are processed in acceptable running time. In particular, our results suggest that ILP is suitable for solving large instances with many modifications.

We believe that better upper bounds will allow even larger instances of (unweighted and weighted) CLUSTER EDITING to be solved exactly in the future. We will make the source code of our reduction and cluster editing tools, as well as the data used in this article publicly available. Furthermore, we plan to implement a web interface for our tools in order to give a large community access to our exact clustering tools and to facilitate comparison and evaluation.

Acknowledgments. We thank Svenja Simon for support with evaluation and implementation. S. Briesemeister gratefully acknowledges financial support from LGFG Promotionsverbund "Pflanzliche Sensorhistidinkinasen" at the University of Tübingen.

References

1. Ben-Dor, A., Shamir, R., Yakhini, Z.: Clustering gene expression patterns. J. Comput. Biol. 6(3-4), 281–297 (1999)
2. Böcker, S., Briesemeister, S., Bui, Q.B.A., Truß, A.: A fixed-parameter approach for weighted cluster editing. In: Proc. of Asia-Pacific Bioinformatics Conference (APBC 2008). Series on Advances in Bioinformatics and Computational Biology, vol. 5, pp. 211–220. Imperial College Press (2008)
3. Böcker, S., Briesemeister, S., Bui, Q.B.A., Truß, A.: Going weighted: Parameterized algorithms for cluster editing (Manuscript) (2008)
4. Charikar, M., Guruswami, V., Wirth, A.: Clustering with qualitative information. J. Comput. Syst. Sci. 71(3), 360–383 (2005)
5. Dehne, F., Langston, M.A., Luo, X., Pitre, S., Shaw, P., Zhang, Y.: The cluster editing problem: Implementations and experiments. In: Bodlaender, H.L., Langston, M.A. (eds.) IWPEC 2006. LNCS, vol. 4169, pp. 13–24. Springer, Heidelberg (2006)

6. Gramm, J., Guo, J., Hüffner, F., Niedermeier, R.: Automated generation of search tree algorithms for hard graph modification problems. Algorithmica 39(4), 321–347 (2004)
7. Gramm, J., Guo, J., Hüffner, F., Niedermeier, R.: Graph-modeled data clustering: Fixed-parameter algorithms for clique generation. Theor. Comput. Syst. 38(4), 373–392 (2005)
8. Grötschel, M., Wakabayashi, Y.: A cutting plane algorithm for a clustering problem. Math. Program. 45, 52–96 (1989)
9. Guo, J.: A more effective linear kernelization for Cluster Editing. In: Chen, B., Paterson, M., Zhang, G. (eds.) ESCAPE 2007. LNCS, vol. 4614, pp. 36–47. Springer, Heidelberg (2007)
10. Kochenberger, G.A., Glover, F., Alidaee, B., Wang, H.: Clustering of microarray data via clique partitioning. J. Comb. Optim. 10(1), 77–92 (2005)
11. Křivánek, M., Morávek, J.: NP-hard problems in hierarchical-tree clustering. Acta Inform. 23(3), 311–323 (1986)
12. Niedermeier, R.: Invitation to Fixed-Parameter Algorithms. Oxford University Press, Oxford (2006)
13. Rahmann, S., Wittkop, T., Baumbach, J., Martin, M., Truß, A., Böcker, S.: Exact and heuristic algorithms for weighted cluster editing. In: Proc. of Computational Systems Bioinformatics (CSB 2007), vol. 6, pp. 391–401 (2007)
14. Shamir, R., Sharan, R., Tsur, D.: Cluster graph modification problems. Discrete Appl. Math. 144(1–2), 173–182 (2004)
15. Sharan, R., Maron-Katz, A., Shamir, R.: CLICK and EXPANDER: a system for clustering and visualizing gene expression data. Bioinformatics 19(14), 1787–1799 (2003)
16. Tatusov, R.L., Fedorova, N.D., Jackson, J.D., Jacobs, A.R., Kiryutin, B., Koonin, E.V., Krylov, D.M., Mazumder, R., Mekhedov, S.L., Nikolskaya, A.N., Rao, B.S., Smirnov, S., Sverdlov, A.V., Vasudevan, S., Wolf, Y.I., Yin, J.J., Natale, D.A.: The COG database: an updated version includes eukaryotes. BMC Bioinformatics 4, 41 (2003)
17. van Zuylen, A., Williamson, D.P.: Deterministic algorithms for rank aggregation and other ranking and clustering problems. In: Proc. of Workshop on Approximation and Online Algorithms (WAOA 2007). Lect. Notes Comput. Sc., vol. 4927, pp. 260–273. Springer, Heidelberg (2008)
18. Wittkop, T., Baumbach, J., Lobo, F., Rahmann, S.: Large scale clustering of protein sequences with FORCE – a layout based heuristic for weighted cluster editing. BMC Bioinformatics 8(1), 396 (2007)

Combining Hierarchical and Goal-Directed Speed-Up Techniques for Dijkstra's Algorithm*

Reinhard Bauer, Daniel Delling, Peter Sanders, Dennis Schieferdecker,
Dominik Schultes, and Dorothea Wagner

Universität Karlsruhe (TH), 76128 Karlsruhe, Germany
{rbauer,delling,sanders,schief,schultes,wagner}@ira.uka.de

Abstract. In [1], basic speed-up techniques for Dijkstra's algorithm
have been combined. The key observation in their work was that it is
most promising to combine *hierarchical* and *goal-directed* speed-up techniques. However, since its publication, impressive progress has been made
in the field of speed-up techniques for Dijkstra's algorithm and huge data
sets have been made available.

Hence, we revisit the systematic combination of speed-up techniques
in this work, which leads to the fastest known algorithms for various
scenarios. Even for road networks, which have been worked on heavily
during the last years, we are able to present an improvement in performance. Moreover, we gain interesting insights into the behavior of
speed-up techniques when combining them.

1 Introduction

Computing shortest paths in a graph $G = (V, E)$ is used in many real-world
applications like route planning in road networks, timetable information for railways, or scheduling for airplanes. In general, Dijkstra's algorithm [2] finds a
shortest path of length $d(s, t)$ between a given source s and target t. Unfortunately, the algorithm is far too slow to be used on huge datasets. Thus, several speed-up techniques have been developed (see [3] for an overview) yielding
faster query times for typical instances, e.g., road or railway networks. In [1],
basic speed-up techniques have been combined systematically. One key observation of their work was that it is most promising to combine hierarchical and
goal-directed techniques. However, since the publication of [1], many powerful
hierarchical speed-up techniques have been developed, goal-directed techniques
have been improved, and huge data sets have been made available to the community. In this work, we revisit the systematic combination of speed-up techniques.

---

* Partially supported by the Future and Emerging Technologies Unit of EC (IST
  priority – 6th FP), under contract no. FP6-021235-2 (project ARRIVAL), and by
  DFG grant SA 933/1-3.

C.C. McGeoch (Ed.): WEA 2008, LNCS 5038, pp. 303–318, 2008.

1.1 Related Work

Since there is an abundance of related work, we decided to concentrate on previous combinations of speed-up techniques and on the approaches that our work is directly based on.

Bidirectional Search executes Dijkstra's algorithm simultaneously forwards from the source s and backwards from the target t. Once some node has been visited from both directions, the shortest path can be derived from the information already gathered [4]. Many more advanced speed-up techniques use bidirectional search as an optional or sometimes even mandatory ingredient.

Hierarchical Approaches try to exploit the hierarchical structure of the given network. In a preprocessing step, a hierarchy is extracted, which can be used to accelerate all subsequent queries.

Reach. Let $R(v) := \max R_{st}(v)$ denote the *reach* of node v, where $R_{st}(v) := \min(d(s, v), d(v, t))$ for all s-t shortest paths including v. Gutman [5] observed that a shortest-path search can be pruned at nodes with a reach too small to get to either source or target from there. The basic approach was considerably strengthened by Goldberg et al. [6], in particular by a clever integration of *shortcuts* [3], i.e., single edges that represent whole paths in the original graph.

Highway-Node Routing [3] computes for a given sequence of node sets $V =: V_0 \supseteq V_1 \supseteq \ldots \supseteq V_L$ a hierarchy of *overlay graphs* [7,8]: the level-ℓ overlay graph consists of the node set V_ℓ and an edge set E_ℓ that ensures the property that all distances between nodes in V_ℓ are equal to the corresponding distances in the underlying graph $G_{\ell-1}$. A bidirectional query algorithm takes advantage of the multi-level overlay graph by never moving downwards in the hierarchy—by that means, the search space size is greatly reduced. The most recent variant of HNR [9], Contraction Hierarchies, obtains a node classification by iteratively contracting the 'least important' node, yielding a hierarchy with up to $|V|$ levels. Moreover, the input graph G is transfered to a search graph G' by storing only edges directing from unimportant to important nodes. As a remarkable result, G' is *smaller* than G yielding a *negative* overhead per node. Finally, by this transformation the query is simply a plain bidirectional Dijkstra operating on G'.

Transit-Node Routing [10] is based on a simple observation intuitively used by humans: When you start from a source node s and drive to somewhere 'far away', you will leave your current location via one of only a few 'important' traffic junctions, called (forward) *access nodes* $\overrightarrow{A}(s)$. An analogous argument applies to the target t, i.e., the target is reached from one of only a few backward access nodes $\overleftarrow{A}(t)$. Moreover, the union of all forward and backward access nodes of all nodes, called *transit-node set* $\mathcal{T}$, is rather small. This implies that for each node the distances to/from its forward/backward access nodes and for each transit-node pair (u, v) the distance between u and v can be stored. For given source and target nodes s and t, the length of the shortest path that passes at least one transit node is given by $d_T(s,t) = \min\{d(s,u) + d(u,v) + d(v,t) \mid u \in \overrightarrow{A}(s), v \in \overleftarrow{A}(t)\}$. As a final ingredient, a *locality filter* $\mathcal{L} : V \times V \rightarrow \{\text{true}, \text{false}\}$ is needed that decides whether given nodes s and t are too close to travel via a transit node. $\mathcal{L}$ has to fulfill the property that $\mathcal{L}(s,t) = \text{false}$ implies $d(s,t) = d_T(s,t)$. Then, the following algorithm can be used to compute the shortest-path length $d(s,t)$:

if $\mathcal{L}(s,t) = \text{false}$ **then** compute and return $d_T(s,t)$; **else** use any other routing algorithm.

Note that for a given source-target pair (s,t), let $a := \max(|\overrightarrow{A}(s)|, |\overleftarrow{A}(t)|)$. For a global query (i.e., $\mathcal{L}(s,t) = \text{false}$), we need $O(a)$ time to lookup all access nodes, $O(a^2)$ to perform the table lookups, and $O(1)$ to check the locality filter.

Goal-Directed Approaches direct the search towards the target t by preferring edges that shorten the distance to t and by excluding edges that cannot possibly belong to a shortest path to t—such decisions are usually made by relying on preprocessed data.

ALT [11] is based on $\underline{A}^*$ search, *Landmarks*, and the $\underline{T}$riangle inequality. After selecting a small number of nodes, called landmarks, for all nodes v, the distances $d(v, \lambda)$ and $d(\lambda, v)$ to and from each landmark λ are precomputed. For nodes v and t, the triangle inequality yields for each landmark λ two lower bounds $d(\lambda, t) - d(\lambda, v) \leq d(v, t)$ and $d(v, \lambda) - d(t, \lambda) \leq d(v, t)$. The maximum of these lower bounds is used during an A^* search. The original ALT approach has fast preprocessing times and provides reasonable speed-ups, but consumes too much space for very large networks. In the subsequent paragraph on "Previous Combinations", we will see that there is a way to reduce the memory consumption by storing landmark distances only for a subset of the nodes.

Arc-Flags. The arc-flag approach, introduced in [12], first computes a partition $\mathcal{C}$ of the graph. A *partition* of V is a family $\mathcal{C} = \{C_0, C_1, \ldots, C_k\}$ of sets $C_i \subseteq V$ such that each node $v \in V$ is contained in exactly one set C_i. An element of a partition is called a *cell*. Next, a *label* is attached to each edge e. A label contains, for each cell $C_i \in \mathcal{C}$, a flag $AF_{C_i}(e)$ which is true if a shortest path to a node in C_i starts with e. A modified Dijkstra then only considers those edges for which the flag of the target node's cell is true. The big advantage of this approach is its easy and fast query algorithm. However, preprocessing is very expensive, either regarding preprocessing time or memory consumption [13].

Previous Combinations. Many speed-up techniques can be combined. In [7], a combination of a special kind of geometric container [14], the separator-based multi-level method [8], and A^* search yields a speed-up of 62 for a railway transportation problem. In [1], combinations of A^* search, bidirectional search, the separator-based multi-level method, and geometric containers are studied: Depending on the graph type, different combinations turn out to be best.

REAL. Goldberg et al. [6] have successfully combined their advanced version of <u>RE</u>ach with landmark-based A^* search (the <u>AL</u>t algorithm), obtaining the REAL algorithm. In the most recent version, they introduce a variant where landmark distances are stored only with the more important nodes, i.e., nodes with high reach values. By this means, the memory consumption can be reduced.

HH* [15] combines highway hierarchies [16] (<u>HH</u>) with landmark-based <u>A*</u> search. Similar to [6], the landmarks are not chosen from the original graph, but for some level k of the highway hierarchy, which reduces the preprocessing time and memory consumption. As a result, the query works in two phases: in an initial phase, a non-goal-directed highway query is performed until all entrance points to level k have been discovered; for the remaining search, the landmark distances are available so that the combined algorithm can be used.

SHARC [17] extends and combines ideas from highway hierarchies (namely, the contraction phase, which produces <u>SH</u>ortcuts) with the <u>ARC</u> flag approach. The result is a fast *unidirectional* query algorithm, which is advantageous in scenarios where bidirectional search is prohibitive. In particular, using an approximative variant allows dealing with time-dependent networks efficiently. Even faster query times can be obtained when a bidirectional variant is applied.

1.2 Our Contributions

In this work, we study a systematic combination of speed-up techniques for Dijkstra's algorithm. However, we observed in [18] that some combinations are more promising than others. Hence, we focus on the most promising ones: adding goal-direction to hierarchical speed-up techniques. By evaluating different inputs and scenarios, we gain interesting insights into the behavior of speed-up techniques when combining them. As a result, we are able to present the fastest known techniques for several scenarios. For sparse graphs, a combination of Highway-Node Routing and Arc-Flags yields excellent speed-ups with low preprocessing effort. The combination is only overtaken by Transit-Node Routing in road networks with travel times, but the gap is almost closed. However, even Transit-Node Routing can be further accelerated by adding goal-direction. Moreover, we introduce a hierarchical ALT algorithm, called CALT, that yields a good performance on denser graphs. Finally, we reveal interesting observations when combining Arc-Flags with Reach.

We start our work on combinations in Section 2 by presenting a generic approach how to improve the performance of basic speed-up techniques in general. The key observation is that we extract an important subgraph, called the *core*, of the input graph and use only the core as input for the preprocessing-routine of the applied speed-up technique. As a result, we derive a two-phase query algorithm, similar to partial landmark REAL or HH*. During phase 1 we use plain Dijkstra to reach the core, while during phase 2, we use a speed-up technique in order to accelerate the search within the core. The full power of this *core-based routing* approach can be unleashed by using a goal-directed technique during phase 2. Our experimental study in Section 5 shows that when using ALT during phase 2, we end in a very robust technique that is superior to plain ALT.

In Section 3, we show how to remedy the crucial drawback of Arc-Flags: its preprocessing effort. Instead of computing arc-flags on the full graph, we use a purely hierarchical method until a specific point during the query. As soon as we have reached an 'important' subgraph, i.e., a high level within the hierarchy, we turn on arc-flags. As a result, we significantely accelerate hierarchical methods like Highway-Node Routing. Our aggressive variant moderately increases preprocessing effort but query performance is almost as good as Transit-Node Routing in road networks: On average, we settle only 45 nodes for computing the distance between two random nodes in a continental road network. The advantage of this combination over Transit-Node Routing is its very low space consumption.

However, we are also able to improve the performance of Transit-Node Routing. In Section 4, we present how to add goal-direction to this approach. As a result, the number of required table lookups can be reduced by a factor of 13, resulting in average query times of less than $2\,\mu s$—more than three million times faster than Dijkstra's algorithm.

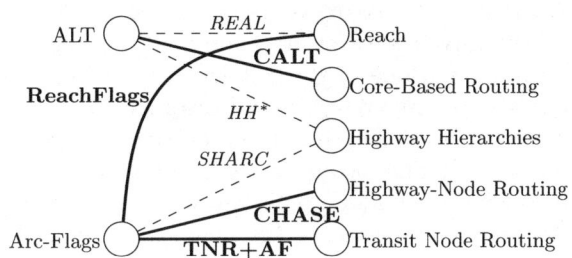

As already mentioned, a few combinations like HH*, REAL, and SHARC have already been published. Hence, Figure 1 provides an overview

Fig. 1. Overview of combinations of speed-up techniques. Speed-up techniques are drawn as nodes (goal-directed techniques on the left, hierarchical on the right). A dashed edge indicates an existing combination, whereas thick edges indicate combinations presented in this work.

over existing combinations already published and those which are presented in this work. Note that all techniques in this work use bidirectional search. Also note that due to space limitations all proofs of correctness are skipped but will be included in the full paper.

2 Core-Based Routing

In this section, we introduce a very easy and powerful approach to generally reduce the preprocessing of the speed-up techniques introduced in Section 1. The central idea is to use contraction [9] to extract an important subgraph and preprocess only this subgraph instead of the full graph.

Preprocessing. At first, the input graph $G = (V, E)$ is contracted to a graph $G_C = (V_C, E_C)$, called the *core*. Note that we could use any contraction routine, that removes nodes from the graph and inserts edges to preserve distances between core nodes. Examples are those from [16,6,17] or the most advanced one from [9]. The key idea of core-based routing is not to use G as input for preprocessing but to use G_C instead. As a result, preprocessing of most techniques can be accelerated as the input can be shrunk. However, sophisticated methods like Highway Hierarchies, REAL, or SHARC already use contraction during preprocessing. Hence, this advantage especially holds for goal-directed techniques like ALT or Arc-Flags. After preprocessing the core, we store the preprocessed data and merge the core and the normal graph to a full graph $G_F = (V, E_F = E \cup E_C)$. Moreover, we mark the core-nodes with a flag.

Query. The s-t query is a modified bidirectional Dijkstra, consisting of two phases and performed on G_F. During phase 1, we search the graph search until all *entrance points* of s and t are found (cf. [15] for details). We identify a superset of those nodes by the following approach. We run a bidirectional Dijkstra rooted at s and t *not* relaxing edges belonging to the core. We add each core node settled by the forward search to a set S (T for the backward search). The first phase terminates if one of the following two conditions hold: (1) either both priority queues are empty or (2) the distance to the closest entry points of s and t is larger than the length of the tentative shortest path. If case (2), the whole query terminates. The second phase is initialized by refilling the queues with the nodes belonging to S and T. As key we use the distances computed during phase 1. Afterwards, we execute the query-algorithm of the applied speed-up technique which terminates according to its stopping condition.

CALT. Although we could use *any* of the speed-up techniques to instantiate our core-based approach we focus on a variant based on ALT due to the following reasons. First of all, ALT works well in *dynamic* scenarios. As contraction seems easy to dynamize, we are optimistic that CALT (Core-ALT) also works well in dynamic scenarios. Second, pure ALT is a very robust technique with respect to the input. Finally, ALT suffers from the critical drawback of high memory consumption—we have to store two distances per node and landmark—which can be reduced by switching to CALT.

On top of the preprocessing of the generic approach, we compute landmarks on the core and store the distances to and from the landmarks for all core nodes. However, ALT needs lower bounds for *all* nodes to the source and target. As we do not store distances from all nodes to the landmarks, we need *proxy nodes*, which were introduced for the partial REAL algorithm in [6]. The method developed there can directly be applied to CALT: The proxy s' of a node s is the core node closest to s. We compute these proxy nodes for a given s–t query during the initialization phase of the first phase of the query. During the second phase we use the landmark information for the core in order to speed-up the query within the core.

3   Hierarchy-Aware Arc-Flags

Two goal-directed techniques have been established during the last years: ALT and Arc-Flags. The advantages of ALT are fast preprocessing and easy adaption to dynamic scenarios, while the latter is superior with respect to query-performance and space consumption. However, preprocessing of Arc-Flags is expensive. The central idea of *Hierarchy-Aware Arc-Flags* is to combine—similar to REAL or HH*—a hierarchical method with Arc-Flags. By computing arc-flags only for a subgraph containing all nodes in high levels of the hierarchy, we are able to reduce preprocessing times. In general, we could use any hierarchical approach but as Contraction Hierarchies (CH) is the hierarchical method with lowest space consumption, we focus on the combination of Contraction Hierarchies and Arc-Flags. However, we also present a combination of Reach and Arc-Flags.

3.1 Contraction Hierarchies + Arc-Flags (CHASE)

As already mentioned in Section 1.1, Contraction Hierarchies is basically a plain bidirected Dijkstra on a search graph constructed during preprocessing. We are able to combine Arc-Flags and Contraction Hierarchies in a very natural way and name it the CHASE-algorithm (Contraction-Hierarchy + Arc-flagS + highway-nodE routing).

Preprocessing. First, we run a complete Contraction Hierarchies preprocessing which assembles the search graph G'. Next, we extract the subgraph H of G' containing the $|V_H|$ nodes of highest levels. The size of V_H is a tuning parameter. Recall thatContraction Hierarchies uses $|V|$ levels with the most important node in level $|V|-1$. We partition H into k cells and compute arc-flags according to [13] for all edges in H. Summarizing, the preprocessing consists of constructing the search graph and computing arc-flags for H.

Query. Basically, the query is a two-phase algorithm. The first phase is a bidirected Dijkstra on G' with the following modification: When settling a node v belonging to H, we do *not* relax any outgoing edge from v. Instead, if v is settled by the forward search, we add v to a node set S, otherwise to T. Phase 1 ends if the search in both directions stops. The search stops in one direction, if either the respective priority queue is empty or if the minimum of the key values in that queue and the distance to the closest entrance point in that direction is equal or larger than the length of the tentative shortest path. The whole search can be stopped after the first phase, if either no entrance points have been found in one direction or if the tentative shortest-path distance is smaller than minimum over all distances to the entrance points and all key values remaining in the queues. Otherwise we switch to phase 2 of the query which we initialize by refilling the queues with the nodes from S and T. As keys we use the distances computed during phase 1. In phase 2, we use a bidirectional Arc-Flags Dijkstra. We identify the set C_S (C_T) of all cells that contain at least one node $u \in S$ ($u \in T$). The forward search only relaxes edges having a true arc-flag for any of the cells C_T. The backward search proceeds analogously. Moreover, we use the CH stopping criterion and the strict alternating strategy for forward and backward search. However, during our experimental study, it turned out that *stall-on-demand* [3], which accelerates pure CH, does not pay off for CHASE. The computational overhead is too high which is not compensated by the slight decrease in search space. So, the resulting query is a plain bidirectional Dijkstra operating on G' with the CH stopping criterion and arc-flags activated on high levels of the hierarchy.

Note that we have a trade-off between performance and preprocessing. If we use bigger subgraphs as input for preprocessing arc-flags, query-performance is better as arc-flags can be used earlier. However, preprocessing time increases as more arc-flags have to be computed.

3.2 Reach + Arc-Flags (ReachFlags)

Similar to CHASE, we can also combine Reach and Arc-Flags, called *Reach-Flags*. However, we slightly alter the preprocessing: Reach-computation according to [6] is a process that iteratively contracts and prunes the input. This iteration can be interpreted as levels of a hierarchy: A node u belongs to level i if u is still part of the graph during iteration step i. With this notion of hierarchy, we are able to preprocess ReachFlags. We first run a complete Reach-preprocessing as described in [6] and assemble the output graph. Next, we extract a subgraph H from the output graph containing all nodes of level $\geq \ell$. Again, we compute arc-flags in H according to [13]. The ReachFlags-query can easily by adapted from the CHASE-query in straight-forward manner. Note that the input parameter ℓ adjusts the size of V_H. Thus, a similar trade-off in performance/preprocessing effort like for CHASE is given.

4   Transit-Node Routing + Arc-Flags (TNR+AF)

Recall that the most time-consuming part of a TNR-query are the table lookups. Hence, we want to further improve the average query times, the first attempt should be to reduce the number of those lookups. This can be done by excluding certain access nodes at the outset, using an idea very similar to the arc-flag approach. We consider the minimal overlay graph $G_{\mathcal{T}} = (\mathcal{T}, E_{\mathcal{T}})$ of G, i.e., the graph with (transit) node set $\mathcal{T}$ and an edge set $E_{\mathcal{T}}$ such that $|E_{\mathcal{T}}|$ is minimal and for each node pair $(s, t) \in \mathcal{T} \times \mathcal{T}$, the distance from s to t in G corresponds to the distance from s to t in $G_{\mathcal{T}}$. We partition this graph $G_{\mathcal{T}}$ into k regions and store for each node $u \in \mathcal{T}$ its region $r(u) \in \{1, \ldots, k\}$. For each node s and each access node $u \in \overrightarrow{A}(s)$, we manage a flag vector $f_{s,u}^{\rightarrow} : \{1, \ldots, k\} \to \{\mathsf{true}, \mathsf{false}\}$ such that $f_{s,u}^{\rightarrow}(x)$ is true iff there is a node $v \in \mathcal{T}$ with $r(v) = x$ such that $d(s, u) + d(u, v)$ is equal to $\min\{d(s, u') + d(u', v) \mid u' \in \overrightarrow{A}(s)\}$. In other words, a flag of an access node u for a particular region x is set to true iff u is useful to get to some transit node in the region x when starting from the node s. Analogous flag vectors $f_{t,u}^{\leftarrow}$ are kept for the backward direction.

Preprocessing. The flag vectors can be precomputed in the following way, again using ideas similar to those used in the preprocessing of the arc-flag approach: Let $B \subseteq \mathcal{T}$ denote the set of border nodes, i.e., nodes that are adjacent to some node in $G_{\mathcal{T}}$ that belongs to a different region. For each node $s \in V$ and each border node $b \in B$, we determine the access nodes $u \in \overrightarrow{A}(s)$ that minimize $d(s, u) + d(u, b)$; we set $f_{s,u}^{\rightarrow}(r(b))$ to true. In addition, $f_{s,u}^{\rightarrow}(r(u))$ is set to true for each $s \in V$ and each access node $u \in \overrightarrow{A}(s)$ since each access node obviously minimizes the distance to itself. An analogous preprocessing step has to be done for the backward direction.

Query. In a query from s to t, we can take advantage of the precomputed flag vectors. First, we consider all backward access nodes of t and build the flag vector f_t such that $f_t(r(u)) = \mathsf{true}$ for each $u \in \overleftarrow{A}(t)$. Second, we consider only forward access nodes u of s with the property that the bitwise AND of $f_{s,u}^{\rightarrow}$ and f_t is not zero; we denote this set by $\overrightarrow{A}'(s)$; during this step, we also build the vector f_s such that $f_s(r(u)) = \mathsf{true}$ for each $u \in \overrightarrow{A}'(s)$. Third, we use f_s to determine the subset $\overleftarrow{A}'(t) \subseteq \overleftarrow{A}(t)$ analogously to the second step. Now, it is sufficient to perform only $|\overrightarrow{A}'(s)| \times |\overleftarrow{A}'(t)|$ table lookups. Note that determining $\overrightarrow{A}'(s)$ and $\overleftarrow{A}'(t)$ is in $O(a)$, in particular operations on the flag vectors can be considered as quite cheap.

Optimizations. Presumably, it is a good idea to just store the bitwise OR of the forward and backward flag vectors in order to keep the memory consumption within reasonable bounds. The preprocessing of the flag vectors can be accelerated by rearranging the columns of the distance table so that all border nodes are stored consecutively, which reduces the number of cache misses.

5   Experiments

In this section, we present an extensive experimental evaluation of our combined speed-up techniques in various scenarios and inputs. Our implementation is written in C++ (using the STL at some points). As priority queue we use a binary heap. The evaluation was done on two similar machines: An AMD Opteron 2218[1] and an Opteron 270[2]. The second machine is used for the combination of Transit-Node Routing and Arc-Flags, the first one for all other experiments. Note that the second machine is roughly 10% faster than the first one due to faster memory. All figures in this paper are based on 10 000 random s-t queries and refer to the scenario that only the lengths of the shortest paths have to be determined, without outputting a complete description of the paths. Efficient techniques for the latter have been published in [15,19].

---

[1] The machine runs SUSE Linux 10.1, is clocked at 2.6 GHz, has 16 GB of RAM and 2 x 1 MB of L2 cache. The DIMACS benchmark on the full US road network with travel time metric takes 6 013.6 s.

[2] SUSE Linux 10.0, 2.0 GHz, 8 GB of RAM, and 2 x 1 MB of L2 cache. The DIMACS benchmark: 5 355.6 s.

5.1   Road Networks

As inputs we use the largest strongly connected component[3] of the road networks of Western Europe, provided by PTV AG for scientific use, and of the US which is taken from the DIMACS Challenge homepage. The former graph has approximately 18 million nodes and 42.6 million edges. The corresponding figures for the USA are 23.9 million and 58.3 million, respectively. In both cases, edge lengths correspond to travel times. For results on the distance metric, see Tab. 4 in Appendix A.

CALT. In [20], we were able to improve query performance of ALT over [11] by improving the organization of landmark data. However, we do not compress landmark information and use a slightly better heuristic for landmark[4] selection. Hence, we report both results. By adding contraction—we use the one from [17] with $c = 3.0$ and $h = 30$—to ALT, we are able to reduce query time to 2.0 ms for Europe and to 4.9 ms for the US. This better performance is due to two facts. On the one hand, we may use more landmarks (we use 64) and on the other hand, the contraction reduces the number of hops of shortest paths. The latter observation is confirmed by the figures of CALT with 16 landmarks. Moreover, the most crucial drawback of ALT—memory consumption—can be reduced to a reasonable amount, even when using 64 landmarks. Still, CALT cannot compete with REAL or pure hierarchical methods, but the main motivation for CALT is its presumably easy dynamization.

CHASE. We report the figures for two variants of CHASE: the *economical* variant computes arc-flags only for a subgraph of 0.5% size of the input while for the *generous* variant, the subgraph H has a size of 5% of the input (with respect to number of nodes). We partition H with SCOTCH [21] into 128 cells.

For Europe, the economical variant only needs 7 additional minutes of preprocessing over pure CH and the preprocessed data is still smaller than the input. Recall that a negative overhead derives from the fact that the search graph is smaller than the input, see Section 1.1. This economical variant is already roughly 4 times faster than pure CH. However, by increasing the size of the subgraph H used as input for arc-flags, we are able to almost close the gap to pure Transit-Node Routing. CHASE is only 5 times slower than TNR (and is even *faster* than the grid-based approach of TNR [19]). However, the preprocessed data is much smaller for CHASE, which makes it more practical in environments with limited memory. Using the distance metric (cf. Tab. 4 in Appendix A), the gap between CHASE and TNR can be reduced even further. Remarkably, both pure Arc-Flags and CH perform much worse on distances than on travel times, whereas the combination CHASE performs—with respect to queries—very similarly on both metrics.

---

[3] For historical reasons, some quoted results are based on the respective original network that contains a few additional nodes that are not connected to the largest strongly connected component.

[4] 16 landmarks are generated by the maxCover algorithm, 64 are generated by avoid [11].

Table 1. Overview of the performance of various speed-up techniques, grouped by (1.) hierarchical methods [Highway Hierarchies (HH), highway-node routing based on HH (HH-HNR) and on Contraction Hierarchies (CH-HNR), Transit-Node Routing (TNR)], (2.) goal-directed methods [landmark-based A^* search (ALT), Arc-Flags (AF)], (3.) previous combinations, and (4.) the new combinations introduced in this paper. The additional overhead is given in bytes per node in comparison to *bidirectional* Dijkstra. Preprocessing times are given in minutes. Query performance is evaluated by the average number of settled nodes and the average running time of 10 000 random queries.

| | | Europe | | | | USA | | | |
| | | PREPRO. | | QUERY | | PREPRO. | | QUERY | |
method		time [min]	overhead [B/node]	#settled nodes	time [ms]	time [min]	overhead [B/node]	#settled nodes	time [ms]
Reach	[6]	83	17	4 643	3.47	44	20	2 317	1.81
HH	[3]	13	48	709	0.61	15	34	925	0.67
HH-HNR	[3]	15	2.4	981	0.85	16	1.6	784	0.45
CH-HNR	[9]	25	-2.7	355	0.18	27	-2.3	278	0.13
TNR	[19]	164	251	N/A	0.0056	205	244	N/A	0.0049
TNR	[9]	112	204	N/A	0.0034	90	220	N/A	0.0030
ALT-a16	[6]	13	70	82 348	160.3	19	89	187 968	400.5
ALT-m16	[20]	85	128	74 669	53.6	103	128	180 804	129.3
AF	[13]	2 156	25	1 593	1.1	1 419	21	5 522	3.3
REAL	[6]	141	36	679	1.11	121	45	540	1.05
HH*	[3]	14	72	511	0.49	18	56	627	0.55
SHARC	[17]	192	20	145	0.091	158	21	350	0.18
CALT-m16	2	16	8	3 017	3.9	26	8	7 079	8.3
CALT-a64	2	14	20	1 394	2.0	21	19	3 240	4.9
CHASE eco	3.1	32	0.0	111	0.044	36	-0.8	127	0.049
CHASE gen	3.1	99	12	45	0.017	228	11	49	0.019
ReachFlags	3.2	229	30	1 168	0.76	318	25	1 636	1.02
TNR+AF	4	229	321	N/A	0.0019	157	263	N/A	0.0017

Size of the Subgraph. The combination of Contraction Hierarchies and Arc-Flags allows a very flexible trade-off between preprocessing and query performance. The bigger the subgraph H used as input for Arc-Flags, the longer preprocessing takes but query performance decreases. Table 2 reports the performance of CHASE for different sizes of H in percentage of the original graph. Recall that 0.5% equals our economical variant, while 5% corresponds to the generous variant.

Two observations are remarkable: the effect of stall-on-demand ($\rightarrow$ Section 3.1) and the size of the subgraphs. While stall-on-demand pays off for pure CH, CHASE does not win from turning on this optimization. The number of settled nodes decreases but due to the overhead query times increase. Another very interesting observation is the influence of the input size for arc-flags. Applying goal-direction on a very high level of the hierarchy speeds up the query significantly. Increasing the size of H to 10% or even 20% yields a much higher preprocessing effort (both space and time) but query performance decreases only

Table 2. Performance of CHASE for Europe with stall-on-demand turned on and off running 10 000 random queries

		0.0%	0.5%	1.0%	2.0%	5.0%	10.0%	20.0%
	size of H							
Prepro.	time [min]	25	31	41	62	99	244	536
	space [Byte/n]	-2.7	0.0	1.9	4.9	12.1	22.2	39.5
Query	# settled	355	86	67	54	43	37	34
(with s-o-d)	time [μs]	180.0	48.5	36.3	29.2	22.8	19.7	17.2
Query	# settled	931	111	78	59	45	39	35
(without s-o-d)	time [μs]	286.3	43.8	30.8	23.1	17.3	14.9	13.0

slightly, compared to 5%. However, our fastest variant settles only 35 nodes on average having query times of 13 μs. Note that for this input, the average shortest path in its contracted form consists of 22 nodes, so only 13 unnecessary nodes are settled on average.

ReachFlags. We use $l = 2$ to determine the sub-graph for arc-flags preprocessing (cf. Section 3.2). We observe that it does not pay off to combine Arc-Flags—instead of landmarks—with REAL. Although query times are slightly faster than REAL, the search space is higher. One reason might be that our choice of parameters for Reach yield an increase in search space by roughly 20% compared to [6]. Still, it seems as if ReachFlags is inferior to CHASE which is mainly due to the good performance of Contraction Hierarchies.

TNR+AF. The fastest variant of Transit-node Routing *without* using flag vectors is presented in [9]; the corresponding figures are quoted in Tab. 1. For this variant, we computed flag vectors according to Section 4 using $k = 48$ regions. This takes, in the case of Europe, about two additional hours and requires 117 additional bytes per node. Then, the average query time is reduced to as little as 1.9 μs, which is an improvement of almost factor 1.8 (factor 2.9 compared to our first publication in [19]) and a speed-up compared to Dijkstra's algorithm of more than factor 3 *million*. The results for the US are even better.

The improved running times result from the reduced number of table accesses: in the case of Europe, on average only 3.1 entries have to be looked up instead of 40.9 when no flag vectors are used. Note that the runtime improvement is considerably less than a factor of 40.9 / 3.1 = 13.2 though. This is due to the fact that the average runtime also includes looking up the access nodes and dealing with local queries.

5.2   Robustness of Combinations

In the last section we focused on the performance of our combinations on road networks. However, existing combinations of goal-directed and hierarchical methods like REAL or SHARC are very robust to the input. Here, we evaluate our most promising combinations—CALT and CHASE—on various other inputs.

Table 3. Performance of bidirectional Dijkstra, ALT, CALT, CH, and economical CHASE on unit disk graphs with different average degree and grid graphs with different number of dimensions. Note that the we use the *aggressive* variant of Contraction Hierarchies, better results may be achieved by better input parameters.

	PREPRO time space [s] [B/n]		QUERY #settled nodes	PREPRO time space [s] [B/n]		QUERY #settled nodes	PREPRO time space [s] [B/n]		QUERY #settled nodes
unit disk	average degree 5			average degree 7			average degree 10		
bidir. Dijkstra	0	0	299 077	0	0	340 801	0	0	325 803
ALT-m16	490	128	10 051	514	128	10 327	566	128	11 704
CALT-m16	34	2	726	166	13	927	658	62	2 523
CALT-a64	32	7	689	135	29	670	511	137	992
CH-HNR	94	-13	236	1 249	-11	1 089	34 274	-4	2 475
CHASE	103	-12	66	1 368	-7	424	34 847	6	1 457
grid	2-dimensional			3-dimensional			4-dimensional		
bidir. Dijkstra	0	0	79 962	0	0	45 269	0	0	21 763
ALT-m16	65	128	2 362	100	128	1 759	133	128	1 335
CALT-m16	113	98	798	202	165	1 057	171	142	1 275
CALT-a64	60	211	458	101	386	557	129	487	774
CH-HNR	70	0	418	13 567	14	2 177	133 734	29	14 501
CHASE	73	2	274	13 585	22	2 836	133 741	32	30 848
railways	Berlin/Brandenburg			Ruhrgebiet			long distance		
bidir. Dijkstra	0	0	1 299 830	0	0	1 134 420	0	0	609 352
ALT-m16	604	128	56 404	556	128	60 004	291	128	30 021
CALT-m16	174	18	4 622	377	32	7 107	158	29	3 335
CALT-a64	123	45	2 830	191	68	4 247	87	63	2 088
CH-HNR	1 636	0	416	2 584	4	546	486	3	376
CHASE	2 008	2	125	2 863	7	244	536	5	229

We use time-expanded timetable networks[5], synthetic unit disk graphs[6] (1 000 000 nodes with an average degree of 5, 7, and 10), and grid graphs (2–4 dimensions with each having 250 000 nodes, edge weights picked uniformly at random between 1 and 1000.). The results can be found in Tab. 3.

For almost all inputs it pays off to combine goal-directed and hierarchical techniques. Moreover, CHASE works very well as long as the graph stays somehow sparse, only on denser graphs like 3- and 4-dimensional grids, preprocessing times increase significantly, which is mainly due to the contraction routine. Especially the last 20% of the graph take a long time to contract.

Concerning CALT, we observe that turning on contraction pays off—in most cases—very well: Preprocessing effort gets less with respect to time and space

---

[5] 3 networks: local traffic of Berlin/Brandenburg (2 599 953 nodes and 3 899 807 edges), local traffic of the Ruhrgebiet (2 277 812 nodes, 3 416 552 edges), long distance connections of Europe (1 192 736 nodes,1 789 088 edges).

[6] We obtain such graphs by arranging nodes uniformly at random on the plane and connecting nodes with a distance below a given threshold. As metric we use the distance according to the embedding.

while query performance improves. However, as soon as the graph gets too dense, e.g. 4-dimensional grids, the gain in performance is achieved by a higher amount of preprocessed data. The reason for this is that contraction works worse on dense graphs, thus the core is bigger. Comparing CALT and CHASE, we observe that CHASE works better or very sparse graphs while CALT yields better performance on denser graphs. So, it seems as if for dense graphs, it is better to stop contraction at some point and use a goal-directed technique on the core of the graph.

6   Conclusion

In this work, we systematically combine hierarchical and goal-directed speed-up techniques. As a result we are able to present the fastest algorithms for several scenarios and inputs. For sparse graphs, CHASE yields excellent speed-ups with low preprocessing effort. The algorithm is only overtaken by Transit-Node Routing in road networks with travel times, but the gap is almost closed. However, even Transit-Node Routing can be further accelerated by adding goal-direction. Finally, we introduce CALT yielding a good performance on denser graphs.

However, our study not only leads to faster algorithms but to interesting insights into the behavior of speed-up techniques in general. By combining goal-directed and hierarchical methods we obtain techniques which are very robust to the input. It seems as if hierarchical approaches work best on sparse graphs but the denser a graph gets, the better goal-directed techniques work. By combining both approaches the influence—with respect to performance—of the type of input fades. Hence, we were able to refine the statement given in [1]: Instead of blindly combining goal-directed and hierarchical techniques, our work suggest that for large networks, it pays off to drop goal-direction on lower levels of the hierarchy. Instead, it is better with respect to preprocessing (and query performance) to use goal-direction *only* on higher levels of the hierarchy.

Regarding future work, it may be interesting how the insight stated above can be used for graphs where hierarchical preprocessing fails. One could think of a technique that runs only a hierarchical query during the first phase and the second phase is only a goal-directed search, similar to CALT. For example, we could stop the construction of a contraction hierarchy at some point and apply Arc-Flags or ALT to the remaining core. We are optimistic that such a technique would even achieve very good results on dense graphs. Another open problem is the dynamization of CALT. We are confident, that CALT is very helpful in scenarios where edge updates occur very frequentely, e.g. dynamic timetable information systems.

Acknowledgments. We would like to thank Riko Jacob for interesting discussions on the combination of Transit-Node Routing and Arc-Flags. Moreover, we thank Robert Geisberger for helping us to use Contraction Hierarchies [9] in our work. He provided his implementation of [9] and some precomputed contraction hierarchies for various networks.

References

1. Holzer, M., Schulz, F., Wagner, D., Willhalm, T.: Combining Speed-up Techniques for Shortest-Path Computations. ACM J. of Exp. Algorithmics 10 (2006)
2. Dijkstra, E.W.: A Note on Two Problems in Connexion with Graphs. Numerische Mathematik 1, 269–271 (1959)
3. Schultes, D.: Route Planning in Road Networks. PhD thesis, Universität Karlsruhe (TH), Fakultät für Informatik (2008)
4. Dantzig, G.B.: Linear Programming and Extensions. Princeton University Press, Princeton (1962)
5. Gutman, R.J.: Reach-Based Routing: A New Approach to Shortest Path Algorithms Optimized for Road Networks. In: Proceedings of the 6th Workshop on Algorithm Engineering and Experiments (ALENEX 2004), pp. 100–111. SIAM (2004)
6. Goldberg, A.V., Kaplan, H., Werneck, R.F.: Better Landmarks Within Reach. In: Demetrescu, C. (ed.) WEA 2007. LNCS, vol. 4525, pp. 38–51. Springer, Heidelberg (2007)
7. Schulz, F., Wagner, D., Weihe, K.: Dijkstra's Algorithm On-Line: An Empirical Case Study from Public Railroad Transport. ACM J. of Exp. Algorithmics 5 (2000)
8. Holzer, M., Schulz, F., Wagner, D.: Engineering Multi-Level Overlay Graphs for Shortest-Path Queries. In: Proceedings of the 8th Workshop on Algorithm Engineering and Experiments (ALENEX 2006), SIAM (2006)
9. Geisberger, R., Sanders, P., Schultes, D., Delling, D.: Contraction Hierarchies: Faster and Simpler Hierarchical Routing in Road Networks. In: Proceedings of the 7th Workshop on Experimental Algorithms (WEA 2008). LNCS, Springer, Heidelberg (2008)
10. Bast, H., Funke, S., Sanders, P., Schultes, D.: Fast Routing in Road Networks with Transit Nodes. Science 316, 566 (2007)
11. Goldberg, A.V., Werneck, R.F.: Computing Point-to-Point Shortest Paths from External Memory. In: Proceedings of the 7th Workshop on Algorithm Engineering and Experiments (ALENEX 2005), pp. 26–40. SIAM (2005)
12. Lauther, U.: An Extremely Fast, Exact Algorithm for Finding Shortest Paths in Static Networks with Geographical Background. In: Geoinformation und Mobilität - von der Forschung zur praktischen Anwendung, vol. 22, pp. 219–230. IfGI prints (2004)
13. Hilger, M.: Accelerating Point-to-Point Shortest Path Computations in Large Scale Networks. Master's thesis, Technische Universität Berlin (2007)
14. Wagner, D., Willhalm, T., Zaroliagis, C.: Geometric Containers for Efficient Shortest-Path Computation. ACM J. of Exp. Algorithmics 10, 1.3 (2005)
15. Delling, D., Sanders, P., Schultes, D., Wagner, D.: Highway Hierarchies Star. In: 9th DIMACS Implementation Challenge - Shortest Paths (2006)
16. Sanders, P., Schultes, D.: Engineering Highway Hierarchies. In: Azar, Y., Erlebach, T. (eds.) ESA 2006. LNCS, vol. 4168, pp. 804–816. Springer, Heidelberg (2006)
17. Bauer, R., Delling, D.: SHARC: Fast and Robust Unidirectional Routing. In: Proceedings of the 10th Workshop on Algorithm Engineering and Experiments (ALENEX 2008), pp. 13–26. SIAM (2008)
18. Schieferdecker, D.: Systematic Combination of Speed-Up Techniques for exact Shortest-Path Queries. Master's thesis, Universität Karlsruhe (TH) (2008)
19. Bast, H., Funke, S., Matijevic, D., Sanders, P., Schultes, D.: In Transit to Constant Shortest-Path Queries in Road Networks. In: Proceedings of the 9th Workshop on Algorithm Engineering and Experiments (ALENEX 2007), pp. 46–59. SIAM (2007)

20. Delling, D., Wagner, D.: Landmark-Based Routing in Dynamic Graphs. In: Demetrescu, C. (ed.) WEA 2007. LNCS, vol. 4525, pp. 52–65. Springer, Heidelberg (2007)
21. Pellegrini, F.: SCOTCH: Static Mapping, Graph, Mesh and Hypergraph Partitioning, and Parallel and Sequential Sparse Matrix Ordering Package (2007)

A   Further Experiments

Table 4. Overview on the performance of prominent speed-up techniques and combinations analogous to Tab. 1 but with travel distances as metric

method		Europe				USA			
		PREPRO.		QUERY		PREPRO.		QUERY	
		time [min]	overhead [B/node]	#settled nodes	time [ms]	time [min]	overhead [B/node]	#settled nodes	time [ms]
Reach	[6]	49	15	7 045	5.53	70	22	7 104	5.97
HH	[3]	32	36	3 261	3.53	38	66	3 512	3.73
CH-HNR	3.1	89	-0.1	1 650	4.19	57	-1.2	953	1.50
TNR	[3]	162	301	N/A	0.038	217	281	N/A	0.086
ALT-a16	[6]	10	70	276 195	530.4	15	89	240 750	430.0
ALT-m16	2	70	128	218 420	127.7	102	128	278 055	166.9
AF	[13]	1 874	33	7 139	5.0	1 311	37	12 209	8.8
REAL	[6]	90	37	583	1.16	138	44	628	1.48
HH*	[3]	33	92	1 449	1.51	40	89	1 372	1.37
SHARC	[17]	156	26	4 462	2.01	-	-	-	-
CALT-m16	2	17	8	6 453	8.1	20	8	9 034	11.0
CALT-a64	2	14	19	2 958	4.2	15	19	4 015	5.6
CHASE eco	3.1	224	7.0	175	0.156	185	2.5	148	0.103
CHASE gen	3.1	1 022	27	67	0.064	1 132	18	63	0.043
ReachFlags	3.2	516	31	5 224	4.05	1 897	27	6 849	4.69

Contraction Hierarchies: Faster and Simpler Hierarchical Routing in Road Networks⋆

Robert Geisberger, Peter Sanders, Dominik Schultes, and Daniel Delling

Universität Karlsruhe (TH), 76128 Karlsruhe, Germany
{robert.geisberger,sanders,schultes,delling}@ira.uka.de

Abstract. We present a route planning technique solely based on the concept of node *contraction*. The nodes are first ordered by 'importance'. A hierarchy is then generated by iteratively *contracting* the least important node. Contracting a node v means replacing shortest paths going through v by *shortcuts*. We obtain a hierarchical query algorithm using bidirectional shortest-path search. The forward search uses only edges leading to more important nodes and the backward search uses only edges coming from more important nodes. For fastest routes in road networks, the graph remains very sparse throughout the contraction process using rather simple heuristics for ordering the nodes. We have five times lower query times than the best previous hierarchical Dijkstra-based speedup techniques and a *negative* space overhead, i.e., the data structure for distance computation needs *less* space than the input graph. CHs can be combined with many other route planning techniques, leading to improved performance for many-to-many routing, transit-node routing, goal-directed routing or mobile and dynamic scenarios.

1 Introduction

Planning optimal routes in road networks has recently attracted considerable interest in algorithm engineering because it is an important application that admits a lot of interesting algorithmic approaches. Many of these techniques exploit the *hierarchical* nature of road networks in some way or another.

Here we present a very simple approach to hierarchical routing. Assume the nodes of a weighted directed graph $G = (V, E)$ are numbered $1..n$ in order of ascending 'importance'. We now construct a hierarchy by *contracting* the nodes in this order. A node v is contracted by removing it from the network in such a way that shortest paths in the remaining *overlay graph* are preserved. This property is achieved by replacing paths of the form $\langle u, v, w \rangle$ by a *shortcut* edge $\langle u, w \rangle$. Note that the shortcut $\langle u, w \rangle$ is only required if $\langle u, v, w \rangle$ is the only shortest path from u to w.

⋆ Partially supported by DFG grant SA 933/1-3, and by the Future and Emerging Technologies Unit of EC (IST priority – 6th FP), under contract no. FP6-021235-2 (project ARRIVAL).

C.C. McGeoch (Ed.): WEA 2008, LNCS 5038, pp. 319–333, 2008.

We shall view the contraction process as a way to add all discovered shortcuts to the edge set E. We obtain a *contraction hierarchy (CH)*. Section 2 gives more details.

In Section 3 we explain how the nodes are ordered. Although 'optimal' node ordering seems a quite difficult problem, already very simple local heuristics turn out to work quite well. The basic idea is to keep the nodes in a priority queue sorted by some estimate of how attractive it is to contract a node. The main ingredient of this heuristic estimate is the *edge difference*: The number of shortcuts introduced when contracting v minus the number of edges incident to v. The intuition behind this is that the contracted graph should have as few edges as possible. Even using only edge difference, quite good CHs are computed. However, further refinements are useful. In particular, it is important to contract nodes 'uniformly'.

For routing, we split the CH (V, E) into an *upward graph* $G_\uparrow := (V, E_\uparrow)$ with $E_\uparrow := \{(u, v) \in E : u < v\}$ and a *downward graph* $G_\downarrow := (V, E_\downarrow)$ with $E_\downarrow := \{(u, v) \in E : u > v\}$. For a shortest path query from s to t, we perform a modified bidirectional Dijkstra shortest path search, consisting of a forward search in $G_\uparrow$ and a backward search in $G_\downarrow$. If, and only if, there exists a shortest s-t-path in the original graph, then both search scopes eventually meet at a node v that has the highest order of all nodes in a shortest s-t-path. More details of the query algorithm are given in Section 4. Applications and refinements like dynamic routing (i.e., edge weights are allowed to change), many-to-many routing, and combinations with other speedup techniques can be found in Section 5. Section 6 shows that in many cases, we get significant improvements over previous techniques for large real world inputs. Lessons learned and possible future improvements are summarized in Section 7.

Related Work

Since there has recently been extensive work on speed-up techniques, we can only give a very abridged overview with emphasis on the directly related techniques beginning with the closest kin. For a more detailed overview we refer to [1,2]. CHs are an extreme case of the hierarchies in highway-node routing (HNR) [3,2] – every node defines its own level of the hierarchy. CHs are nevertheless a new approach in the sense that the node ordering and hierarchy construction algorithms used in [3,2] are only efficient for a small number of geometrically shrinking levels. We also give a faster and more space efficient query algorithm using $G_\uparrow$ and $G_\downarrow$.

The node ordering in highway-node routing uses levels computed by *highway hierarchies* (HHs) [4,5,2]. Our original motivation for CHs was to simplify HNR by obviating the need for another (more complicated) speedup technique (HHs) for node ordering. HHs are constructed by alternating between two subroutines: *Edge reduction* is a sophisticated and relatively costly routine that only keeps edges required 'in the middle' of 'long-distance' paths. *Node reduction* contracts nodes. In the original paper for undirected HHs [5], node reduction only contracted nodes of degrees one and two, i.e., it removed attached trees and multihop paths. We originally viewed node contraction as a mere helper for the main work-horse edge reduction. For directed graphs [5], we needed a more general criterion which nodes should be contracted away. It turned out that the edge difference is a good way to estimate the cost of contracting a node v. In [6,7] this method is further refined to use a priority queue and to avoid parallel edges. All previous approaches to contraction had in common that the average degree of the nodes in the overlay graph would eventually explode. So it looked like an additional technique such as edge reduction or reaches would be a necessary ingredient of any high-performance hierarchical routing method. Perhaps the most important result of CHs is that using *only* (a more sophisticated) node contraction, we get very good performance.

The fastest speedup technique so far, *transit-node routing* [8,2], offers a factor up to 40 times better query times than CHs. However, it needs considerably higher preprocessing time and space, is less amenable to dynamization, and, most importantly it relies on another hierarchical speedup technique for its pre-processing. We have preliminary evidence that using CHs for this purpose leads to improved performance.

Finally, there is an entirely different family of speedup techniques based on goal-directed routing. Combination of CHs with goal-directed routing is the subject of another paper [9] that systematically studies such combinations.

2   Contraction

Recall from the introduction that when contracting node v, we are dealing with an overlay graph $G' = (V', E')$ with $V' = v..n$ and an edge set E' that preserves shortest path distances wrt the input graph. In G', we face the following many-to-many shortest-path problem: For each source node $u \in v + 1..n$ with $(u, v) \in E'$ and each target node $w \in v + 1..n$ with $(v, w) \in E'$, we want to compare the shortest-path distance $d(u, w)$ with the shortcut length $c(u, v) + c(v, w)$ in order to decide whether the shortcut is really needed. A simple way to implement this is to perform a forward shortest-path search in the current overlay graph G' from each source, ignoring node v, until all targets have been found. We can also stop the search from u when it has reached distance $d(u, v) + \max \{c(v, w) : (v, w) \in E'\}$.

Our actual implementation uses a simple, asymmetric form of bidirectional search inspired by [10]: For each target node w we perform a single-hop backward search. For each edge $(x, w) \in E'$ we store a *bucket entry* $(c(x, w), w)$ with node x. This way, forward search from u can be limited to distance

$$c(u, v) + \max_{w:(v,w)\in E'} c(v, w) - \min_{x:(x,w)\in E'} c(x, w) \ .$$

When reaching a node x, we scan its bucket entries. For each entry (C, w), we can infer that there is a path from u to w of length $d(u, x) + C$.

Since exact shortest path search for contraction can be rather expensive, we have implemented two ways to limit the range of searches: We can limit the number of hops (edges) used in any path $\langle u, \ldots, w \rangle$, and we can limit the total search space size of a forward search. Note that this has no influence on the correctness of subsequent queries in the CH as long as we make sure to always insert a shortcut (u, w) when we have not found a path from u to w witnessing that the shortcut is unnecessary. Also note that for hop limit two, our bidirectional approach obviates a full fledged Dijkstra search. It suffices to scan the edges leaving a source node u.

Let us now focus the discussion on the hop limit. We get a tradeoff between fast contraction 'now' for small hop limits and a more sparse graph with better query time and possible easier contraction 'later' for a large hop limit. In our experiments it turned out, that it makes sense to start with a hop limit as small as one and to later increase it. We switch from one hop limit to the next when the average degree of the overlay graph G' exceeds a specified bound.

3   Node Ordering

As already mentioned in the introduction, our basic approach uses a priority queue whose minimum element contains the node looking most attractive to be contracted next. The priority used is a linear combination of several terms. In addition to the single terms used, the linear coefficients of the different terms are important, some of them can be found in Section 6. In this section we focus on different possible terms. One difficulty with this approach is that when node v is contracted, this might affect the priorities of other nodes. We use several techniques to handle this problem:

- We use *lazy update*, i.e., before actually contracting v, we update its priority. If it now exceeds the priority of the second largest element v', we reinsert v and continue with v'. This process is repeated until a consistent minimum is found. Note that (at least wrt the result of node ordering) lazy update obviates immediate updates when a priority *increases*.
- We recompute the priority of the neighbors of v.
- We periodically reevaluate all priorities and rebuild the priority queue.

Edge Difference. Arguably the most important term is the edge difference. For computing it, node ordering uses the same heuristics for limiting search spaces as are later used in the actual contraction.[1]

Uniformity. Using only the edge difference, one can get quite slow routing. For example, if the the input graph is a path, contraction would produce a linear hierarchy where most queries would again follow paths of linear length. In contrast, if we iteratively contract maximal independent sets, we would get a hierarchy where any query is finished in logarithmic time.

---

[1] Updating neighbors of contracted nodes and lazy update 'almost' suffice to keep the priorities up to date wrt the edge difference. However, with some highly constructed example, not all priorities are updated in time when the search horizon is limited.

More generally, it seems to be a good idea, to contract nodes everywhere in the graph in a uniform way, rather than keep contracting nodes in a small region. We have tried several heuristics for choosing nodes uniformly out of which we present the two most successful ones. For all measures used here, a large value means that the node is contracted late.

Deleted Neighbors: We count the number of neighbors that have already been contracted. This includes neighbors reached via shortcuts. Obviously, this quantity can be maintained correctly by either lazy update or by updating the neighbors of a contracted node. This heuristics is very simple and can be computed efficiently.

Voronoi Regions: Define the Voronoi-Region $R(v)$ of a node v in an overlay graph as the set of nodes in the input graph that are closer to v than to any other node in the overlay graph. We use the square root of the size of the Voronoi-region as a term in the priority function. By preferably contracting small Voronoi regions, we can hope that the nodes of the overlay graph are spread uniformly over the network. When v is contracted, its neighboring Voronoi regions will 'eat up' $R(v)$. The necessary computations can be made using $O(|R(v)|)$ steps of Dijkstra's algorithm [11]. If we always contract Voronoi regions of size at most a constant times the average region size, we can easily show that the total number of Dijkstra-steps for maintaining the size of the Voronoi regions is $O(n \log n)$, i.e., computing Voronoi regions is reasonably efficient. Since Voronoi regions can only grow, lazy update ensures that the priority queue works correctly wrt this term of the priority function.

There are a number of further, optional parameters of the priority function that turn out to further improve the hierarchy at the cost of increased time for node ordering.

Cost of contraction. A time consuming part of the contraction are the forward shortest-path searches to decide the necessity of shortcuts. So for example, we can use the sum of these search space sizes as a priority term. Note that this quantity can change beyond the direct neighborhood of the contracted node, i.e., our update rules are only heuristics.

Cost of queries. One can try to estimate how contracting nodes affects the size of query search spaces. We have implemented the following simple estimate $Q(v)$ that can be shown to be an upper bound for the number of hops of a path $\langle s, \ldots, v \rangle$ explored during a query: Initially, $Q(v) = 0$. When v is contracted then for each neighbor u of v, $Q(u) := \max(Q(u), Q(v) + 1)$.

Global measures. We can prefer contracting globally unimportant nodes based on some path based centrality measure such as (approximate) betweenness [12] or reach [13,6].

Generally speaking, one can come up with many heuristic terms. But one gets an inflation of tuning parameters. Therefore, in the experiments we try to keep the number of actually used terms small, we use the same set of parameters for different inputs, and we make some sensitivity analysis to find out how robust the parameter choices are.

4   Query

In the introduction we have already outlined the basic approach which we shall now describe in more detail. An algorithm that already works quite well performs complete Dijkstra searches from s in $G_\uparrow$ and from t in $G_\downarrow$. We have

Lemma 1. $d(s,t) = \min\{d(s,v) + d(v,t) : v \text{ is settled in both searches}\}.$

Proof. We only give a proof outline for self-containedness since the CH-query is a special case of the HNR-query for which a detailed yet simple correctness proof is given in [2]. In particular, here we only consider the case where shortest paths are unique.

Let v denote the largest[2] node on the shortest path P from s to t. We first claim that the sequence of prefix maxima[3] of P forms the shortest path from s to v in the upward graph $G_\uparrow$. If $s = v$ there is nothing to prove. Otherwise, consider any pair (u,w) of subsequent prefix maxima in P and the overlay graph $G' = (u..n, E')$ existing at some point during contraction. Since the shortest path from u to w uses only interior nodes smaller than u, and by definition of the properties of an overlay graph, $(u,w) \in E'$ and $c(u,w) = d(u,w)$. Moreover, $u < w$ and hence $(u,w) \in G_\uparrow$. Analogously, the sequence of suffix maxima of P forms the shortest path from v to t in the downward graph.   □

There are two refinements to the complete search algorithm (that are also analogous to the HNR-query algorithm [3,2]). The query alternates between forward and backward search. Whenever we settle a node in one direction that is already settled in the other direction, we get a new candidate for a shortest path. Search is aborted in one direction if the smallest element in the queue is at least as large as the best candidate path found so far. This does not affect correctness, since additional settled nodes in this direction cannot possibly contribute to better solutions.

We also prune the search space using the *stall-on-demand* technique: Before a node v is settled at distance $d(v)$ in the forward search, it uses the information available in $G_\downarrow$ to inspect downward edges (w,v) with $w > v$. If $d(w)+c(w,v) < d(v)$, then the search can be stopped (*stalled*) at v with stalling distance $d(w)+c(w,v)$ since the computed distance to v is suboptimal so that a continuation of the search from v would be futile. Such stalled nodes are settled but their incident edges are not relaxed, leading to a considerably smaller search space. Moreover, stalling can propagate to further nodes x in the neighborhood of v, if the path over w in G to x is shorter than the currently found path to x in

---

[2] Recall that nodes are considered to be numbered during node ordering.
[3] i.e., the sequence of nodes u_i on $P = \langle s = u_1, u_2, \ldots, u_k = t \rangle$ with the property that $u_i > \max\{u_1, u_2, \ldots, u_{i-1}\}$.

$G_\uparrow$. We perform a local BFS from v using the edges available in $G_\uparrow$ or $G_\downarrow$.[4] The search stops at nodes that are not being stalled. To ensure correctness, we unstall a node x if a shorter path in $G_\uparrow$ to x than the current one in $G_\uparrow$ is found. Stall-on-demand is also applied to the backward search in the same way.

The graphs $G_\uparrow$ and $G_\downarrow$ can be stored in one data structure, using two direction flags for each edge to indicate whether it belongs to $G_\uparrow$ or $G_\downarrow$. Irrespective of the direction flags, each edge (u, v) is stored only once, namely at the smaller node, which complies with the requirements of both forward and backward search (including the stall-on-demand technique). In particular, this also applies to undirected edges $\{u, v\}$ with the same weight in both directions. In contrast, an efficient implementation of Dijkstra's (even unidirectional) algorithm needs to store such undirected edges $\{u, v\}$ both at u and v. This is the reason why we may need less space than Dijkstra's algorithm for the original graph, even though we have to insert shortcuts.

Outputting Paths. As all routing techniques that use shortcuts, we need a way to unpack them in order to obtain a shortest path in the input graph. This is particularly simple for CHs since each shortcut (u, w) bypasses exactly one node v. We therefore obtain a simple recursive unpacking routine. In order to implement this efficiently, we need to store v together with the shortcut somewhere. Note that this information is *not* easily obtained just from $G_\uparrow$ or $G_\downarrow$, i.e., our observation that we may need less space than the input graph only holds when path unpacking is not required.

5   Applications

Changing all Edge Weights. In CHs we can distinguish between two main phases of preprocessing, node ordering and hierarchy construction. Similar to highway-node routing, we do not have to redo node ordering when the edge weights change – for example when we switch from driving times for a fast car to a slow truck. Hierarchy construction ensures correctness for *all* node orderings. We will see that the resulting hierarchies are almost as good as hierarchies where node ordering has been repeated. The intuition behind this is that most important nodes remain important even if the actual edge weights change – both sports cars and trucks are fastest on the motorway.

Changing some Edge Weights. Since CHs are a special case of HNR [3,2], we can also adopt the successful approaches used there for routing in presence of some changed edges (e.g., due to traffic jams).

Many-to-Many Routing. In [10] we developed an algorithm based on highway hierarchies that finds all shortest path distances between a set S of source nodes and a set T of target nodes. The idea is to perform only $|T|$ backward searches,

---

[4] We also have a version that additionally exploits the parent pointers of the shortest path tree. This slightly decreases search space but slightly *increases* query time.

store the resulting search spaces appropriately and then to perform $|S|$ forward searches that use the stored information on the backward searches to find the shortest path distances. As explained in [2], this works for a large family of non-goal-directed hierarchical routing techniques including highway-node routing and reach-based routing [13,6]. CHs are particularly well suited for many-to-many routing because they have very small search spaces and because for the backward search spaces *we only need to store nodes that are not stalled.*

Distance Oracles for Replacing Large Distance Tables. CH search-spaces are so small that we can drop the distance tables computed by many-to-many routing and instead store the search spaces from S and T as arrays of node-distance pairs sorted by node-id. Then an *s-t* query amounts to intersecting the search spaces for s and t and computing the minimum resulting distance. This intersection operation is similar to binary merging and thus runs very fast and cache efficiently.

Transit-Node Routing. Transit-node routing [8,2] is currently the fastest static routing technique available. Its main disadvantage compared to simpler techniques is that it needs considerably more preprocessing time. The preprocessing for transit-node routing is essentially a generalization of many-to-many routing. Hence, we can also do preprocessing using CHs and expect to obtain an improvement. We can use the nodes designated as most important by node ordering to define the sets of transit nodes. The edge difference criterion used by node ordering might help to identify transit-node sets that imply small sets of access nodes.

Combination with Other Speedup Techniques. There are interesting synergies between hierarchical speedup techniques and goal-directed methods such as landmark A^* [6] or arc flags [14,15]. Goal-directed techniques become cheaper in terms of preprocessing time and space if they are only applied to a *core* obtained after some contraction [16,17,6,7]. Since CHs are a fast, flexible, effective, and very fine-grained approach to this contraction, they seem best suited for this. The resulting overall query time is often better than any of the techniques alone. For example, an integration of CHs and arc-flags is so fast that it almost achieves the query times of transit-node routing using less space [9]. Another interesting example is SHARC-routing [7] which applies a sophisticated, multilevel variant of arc-flags to an network enriched with shortcuts. This has the advantage that it yields a unidirectional, very simple query algorithm that takes hierarchy into account indirectly via the arc flags.

Perhaps most importantly, not all graph families are as well behaved as road networks with travel time weights with respect to contraction. So it sometimes seems to be the best idea to stop contraction at some point and solely rely on goal-directed techniques for the core [9].

Node contraction started out as an ingredient of highway hierarchies (HHs). It would be interesting to see how good HHs would perform if we would reintegrate CHs into HHs. We could expect a more sparse network in the upper levels but also a more complicated, less focused query algorithm. Our guess would be that for road networks, we cannot expect an additional improvement but perhaps we should keep this approach in mind for network where contraction does not work so well.

Similarly, we could integrate CHs with reach-based routing [13,6]. CHs could contribute the shortcuts to be used, possibly simplifying the reach approximations during preprocessing. During the query, we could use reach values to prune the search additionally.

Implementation on Mobile Devices. Due to its small memory overhead and search space, CHs are a good starting point for route planning on mobile devices. This is the subject of a separate paper [18].

6   Experiments

Environment. Experiments have been done on one core of a single AMD Opteron Processor 270 clocked at 2.0 GHz with 8 GB main memory and 2×1 MB L2 cache, running SuSE Linux 10.3 (kernel 2.6.22). The program was compiled by the GNU C++ compiler 4.2.1 using optimization level 3.

Test Instances. Our experiments in this section have been done on a road network of Western Europe[5] with 18 029 721 nodes and 42 199 587 directed edges, which has been made available for scientific use by the company PTV AG. For each edge, its length and one out of 13 road categories (e.g., motorway, national road, regional road, urban street) is provided so that an expected travel time can be derived, which we use as edge weight. Results for other test instances can be found in the full paper.

Different Variants. Although the basic idea of CHs is simple, we have many tuning parameters that should be set carefully and we should verify that these choices are robust in the sense that they work reasonably well for different instances. Therefore, we build up the system incrementally. Tab. 1 shows the most fundamental performance parameters for a number of increasingly sophisticated variants. For comparison, we add the times for the fastest variant of highway-node routing (HNR) from [3] using the same system environment. Note that this version of HNR outperforms all previous speedup techniques with comparable preprocessing time so that focusing on HNR is meaningful.

Already using only the edge difference we obtain query times better than HNR. However, the preprocessing time and space is quite large. Just adding the uniformity parameter based on number of deleted neighbors (Line ED), we obtain more than four times better query time than HNR. The time for hierarchy construction becomes better than HNR once we take the search space size into

---

[5] Austria, Belgium, Denmark, France, Germany, Italy, Luxembourg, the Netherlands, Norway, Portugal, Spain, Sweden, Switzerland, and the UK.

Table 1. Performance of various node ordering heuristics. Terms of the priority function: E=edge difference, D=deleted neighbors, S=search space size, W=relative betweenness, V=$\sqrt{\text{Voronoi region size}}$, L=limit[8] search space on weight calculation, Q=upper bound on edges in search paths. Digits denote hop limits for testing short-cuts. Space overhead is wrt an adjacency array for *bidirectional* Dijkstra that stores each directed edge at both endpoints. The bottom line shows the performance for highway-node routing using the code from [3].

method	node ordering [s]	hierarchy construction [s]	query [μs]	nodes settled	non-stalled nodes	edges relaxed	space overhead [B/node]
E	13010	1739	670	1791	1127	4999	-1.6
ED	7746	1062	183	403	236	1454	-2.3
ES	5355	123	245	614	366	1803	**-3.5**
ESL	1158	123	292	758	465	2169	**-3.5**
EDL	2071	576	187	418	243	1483	-2.3
EDSL	1414	165	175	399	228	1335	-2.6
ED5	634	98	224	470	250	1674	-1.6
EDS5	652	99	213	462	256	1651	-2.1
EDS1235	**545**	**57**	223	459	234	1638	0.6
EDSQ1235	591	64	211	440	236	1621	1.0
EDSQL	1648	199	173	385	220	1378	-2.1
EVSQL	1627	170	159	368	209	1181	-2.7
EDSQWL	1629	199	163	372	218	1293	-2.5
EVSQWL	1734	180	**154**	**359**	**208**	**1159**	-3.0
HNR	594	203	802	957	630	7561	9.5

account (letter S). This also improves node ordering if we limit the size of a local search (letter L).

To improve the preprocessing times, it helps to limit the number of hops in the searches during preprocessing and to take search space sizes for contraction into account. Figure 1 shows the development of the average degree during node contraction for different hop limits. We see that for hop limits below four, the average degree eventually explodes. We choose limits for the average degree that switch to a larger hop limit sufficiently before this explosion.[6] Interestingly, this also further improves query time. The algorithm in Line EDS1235 of Tab. 1 outperforms HNR in all respects and with a wide margin with respect to query time and hierarchy construction[7] time. As explained in Section 5, the latter time is particularly interesting when we want to exchange the edge weight function. We use this variant as our main *economical*[8] variant for further experiments.

---

[6] 1 → 2 hops @ degree 3.3, 2 → 3 @ 10, 3 → 5 @ 10. After switching to hop limit 3, we remove all edges e for which there is a witness with at most 3 edges that e is not a shortest path. This reduces the average degree and leaves some time before we have to switch to hop limit 5.

[7] There is a version of HNR in [3] with about two times faster hierarchy construction but with slower queries and more space consumption.

[8] Coefficients for priority: E=190, D=120, S=1.

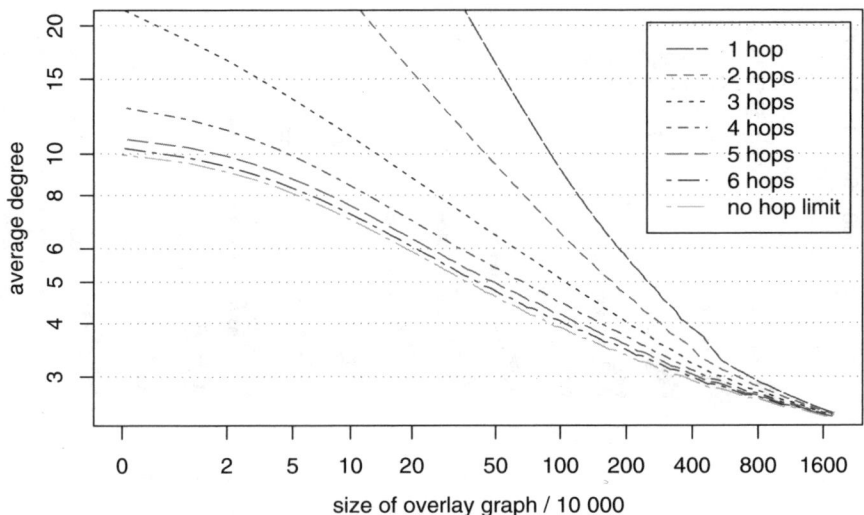

Fig. 1. Average degree development for different hop limits

By investing more preprocessing time, we can further improve the query performance. We abandon hop limitations and take the path-length estimate $Q(v)$ into account. The resulting algorithm, Line EVSQL in Tab. 1, is used as our *aggressive*[9] variant for further experiments. Using betweenness[10] approximations (letter W) can improve the query time by additional 3%.[11] It is interesting to compare different indicators for query performance between aggressive CH and HNR. CHs are 5 times faster although the number of settled nodes is only 2.6 times smaller. This is in part due to a simpler data structure[12] and in part due to a far larger improvement (factor 6.4) wrt the number of relaxed edges. For many-to-many routing, we are mostly interested in the number of non-stalled nodes, which make the bucket-scan operations more expensive. In this respect, CHs are a factor 3 better.

Local Queries. Since random queries are unrealistic for large graphs, Fig. 2 shows the distributions of query times for various degrees of locality [4]. We see a uniform improvement over HNR and small fluctuations in query time. This is further underlined in Fig. 3 where we give upper bounds for the search space size of *all* $n \times n$ possible queries (see [5] for the algorithm). We see a superexponential decay of the probability to observe a certain search-space size and maximal search-space size bound less than 2.5 times the size of the average actual search-space sizes (see also Tab. 1).

---

[9] Coefficients for priority: E=190, V=60, S=1, Q=145, L=1000.

[10] The execution times for betweenness approximation [12] are not included in Tab. 1.

[11] Preliminary experiments with reach-approximations were not successful.

[12] The HNR implementation from [3] has to compare level information to find out which edges should be relaxed.

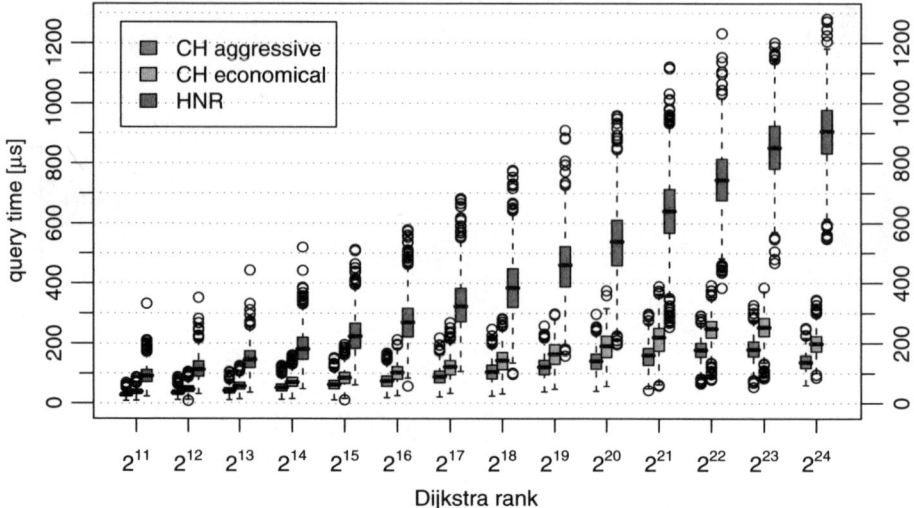

Fig. 2. Local queries, box-and-whisker plot [19]: each box spreads from the lower to the upper quartile and contains the median, the whiskers extend to the minimum and maximum value omitting outliers, which are plotted individually. The queries generated for x-value r are random s–t-queries under the constraint that t is the r-th node visited by Dijkstra's algorithm (see also [4]).

Unpacking Paths needs an average of $317\,\mu s$ for the aggressive variant and $332\,\mu s$ for the economical variant. The difference between the two variants is bigger for the space overhead which is 5.8 B/node and 10.8 respectively. Among the path unpacking times we have seen, this is only outperformed by the fastest variant for highway hierarchies in [5] that explicitly stores completely unpacked representations of the most important shortcuts. Note that this optimization works for any shortcut-based speedup technique including CHs.

Many-to-Many Routing for a random 10000×10000 table using the aggressive variant needs 10.2 s. This is about six times faster than the highway-hierarchy-based code from [10] and more than twice as fast as the HNR-based implementation from [2]. Our current implementation of many-to-many routing does not (yet) use the asymmetry between forward and backward search that has proved useful in [10,2]. Hence, we can expect further improvements.

Exchanging the Edge Weight Function. The table below shows the hierarchy construction time and query time using our economical variant for different speed profiles which come from the company PTV (see also [3]). The times in brackets refer to the case when node ordering was done with the same speed profile and the main times are for the case that node ordering was done for our default speed profile.

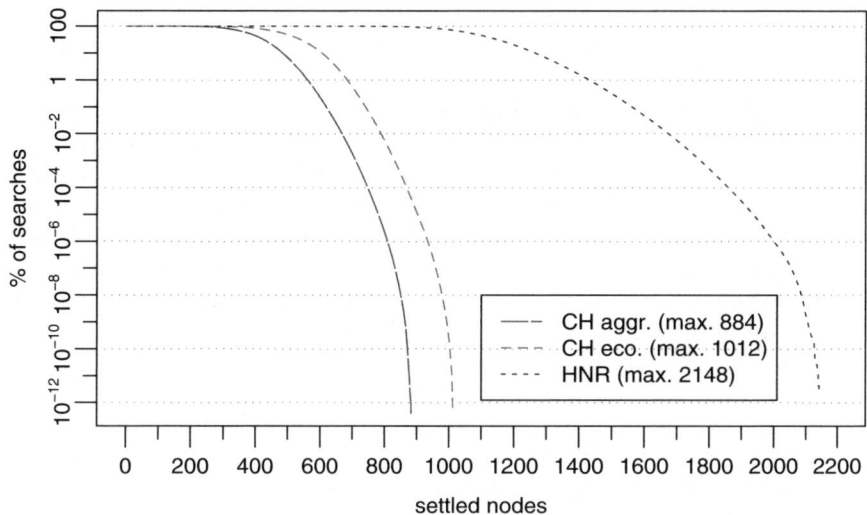

Fig. 3. Upper bound on search space in settled nodes for the worst percentages of queries

	default	fast car	slow car	slow truck
hier. construction [s]	57	80 (62)	82 (63)	88 (65)
query [μs]	223	208 (211)	232 (243)	294 (291)

We can see that preprocessing time goes up by about 30 %. Query times are about the same. Query performance decreases with the speed of the vehicle since the hierarchy induced by fast streets gets less pronounced.

Transit-Node Routing. We used the node ordering with the aggressive variant of CHs to determine the transit-node sets for the implementation from [2]. As we hoped for, this resulted in a reduced number of access nodes, which in turn results in better query time ($4.3 \rightarrow 3.4\,\mu s$) and lower space consumption ($247 \rightarrow 204$ Byte/node), compared to [2]. Preliminary experiments suggest that we get further improvements with an additional term for node ordering that takes into account the number of edges of the input graph that make up a shortcut. We have not yet implemented a CH-based preprocessing so that it is too early to judge the effect of CHs on preprocessing time. It is quite likely however, that we will also see an improvement in preprocessing time.

7   Conclusions

CHs are a simple and efficient basis for many hierarchical routing methods in road networks. The experiments in [9] suggest that CHs also work well for other sparse networks with high locality such as transportation networks, or sparse unit-disk graphs. For more dense networks, CHs can be used for an initial contraction phase whereas a goal-directed technique is applied to the resulting core network.

Several further improvements might be possible. The performance of node ordering is so far only slightly better than the HH based method used in [3] for HNR. One reason is that we perform many similar searches that might be saved if we would reuse search spaces. The main problem with reuse is that storing search spaces would cost a lot of space. But perhaps we can partition large networks into smaller networks; perform the node ordering separately for each subnetwork; and only then merge the pieces into a global order. To a lesser extend such an optimization might also accelerate hierarchy construction. As a side effect we might also obtain a way to update the search space sizes of all nodes affected by a node contraction.

Although we have established that uniformity is important for good node ordering, it is not so clear whether the two uniformity measures we have introduced are the final word. In particular, the right measure might depend on the application. For example, our current code for transit-node routing uses a geometric locality filter and hence it might be good if the uniformity measure would take geometry into account.

We have already demonstrated that CHs yield improved preprocessing times when changing the entire cost function. We still have to try how well the dynamization techniques for changing few edge weights from [3,2] translate.

Last but not least, we are now developing a method for fast routing in road networks with time-dependent edge weights. We hope that the simplicity and efficiency of CHs will give us a good starting point for this challenging task. The good performance of CHs for (unrolled) transportation networks observed in [9] may be an indicator that this will work well.

References

1. Sanders, P., Schultes, D.: Engineering fast route planning algorithms. In: Demetrescu, C. (ed.) WEA 2007. LNCS, vol. 4525, pp. 23–36. Springer, Heidelberg (2007)
2. Schultes, D.: Route Planning in Road Networks. PhD thesis (2008)
3. Schultes, D., Sanders, P.: Dynamic highway-node routing. In: Demetrescu, C. (ed.) WEA 2007. LNCS, vol. 4525, pp. 66–79. Springer, Heidelberg (2007)
4. Sanders, P., Schultes, D.: Highway hierarchies hasten exact shortest path queries. In: Brodal, G.S., Leonardi, S. (eds.) ESA 2005. LNCS, vol. 3669, pp. 568–579. Springer, Heidelberg (2005)
5. Sanders, P., Schultes, D.: Engineering highway hierarchies. In: Azar, Y., Erlebach, T. (eds.) ESA 2006. LNCS, vol. 4168, pp. 804–816. Springer, Heidelberg (2006)
6. Goldberg, A.V., Kaplan, H., Werneck, R.F.: Better landmarks within reach. In: Demetrescu, C. (ed.) WEA 2007. LNCS, vol. 4525, pp. 38–51. Springer, Heidelberg (2007)
7. Bauer, R., Delling, D.: SHARC: Fast and robust unidirectional routing. In: Workshop on Algorithm Engineering and Experiments (ALENEX) (2008)
8. Bast, H., Funke, S., Sanders, P., Schultes, D.: Fast routing in road networks with transit nodes. Science 316(5824), 566 (2007)

9. Bauer, R., Delling, D., Sanders, P., Schieferdecker, D., Schultes, D., Wagner, D.: Combining hierarchical and goal-directed speed-up techniques for Dijkstra's algorithm. In: WEA 2008. LNCS, vol. 5038, Springer, Heidelberg (2008)
10. Knopp, S., Sanders, P., Schultes, D., Schulz, F., Wagner, D.: Computing many-to-many shortest paths using highway hierarchies. In: Workshop on Algorithm Engineering and Experiments (ALENEX) (2007)
11. Maue, J., Sanders, P., Matijevic, D.: Goal directed shortest path queries using Precomputed Cluster Distances. In: Àlvarez, C., Serna, M.J. (eds.) WEA 2006. LNCS, vol. 4007, pp. 316–328. Springer, Heidelberg (2006)
12. Geisberger, R., Sanders, P., Schultes, D.: Better approximation of betweenness centrality. In: Workshop on Algorithm Engineering and Experiments (ALENEX) (2008)
13. Gutman, R.: Reach-based routing: A new approach to shortest path algorithms optimized for road networks. In: Workshop on Algorithm Engineering and Experiments (ALENEX)., pp. 100–111 (2004)
14. Lauther, U.: An extremely fast, exact algorithm for finding shortest paths in static networks with geographical background. In: Geoinformation und Mobilität – von der Forschung zur praktischen Anwendung, vol. 22, pp. 219–230. IfGI prints, Institut für Geoinformatik, Münster (2004)
15. Köhler, E., Möhring, R.H., Schilling, H.: Acceleration of shortest path and constrained shortest path computation. In: WEA 2005. LNCS, vol. 3503, Springer, Heidelberg (2005)
16. Delling, D., Sanders, P., Schultes, D., Wagner, D.: Highway hierarchies star. In: 9th DIMACS Implementation Challenge [20] (2006)
17. Goldberg, A.V., Kaplan, H., Werneck, R.F.: Better landmarks within reach. In: 9th DIMACS Implementation Challenge [20] (2006)
18. Sanders, P., Schultes, D., Vetter, C.: Mobile Route Planning (2008) in preparation, http://algo2.iti.uka.de/schultes/hwy/
19. R Development Core Team: R: A Language and Environment for Statistical Computing (2004), http://www.r-project.org
20. 9th DIMACS Implementation Challenge: Shortest Paths (2006), http://www.dis.uniroma1.it/~challenge9/

Bidirectional A^* Search for Time-Dependent Fast Paths⋆

Giacomo Nannicini[1,2], Daniel Delling[3], Leo Liberti[1], and Dominik Schultes[3]

[1] LIX, École Polytechnique, F-91128 Palaiseau, France
{giacomon,liberti}@lix.polytechnique.fr
[2] Mediamobile, 10 rue d'Oradour sur Glane, Paris, France
giacomo.nannicini@v-trafic.com
[3] Universität Karlsruhe (TH), 76128 Karlsruhe, Germany
{delling,schultes}@ira.uka.de

Abstract. The computation of point-to-point shortest paths on time-dependent road networks has many practical applications, but there have been very few works that propose efficient algorithms for large graphs. One of the difficulties of route planning on time-dependent graphs is that we do not know the exact arrival time at the destination, hence applying bidirectional search is not straightforward; we propose a novel approach based on A^* with landmarks (ALT) that starts a search from both the source and the destination node, where the backward search is used to bound the set of nodes that have to be explored by the forward search. Extensive computational results show that this approach is very effective in practice if we are willing to accept a small approximation factor, resulting in a speed-up of several times with respect to Dijkstra's algorithm while finding only slightly suboptimal solutions.

1 Introduction

We consider the TIME-DEPENDENT SHORTEST PATH PROBLEM (TDSPP): given a directed graph $G = (V, A)$, a source node $s \in V$, a destination node $t \in V$, an interval of time instants T, a departure time $\tau_0 \in T$ and a time-dependent arc weight function $c : A \times T \to \mathbb{R}_+$, find a path $p = (s = v_1, \ldots, v_k = t)$ in G such that its *time-dependent cost* $\gamma_{\tau_0}(p)$, defined recursively as follows:

$$\gamma_{\tau_0}(v_1, v_2) = c(v_1, v_2, \tau_0) \tag{1}$$

$$\gamma_{\tau_0}(v_1, \ldots, v_i) = \gamma_{\tau_0}(v_1, \ldots, v_{i-1}) + c(v_{i-1}, v_i, \tau_0 + \gamma_{\tau_0}(v_1, \ldots, v_{i-1})) \tag{2}$$

for all $2 \leq i \leq k$, is minimum. We also consider a function $\lambda : A \to \mathbb{R}_+$ which has the following property:

$$\forall (u, v) \in A, \tau \in T \quad (\lambda(u, v) \leq c(u, v, \tau)).$$

⋆ Partially supported by the Future and Emerging Technologies Unit of EC (IST priority – 6th FP), under contract no. FP6-02123502 (project ARRIVAL), and by DFG grant SA 933/1-3.

C.C. McGeoch (Ed.): WEA 2008, LNCS 5038, pp. 334–346, 2008.
© Springer-Verlag Berlin Heidelberg 2008

In other words, $\lambda(u,v)$ is a lower bound on the travelling time of arc (u,v) for all time instants in T. In practice, this can easily be computed, given an arc length and the maximum allowed speed on that arc. We naturally extend λ to be defined on paths, i.e. $\lambda(p) = \sum_{(v_i,v_j) \in p} \lambda(v_i, v_j)$.

In this paper, we propose a novel algorithm for the TDSPP based on a bidirectional A^* algorithm. Since the arrival time is not known in advance (so c cannot be evaluated on the arcs adjacent to the destination node), our backward search occurs on the graph weighted by the lower bounding function λ. This is used for bounding the set of nodes that will be explored by the forward search.

Many ideas have been proposed for the computation of point-to-point shortest paths on static graphs (see [1,2] for a review), and there are algorithms capable of finding the solution in a matter of a few microseconds [3]; adaptations of those ideas for dynamic scenarios, i.e. where arc costs are updated at regular intervals, have been tested as well (see [4,5,6]).

Much less work has been undertaken on the time-dependent variant of the shortest paths problem; this problem has been first addressed by [7] (a good review of this paper can be found in [8], p. 407): Dijkstra's algorithm [9] is extended to the dynamic case through a recursion formula based on the assumption that the network $G = (V, A)$ has the FIFO property. The FIFO property is also called the *non-overtaking property*, because it basically states that if A leaves u at time τ_0 and B at time $\tau_1 > \tau_0$, B cannot arrive at v before A using the arc (u,v). The FIFO assumption is usually necessary in order to mantain an acceptable level of complexity: the TDSPP in FIFO networks is polynomially solvable [10], while it is NP-hard in non-FIFO networks [11]. Given source and destination nodes s and t, the problem of maximizing the departure time from node s with a given arrival time at node t is equivalent to the TDSPP (see [12]).

Goal-directed search, also called A^* [13], has been adapted to work on all the previously described scenarios; an efficient version for the static case has been presented in [14], and then developed and improved in [15]. Those ideas have been used in [4] on dynamic graphs as well, while the time-dependent case on graphs with the FIFO property has been addressed in [16] and [4].

Moreover, the recently developed SHARC-algorithm [17] allows fast *unidirectional* shortest-path calculations in large scale networks. Due to its unidirectional nature, it can easily be used in a time-dependent scenario. However, in that case SHARC cannot guarantee to find the optimal solution.

The rest of this paper is organised as follows. In Section 2 we describe A^* search and the ALT algorithm, which are needed for our method. In Section 3 we describe the foundations of our idea, and present an adaptation of the ALT algorithm based on it. In Section 4 we formally prove our method's correctness. In Section 5 we propose some improvements. In Section 6 we discuss computational experiments and provide computational results.

2 A^* with Landmarks

A^* is an algorithm for goal-directed search, similar to Dijkstra's algorithm, but which adds a potential function to the priority key of each node in the queue. The

A^* algorithm on static graphs can be described as follows. The potential function on a node v is an estimate of the distance to reach the target from v; A^* then follows the same procedure as Dijkstra's algorithm, but the use of this potential function has the effect of giving priority to nodes that are (supposedly) closer to target node t. If the potential function π is such that $\pi(v) \leq d(v,t) \, \forall v \in V$, where $d(v,t)$ is the distance from v to t, then A^* always finds shortest paths. A^* is guaranteed to explore no more nodes than Dijkstra's algorithm: if $\pi(v)$ is a good approximation from below of the distance to target, A^* efficiently drives the search towards the destination node, and it explores considerably fewer nodes than Dijkstra's algorithm; if $\pi(v) = 0 \, \forall v \in V$, A^* behaves exactly like Dijkstra's algorithm. In [18] it is shown that A^* is equivalent to Dijkstra's algorithm on a graph with reduced costs, i.e. $w_\pi(u,v) = w(u,v) - \pi(u) + \pi(v)$.

One way to compute the potential function, instead of using Euclidean distances, is to use the concept of *landmarks*. Landmarks have first been proposed in [14]; they are a preprocessing technique which is based on the triangular inequality. The basic principle is as follows: suppose we have selected a set $L \subset V$ of landmarks, and we have precomputed distances $d(v, \ell), d(\ell, v) \, \forall v \in V, \ell \in L$; the following triangle inequalities hold: $d(u,t) + d(t,\ell) \geq d(u,\ell)$ and $d(\ell, u) + d(u,t) \geq d(\ell, t)$. Therefore $\pi_t(u) = \max_{\ell \in L}\{d(u,\ell) - d(t,\ell), d(\ell,t) - d(\ell,u)\}$ is a lower bound for the distance $d(u,t)$, and it can be used as a potential function which preserves optimal paths. On a static graph (i.e. non time-dependent), bidirectional search can be implemented, using some care in modifying the potential function so that it is consistent for the forward and backward search (see [15]); the consistency condition states that $w_{\pi_f}(u,v)$ in G is equal to $w_{\pi_b}(v,u)$ in the reverse graph $\overline{G}$, where π_f and π_b are the potential functions for the forward and the backward search, respectively. Bidirectional A^* with the potential function described above is called ALT. It is straightforward to note that, if arc costs can only increase with respect to their original value, the potential function associated with landmarks is still a valid lower bound, even on a time-dependent graph; in [4] this idea is applied to a real road network in order to analyse the algorithm's performances, but with a unidirectional search.

The choice of landmarks has a great impact on the size of the search space, as it severely affects the quality of the potential function. Several selection strategies exist, although none of them is optimal with respect to random queries, i.e., is guaranteed to yield the smaller search space for random source-destination pairs. The best known heuristics are *avoid* and *maxCover* [19].

3   Bidirectional Search on Time-Dependent Graphs

Our algorithm is based on restricting the scope of a time-dependent A^* search from the source using a set of nodes defined by a time-*independent* A^* search from the destination, i.e. the backward search is a reverse search in G_λ, which corresponds to the graph G weighted by the lower bounding function λ.

Given a graph $G = (V, A)$ and source and destination vertices $s, t \in V$, the algorithm for computing the shortest time-dependent cost path p^* works in three phases.

1. A bidirectional A^* search occurs on G, where the forward search is run on the graph weighted by c with the path cost defined by (1)-(2), and the backward search is run on the graph weighted by the lower bounding function λ. All nodes settled by the backward search are included in a set M. Phase 1 terminates as soon as the two search scopes meet.
2. Suppose that $v \in V$ is the first vertex in the intersection of the heaps of the forward and backward search; then the time dependent cost $\mu = \gamma_{\tau_0}(p_v)$ of the path p_v going from s to t passing through v is an upper bound to $\gamma_{\tau_0}(p^*)$. In the second phase, both search scopes are allowed to proceed until the backward search queue only contains nodes whose associated key exceeds μ. In other words: let β be the key of the minimum element of the backward search queue; phase 2 terminates as soon as $\beta > \mu$. Again, all nodes settled by the backward search are included in M.
3. Only the forward search continues, with the additional constraint that only nodes in M can be explored. The forward search terminates when t is settled.

The pseudocode for this algorithm is given in Algorithm 1.

4   Correctness

We denote by $d(u, v, \tau)$ the length of the shortest path from u to v with departure time τ, and by $d_\lambda(u, v)$ the length of the shortest path from u to v on the graph G_λ. We have the following theorems.

Theorem 4.1. *Algorithm 1 computes the shortest time-dependent path from s to t for a given departure time τ_0.*

Proof. The forward search of Algorithm 1 is exactly the same as the unidirectional version of the A^* algorithm during the first 2 phases, and thus it is correct; we have to prove that the restriction applied during phase 3 does not interfere with the correctness of the A^* algorithm.

Let μ be an upper bound on the cost of the shortest path; in particular, this can be the cost $\gamma_{\tau_0}(p_v)$ of the $s \to t$ path passing through the first meeting point v of the forward and backward search. Let β be the smallest key of the backward search priority queue at the end of phase 2. Suppose that Algorithm 1 is not correct, i.e. it computes a sub-optimal path. Let p^* be the shortest path from s to t with departure time τ_0, and let u be the first node on p^* which is not explored by the forward search; by phase 3, this implies that $u \notin M$, i.e. u has not been settled by the backward search during the first 2 phases of Algorithm 1. Hence, we have that $\beta \leq \pi_b(u) + d_\lambda(u, t)$; then we have the chain $\gamma_{\tau_0}(p^*) \leq \mu < \beta \leq \pi_b(u) + d_\lambda(u, t) \leq d_\lambda(s, u) + d_\lambda(u, t) \leq d(s, u, \tau_0) + d(u, t, d(s, u, \tau_0)) = \gamma_{\tau_0}(p^*)$, which is a contradiction.   □

Algorithm 1. Compute the shortest time-dependent path from s to t with departure time τ_0

```
1:  $\overrightarrow{Q}$.insert(s,0); $\overleftarrow{Q}$.insert(t,0); M := ∅; μ := +∞; done := false; phase := 1.
2:  while ¬done do
3:      if (phase = 1) ∨ (phase = 2) then
4:          ↔∈ {→, ←}
5:      else
6:          ↔:=→
7:      u := $\overleftrightarrow{Q}$.extractMin()
8:      if (u = t) ∧ (↔=→) then
9:          done := true
10:         continue
11:     if (phase = 1) ∧ (u.dist→ + u.dist← < ∞) then
12:         μ := u.dist→ + u.dist←
13:         phase := 2
14:     if (phase = 2) ∧ (↔=←) ∧ (μ < u.key←) then
15:         phase := 3
16:         continue
17:     for all arcs (u,v) ∈ $\overleftrightarrow{A}$ do
18:         if ↔=← then
19:             M.insert(u)
20:         else if (phase = 3) ∧ (v ∉ M) then
21:             continue;
22:         if (v ∈ $\overleftrightarrow{Q}$) then
23:             if u.dist↔ + c(u,v,u.dist↔) < v.dist↔ then
24:                 $\overleftrightarrow{Q}$.decreaseKey(v, u.dist↔ + c(u,v,u.dist↔) + $\overleftrightarrow{π}$(v))
25:         else
26:             $\overleftrightarrow{Q}$.insert(v, u.dist↔ + c(u,v,u.dist↔) + $\overleftrightarrow{π}$(v))
27: return t.dist→
```

Theorem 4.2. *Let p^* be the shortest path from s to t. If the condition to switch to phase 3 is $\mu < K\beta$ for a fixed parameter K, then Algorithm 1 computes a path p from s to t such that $\gamma_{\tau_0}(p) \leq K\gamma_{\tau_0}(p^*)$ for a given departure time τ_0.*

5 Improvements

The basic version of the algorithm can be enhanced by making use of the following results.

Proposition 5.1. *During phase 2 the backward search does not need to explore nodes that have already been settled by the forward search.*

We can take advantage of the fact that the backward search is used only to bound the set of nodes explored by the forward search, i.e. the backward search does not have to compute shortest paths. This means that we can tighten the bounds used by the backward search, as long as they are still valid lower bounds, even if doing so would result in an A^* backward search that computes suboptimal distances.

Proposition 5.2. *At a given iteration, let v be the last node settled by the forward search. Then, for each node w which has not been settled by the forward search, $d(s, v, \tau_0) + \pi_f(v) - \pi_f(w) \leq d(s, w, \tau_0)$.*

Let v be as in Prop. 5.2, and w a node which has not been settled by the forward search. Prop. 5.2 suggests that we can use

$$\pi_b^*(w) = \max\{\pi_b(w), d(s, v, \tau_0) + \pi_f(v) - \pi_f(w)\} \tag{3}$$

as a lower bound to $d(s, w, \tau_0)$ during the backward search. However, to prove the algorithm's correctness when using π_b^* we must assume that the node v used in (3) is fixed at each backward search iteration. Thus, we do the following: we set up 10 checkpoints during the query; when a checkpoint is reached, the node v used to compute (3) is updated, and the backward search queue is flushed and filled again using the updated π_b^*. This is enough to guarantee correctness. The checkpoints are computed comparing the initial lower bound $\pi_f(t)$ and the current distance from the source node, both for the forward search.

6 Experiments

In this section, we present an extensive experimental evaluation of our time-dependent ALT algorithm. Our implementation is written in C++ using solely the STL. As priority queue we use a binary heap. Our tests were executed on one core of an AMD Opteron 2218 running SUSE Linux 10.1. The machine is clocked at 2.6 GHz, has 16 GB of RAM and 2 x 1 MB of L2 cache. The program was compiled with GCC 4.1, using optimization level 3.

Unless otherwise stated, we use 16 maxcover landmarks [14], computed on the input graph using the lower bounding function λ to weight edges, and we use (3) as potential function for the backward search, with 10 checkpoints (see Section 5). When performing random s-t queries, the source s, target t, and the starting time τ_0 are picked uniformly at random and results are based on 10 000 queries.

Inputs. We tested our algorithm on two different road networks: the road network of Western Europe provided by PTV AG for scientific use, which has approximately 18 million vertices and 42.6 million arcs, and the road network of the US, taken from the TIGER/Line Files, with 23.9 million vertices and 58.3 million arcs. A travelling time in uncongested traffic situation was assigned to each arc using that arc's category (13 categories for Europe, 4 for US) to determine the travel speed.

Modeling Traffic. Unfortunately, we are not aware of a *large* publicly available real-world road network with time-dependent arc costs. Therefore, we have to use artificially generated costs. In order to model the time-dependent costs on each arc, we developed an heuristic algorithm, based on statistics gathered using real-world data on a limited-size road network; we used piecewise linear cost functions, with one breakpoint for each hour over a day. Arc costs are generated assigning, at each node, several random values that represent peak hour (i.e. hour with maximum traffic increase), duration and speed of traffic increase/decrease for a traffic jam; for each node, two traffic jams are generated, one in the morning and one in the afternoon. Then, for each arc in a node's arc star, a *speed profile* is generated, using the traffic jam's characteristics of the corresponding node, and assigning a random increase factor between 1.5 and 3 to represent that arc's slowdown during peak hours with respect to uncongested hours. We do not assign speed profile to arcs that have both endpoints at nodes with level 0 in a pre-constructed Highway Hierarchy [20], and as a result those arcs will have the same travelling time value throughout the day; for all other arcs, we use the traffic jam values associated with the endpoint with smallest ID. The breakpoints of these speed profiles are stored in memory as a multiplication factor with respect to the speed in uncongested hours, which allows us to use only 7 bits for each breakpoint. We assume that all roads are uncongested between 11PM and 4AM, so that we do not need to store the corresponding breakpoints; as a result, we store all breakpoints using 16 additional bytes per edge. The travelling time of an arc at time τ is computed via linear interpolation of the two breakpoints that precede and follow τ.

This method was developed to ensure spatial coherency between traffic increases, i.e. if a certain arc is congested at a given time, then it is likely that adjacent arcs will be congested too. This is a basic principle of traffic analysis [21].

Random Queries. Table 1 reports the results of our bidirectional ALT variant on time-dependent networks for different approximation values K using the European and the US road network as input. For the European road network, preprocessing takes approximately 75 minutes and produces 128 *additional* bytes per node (for each node we have to store distances to and from all landmarks); for the US road network, the corresponding figures are 92 minutes and 128 bytes per node. For comparison, we also report the results on the same road network for the time-dependent versions of Dijkstra, unidirectional ALT, and the SHARC algorithm [17].

As the performed ALT-queries compute approximated results instead of optimal solutions, we record three different statistics to characterize the solution quality: error rate, average relative error, maximum relative error. By *error rate* we denote the percentage of computed suboptimal paths over the total number of queries. By *relative error* on a particular query we denote the relative percentage increase of the approximated solution over the optimum, computed as $\omega/\omega^* - 1$, where ω is the cost of the approximated solution computed by our algorithm and ω^* is the cost of the optimum computed by Dijkstra's algorithm. We report *average* and *maximum* values of this quantity over the set of all queries. We also report the number of nodes settled at the *end* of each phase of our algorithm, denoting them with the labels phase 1, phase 2 and phase 3.

Table 1. Performance of the time-dependent versions of Dijkstra, unidirectional ALT, SHARC, and our bidirectional approach. For SHARC, we use approximation values of 1.001 and 1.002 (cf. [17] for details).

input	method	K	ERROR rate	relative avg	max	QUERY # settled nodes phase 1	phase 2	phase 3	time [ms]
EUR	Dijkstra	-	0.0%	0.000%	0.00%	-	-	8 908 300	6 325.8
	uni-ALT	-	0.0%	0.000%	0.00%	-	-	2 192 010	1 775.8
	1.001-SHARC	-	57.1%	0.686%	34.31%	-	-	140 945	60.3
	1.002-SHARC	-	42.8%	0.583%	34.31%	-	-	930 251	491.4
	ALT	1.00	0.0%	0.000%	0.00%	125 068	2 784 540	3 117 160	3 399.3
		1.02	1.0%	0.003%	1.13%	125 068	2 154 900	2 560 370	2 723.3
		1.05	4.0%	0.029%	4.93%	125 068	1 333 220	1 671 630	1 703.6
		1.10	18.7%	0.203%	8.10%	125 068	549 916	719 769	665.1
		1.13	30.5%	0.366%	12.63%	125 068	340 787	447 681	385.5
		1.15	36.4%	0.467%	13.00%	125 068	265 328	348 325	287.3
		1.20	44.7%	0.652%	18.19%	125 068	183 899	241 241	185.3
		1.30	48.2%	0.804%	23.63%	125 068	141 358	186 267	134.6
		1.50	48.8%	0.844%	25.70%	125 068	130 144	172 157	121.9
		2.00	48.9%	0.886%	48.86%	125 068	125 071	165 650	115.7
USA	Dijkstra	-	0.0%	0.000%	0.00%	-	-	12 435 900	8 020.6
	uni-ALT	-	0.0%	0.000%	0.00%	-	-	2 908 170	2 403.9
	ALT	1.00	0.0%	0.000%	0.00%	272 790	4 091 050	4 564 030	4 534.2
		1.10	21.5%	0.135%	7.02%	272 790	633 758	829 176	656.3
		1.15	54.4%	0.402%	9.98%	272 790	312 575	405 699	289.6
		1.20	62.0%	0.482%	9.98%	272 790	278 345	359 190	251.1
		1.50	64.8%	0.506%	13.63%	272 790	272 790	351 865	247.5
		2.00	64.8%	0.506%	16.00%	272 790	272 791	351 854	246.8

As expected, we observe a clear trade-off between the quality of the computed solution and query performance. If we are willing to accept an approximation factor of $K = 2.0$, on the European road network queries are on average 55 times faster than Dijkstra's algorithm, but almost 50% of the computed paths will be suboptimal and, although the average relative error is still small, in the worst case the approximated solution has a cost which is 50% larger than the optimal value. The reason for this poor solution quality is that, for such high approximation values, phase 2 is very short. As a consequence, nodes in the middle of the shortest path are not explored by our approach, and the meeting point of the two search scopes is far from being the optimal one. However, by decreasing the value of the approximation constant K we are able to obtain solutions that are very close to the optimum, and performance is significantly better than for unidirectional ALT or Dijkstra. In our experiments, it seems as if the best trade-off between quality and performance is achieved with an approximation value of $K = 1.15$, which yields average query times smaller than 300 ms on both road neworks with a maximum recorded relative error of 13% (on the European road network, while the corresponding figure is 9.98% for the

US instance). By decreasing K to values < 1.05 it does not pay off to use the bidirectional variant any more, as the unidirectional variant of ALT is faster and is always correct.

Comparing results for $K > 1.15$ for the US with those for Europe, we observe that number of queries that return suboptimal paths increases, but the average and maximum error rates are smaller than the corresponding values on the European road network with the same values of K. Moreover, the speed-ups of our algorithm with respect to plain Dijkstra are lower on the US instance: the maximum recorded speed-up (for $K = 2.0$) is only of a factor 33. This behaviour has also been observed in the static scenario [4]. However, with $K = 1.15$, which is a good trade-off between quality and speed, query performance is very similar on both networks.

An interesting observation is that for $K = 2.0$ switching from a static to a time-dependent scenario increases query times only of a factor of ≈ 2: on the European road network, in a static scenario, ALT-16 has query times of 53.6 ms (see [4]), while our time-dependent variant yields query times of 115 ms. We also note that for our bidirectional search there is an additional overhead which increases the time spent per node with respect to unidirectional ALT: on the European road network, using an approximation factor of $K = 1.05$ yields similar query times to unidirectional ALT, but the number of nodes settled by the bidirectional approach is almost 30% smaller. We suppose that this is due to the following facts: in the bidirectional approach, one has to check at each iteration if the current node has been settled in the opposite direction, and during phase 2 the upper bound has to be updated from time to time. The cost of these operations, added to the phase-switch checks, is probably not negligible.

Comparing the time-dependent variant of SHARC with our approach, we observe that SHARC with an approximation value of 1.001 settles as many nodes as ALT with $K = 2.0$. However, query performance is better for SHARC due to its small computational overhead. By increasing the approximation value, computational times are slowed by almost one order of magnitude, but the solution quality merely improves. The reason for this poor performance is that SHARC uses a contraction routine which cannot bypass nodes incident to time-dependent edges. As in our scenario about half of the edges are time-dependent, the preprocessing of SHARC takes quite long (≈ 12 hours) and query performance is poor. Summarizing, ALT seems to work much better in a time-dependent scenario.

Local Queries. For random queries, our bidirectional ALT algorithm (with $K = 1.15$) is roughly 6.7 times faster than unidirectional ALT on average. In order to gain insight whether this speed-up derives from small or large distance queries, Fig. 1 reports the query times with respect to the Dijkstra rank[1]. These values were gathered on the European road network instance. Note that we use a logarithmic scale due to the fluctuating query times of bidirectional ALT. Comparing both ALT version, we observe that switching from uni- to bidirectional queries

---

[1] For an s-t query, the Dijkstra rank of node t is the number of nodes settled before t is settled. Thus, it is some kind of distance measure.

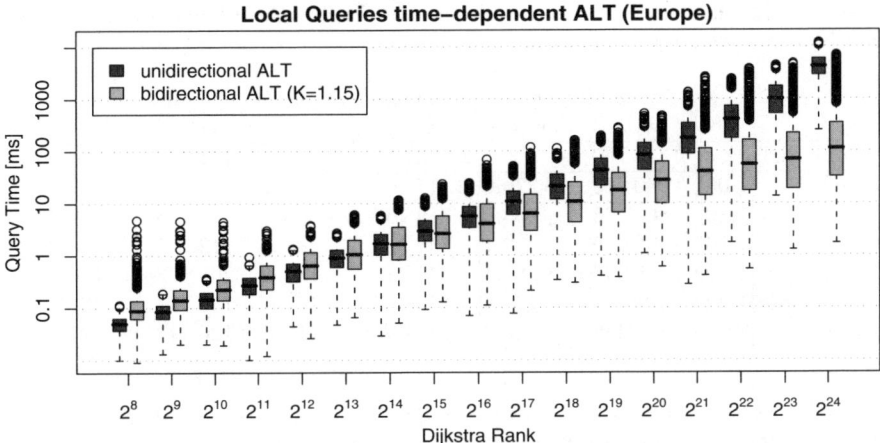

Fig. 1. Comparison of uni- and bidirectional ALT using the Dijkstra rank methodology [20]. The results are represented as box-and-whisker plot: each box spreads from the lower to the upper quartile and contains the median, the whiskers extend to the minimum and maximum value omitting outliers, which are plotted individually.

pays off especially for long-distance queries. This is not surprising, because for small distances the overhead for bidirectional routing is not counterbalanced by a significant decrease in the number of explored nodes: unidirectional ALT is faster for local queries. For ranks of 2^{24}, the median of the bidirectional variant is almost 2 orders of magnitude lower than for the unidirectional variant. Another interesting observation is the fact that some outliers of bidirectional ALT are almost as slow as the unidirectional variant.

Number of Landmarks. In static scenarios, query times of bidirectional ALT can be significantly reduced by increasing the number of landmarks to 32 or even 64 (see [4]). In order to check whether this also holds for our time-dependent variant, we recorded our algorithm's performance using different numbers of landmarks. Tab. 2 reports those results on the European road network. We evaluate 8 maxcover landmarks (yielding a preprocessing effort of 33 minutes and an overhead of 64 bytes per node), 16 maxcover landmarks (75 minutes, 128 bytes per node) and 32 avoid landmarks (29 minutes, 256 bytes per node). Note that we do not report error rates here, as it turned out that the number of landmarks has almost no impact on the quality of the computed paths. Surprisingly, the number of landmarks has a very small influence on the performance of time-dependent ALT. Even worse, increasing the number of landmarks even yields larger average query times for unidirectional ALT and for bidirectional ALT with low K-values. This is due the fact that the search space decreases only slightly, but the additional overhead for accessing landmarks increases when there are more landmarks to take into account. However, when increasing K, a larger number of landmarks yields faster query times: with $K = 2.0$ and 32

Table 2. Performance of uni- and bidirectional ALT with different number of landmarks in a time-dependent scenario

	K	8 landmarks		16 landmarks		32 landmarks	
		# settled	time [ms]	# settled	time [ms]	# settled	time [ms]
uni-ALT	-	2 321 760	1 739.8	2 192 010	1 775.8	2 111 090	1 868.5
ALT	1.00	3 240 210	3 270.6	3 117 160	3 399.3	3 043 490	3 465.1
	1.10	863 526	736.5	719 769	665.1	681 836	669.7
	1.15	495 649	382.1	348 325	287.3	312 695	280.0
	1.20	389 096	286.3	241 241	185.3	204 877	170.1
	1.50	320 026	228.4	172 157	121.9	133 547	98.3
	2.00	313 448	222.2	165 650	115.7	126 847	91.1

landmarks we are able to perform time-dependent queries 70 times faster than plain Dijkstra, but the solution quality in this case is as poor as in the 16 landmarks case. Summarizing, for $K > 1.10$ increasing the number of landmarks has a positive effect on computational times, although switching from 16 to 32 landmarks does not yield the same benefits as from 8 to 16, and thus in our experiments is not worth the extra memory. On the other hand, for $K \leq 1.10$ and for unidirectional ALT increasing the number of landmarks has a negative effect on computational times, and thus is never a good choice in our experiments.

7   Conclusion and Future Work

We have presented an algorithm which applies bidirectional search on a time-dependent road network, where the backward search is used to bound the set of nodes that have to be explored by the forward search; this algorithm is based on the ALT variant of the A^* algorithm. We have discussed related theoretical issues, and we proved the algorithm's correctness. Extensive computational experiments show that this algorithm is very effective in practice if we are willing to accept a small approximation factor: the exact version of our algorithm is slower than unidirectional ALT, but if we can accept a decrease of the solution quality of a few percentage points with respect to the optimum then our algorithm is several times faster. For practical applications, this is usually a good compromise. We have compared our algorithm to existing methods, showing that this approach for bidirectional search is able to significantly decrease computational times.

Future research will include the possibility of an initial contraction phase for a time-dependent graph, which would be useful for several purposes, and algorithm engineering issues such as the balancing of the forward and backward search, and the update of the available upper bound on the optimal solution cost. The idea of bidirectional routing on time-dependent graphs, using a time-dependent forward search and a time-independent backward search, may be applied to other static routing algorithms, in order to generalize them in a time-dependent scenario.

Acknowledgements

We would like to acknowledge Mediamobile's financial support, and thank Daniele Pretolani for providing useful discussions.

References

1. Wagner, D., Willhalm, T.: Speed-up techniques for shortest-path computations. In: Thomas, W., Weil, P. (eds.) STACS 2007. LNCS, vol. 4393, pp. 23–36. Springer, Heidelberg (2007)
2. Sanders, P., Schultes, D.: Engineering fast route planning algorithms. In: Demetrescu, C. (ed.) WEA 2007. LNCS, vol. 4525, pp. 23–36. Springer, Heidelberg (2007)
3. Bast, H., Funke, S., Sanders, P., Schultes, D.: Fast routing in road networks with transit nodes. Science 316(5824), 566 (2007)
4. Delling, D., Wagner, D.: Landmark-based routing in dynamic graphs. In: Demetrescu, C. (ed.) WEA 2007. LNCS, vol. 4525, Springer, Heidelberg (2007)
5. Sanders, P., Schultes, D.: Dynamic highway-node routing. In: Demetrescu, C. (ed.) WEA 2007. LNCS, vol. 4525, pp. 66–79. Springer, Heidelberg (2007)
6. Wagner, D., Willhalm, T., Zaroliagis, C.: Geometric containers for efficient shortest-path computation. ACM Journal of Experimental Algorithmics 10, 1–30 (2005)
7. Cooke, K., Halsey, E.: The shortest route through a network with time-dependent internodal transit times. Journal of Mathematical Analysis and Applications 14, 493–498 (1966)
8. Dreyfus, S.: An appraisal of some shortest-path algorithms. Operations Research 17(3), 395–412 (1969)
9. Dijkstra, E.: A note on two problems in connexion with graphs. Numerische Mathematik 1, 269–271 (1959)
10. Kaufman, D.E., Smith, R.L.: Fastest paths in time-dependent networks for intelligent vehicle-highway systems application. Journal of Intelligent Transportation Systems 1(1), 1–11 (1993)
11. Orda, A., Rom, R.: Shortest-path and minimum delay algorithms in networks with time-dependent edge-length. Journal of the ACM 37(3), 607–625 (1990)
12. Daganzo, C.: Reversibility of time-dependent shortest path problem. Technical report, Institute of Transportation Studies, University of California, Berkeley (1998)
13. Hart, E., Nilsson, N., Raphael, B.: A formal basis for the heuristic determination of minimum cost paths. IEEE Transactions on Systems, Science and Cybernetics SSC-4(2), 100–107 (1968)
14. Goldberg, A., Harrelson, C.: Computing the shortest path: A^* meets graph theory. In: Proceedings of the 16th Annual ACM-SIAM Symposium on Discrete Algorithms (SODA 2005), SIAM (2005)
15. Goldberg, A., Kaplan, H., Werneck, R.: Reach for A^*: Efficient point-to-point shortest path algorithms. In: Demetrescu, C., Sedgewick, R., Tamassia, R. (eds.) Proceedings of the 7th Workshop on Algorithm Engineering and Experimentation (ALENEX 2005), SIAM (2005)
16. Chabini, I., Shan, L.: Adaptations of the A^* algorithm for the computation of fastest paths in deterministic discrete-time dynamic networks. IEEE Transactions on Intelligent Transportation Systems 3(1), 60–74 (2002)

17. Bauer, R., Delling, D.: SHARC: Fast and Robust Unidirectional Routing. In: Proceedings of the 10th Workshop on Algorithm Engineering and Experiments (ALENEX 2008), SIAM (to appear, 2008)

18. Ikeda, T., Tsu, M., Imai, H., Nishimura, S., Shimoura, H., Hashimoto, T., Tenmoku, K., Mitoh, K.: A fast algorithm for finding better routes by ai search techniques. In: Proceedings for the IEEE Vehicle Navigation and Information Systems Conference, pp. 291–296 (2004)

19. Goldberg, A., Werneck, R.: An efficient external memory shortest path algorithm. In: Demetrescu, C., Sedgewick, R., Tamassia, R. (eds.) Proceedings of the 7th Workshop on Algorithm Engineering and Experimentation (ALENEX 2005), pp. 26–40. SIAM (2005)

20. Sanders, P., Schultes, D.: Highway hierarchies hasten exact shortest path queries. In: Brodal, G.S., Leonardi, S. (eds.) ESA 2005. LNCS, vol. 3669, pp. 568–579. Springer, Heidelberg (2005)

21. Kerner, B.S.: The Physics of Traffic. Springer, Berlin (2004)

Multi-criteria Shortest Paths in Time-Dependent Train Networks

Yann Disser[1], Matthias Müller–Hannemann[2], and Mathias Schnee[1]

[1] Technische Universität Darmstadt, Department of Computer Science,
Hochschulstraße 10, 64289 Darmstadt, Germany
{disser,schnee}@algo.informatik.tu-darmstadt.de
[2] Martin-Luther-Universität Halle-Wittenberg, Department of Computer Science,
Von-Seckendorff-Platz 1, 06120 Halle, Germany
muellerh@informatik.uni-halle.de

Abstract. We study the problem of finding all Pareto-optimal solutions in a multi-criteria setting of the shortest path problem in time-dependent graphs. This has important applications in timetable information systems for train schedules. We present a new prototype to solve this problem in a fully realistic scenario based on a multi-criteria generalization of Dijkstra's algorithm. As optimization criteria we use travel time and number of train changes, as well as a new criterion "reliability of transfers".

The performance of the prototype and various speed-up techniques are analyzed experimentally on a large set of real test instances. In comparison with a base-line implementation, our prototype achieves significant speed-up factors of 20 with respect to the number of label creations and of 138 with respect to label insertions into the priority queue. We also compare our prototype with a time-expanded graph model.

Keywords: shortest paths, time-dependent graphs, multi-criteria optimization, speed-up techniques, case study.

1 Introduction

In peak times the timetable information system of the German railway company Deutsche Bahn AG calculates over 1,600,000 connections per hour [1]. This demonstrates the importance that such systems have gained. It is obvious that efficient algorithms have to be used in order to cope with that large a demand. To achieve this kind of efficiency, the system currently in use by Deutsche Bahn AG applies rather restrictive heuristics. Therefore, optimality of the gained results cannot be guaranteed. Moreover, commercial systems usually apply single-criteria algorithms optimizing travel time only.

In recent years, several research efforts have demonstrated that *exact* single-criteria shortest path queries in train networks can be performed very efficiently due to powerful speed-up techniques. Multi-criteria shortest path search is much more challenging. Given two paths p and q, we say that p *dominates* q if and only if there is at least one criterion for which p has a better value than q and

C.C. McGeoch (Ed.): WEA 2008, LNCS 5038, pp. 347–361, 2008.

there is no criterion for which p has a worse value than q. A path is called *Pareto-optimal* if it is not dominated by any other path. Here, the usual goal is to find *all* Pareto-optimal solutions. In theory, there can be exponentially many Pareto-optima in the worst case, although in practice only a few are observed in a realistic setting [2]. But in contrast to single-criteria search, one cannot abort the search after finding a first optimal solution. In fact, even after finding all Pareto-optima, search algorithms require a substantial amount of time to find a certificate that no further solutions exist.

Related work. Two main approaches have been proposed for modeling timetable information as a shortest path problem: the *time-expanded* [3,4,5], and the *time-dependent* approach [6,7,8,9,10,11,5]. These models and algorithms are described in detail in a recent survey [12].

Pyrga et al. [5] have presented an extensive computational study comparing the time-expanded and the time-dependent graph for the earliest arrival problem. They also consider a bicritera search for all Pareto-optima with respect to travel time and number of transfers. However, in the time-dependent version they heavily exploit the fact that the number of transfers in Pareto-optimal solutions is usually fairly small. More specifically, Pyrga et al. reduce the bicriteria search to a sequence of single-criteria problems with a bounded number of transfers. They start by obtaining the lexicographically smallest optimum for the combination of earliest arrival and number of transfers. If this optimum uses T transfers, another T searches are performed bounding the number of transfers by $T - 1, \ldots, 0$. By excluding dominated results from all the obtained ones, all Pareto-optima are computed in this particular bicriteria scenario. This trick is neither well-suited for more than two criteria nor for criteria which may attain a large range of values.

The second and third author have designed a fully realistic multi-criteria prototype MOTIS (multi-criteria timetable information system) which is capable of answering queries in about one second on standard PCs [13]. The MOTIS system is currently based on a time-expanded graph due to the fact that it is much easier to model all side constraints arising in practice in this framework. However, the major drawback of time-expanded graphs in comparison to time-dependent models is the higher space consumption, in particular if highly-periodically operating regional mass transit has to be included. In addition, the time-dependent graph model is easier to adapt in case of dynamic graph changes due to train delays. This motivates our investigation of the time-dependent graph model in this paper.

Only a few months ago Bauer et al. [14] have presented an experimental study on speed-up techniques for timetable information systems. They observed that many of the recently developed speed-up techniques are much slower on graphs derived from timetable information than on road networks. Moreover, many single-criteria speed-up techniques rely on a simultaneous bidirected search from source and target. Such techniques are not applicable in train search applications since we only know the target station but not the time at which an arrival can be expected. A recently developed technique is the unidirectional routing algorithm

SHARC [15]. A time-dependent version of SHARC yields only approximations, but works well on road networks.

Our Contribution and Overview. To the best of our knowledge, no complete, realistic system has been built for exact multi-criteria search of all Pareto-optimal solutions in the time-dependent graph model. In [5], Pyrga et al. treat constant transfer times and traffic days, but other aspects of real timetables like foot-paths and special transfer rules are not considered. In this paper we describe a first prototype for multi-criteria search of all Pareto-optima within a fully featured, real timetable. Its search results are guaranteed to be optimal. We provide an extensive computational study showing the impact of several speed-up techniques. Even though the number of possible speed-up techniques is restricted severely in order to guarantee the optimality of all search results, the performance of our prototype is already comparable to time-expanded systems, but consumes much less space.

Most previous research (in particular [5]) concentrates on the earliest arrival problem from a given point in time. But here we focus on a many-source shortest path version because in a pre-trip search for train connections a user usually wants to specify a *time interval* in which his journey should start. This implies that we have to perform a simultaneous search from multiple starting times. In a time-expanded graph model this can be handled very easily: One simply adds a "super-source" and edges of length zero to all start events, thereby reducing the search to a single-source search. In time-dependent graphs, however, solving the many-source shortest path problem is more subtle if travel time is used as an optimization criterion. Consider two subpaths from the source to some intermediate node. Then, path p_1 with start time s_1 and travel time t_1 dominates another path p_2 with start time s_2 and travel time t_2 with respect to travel time only if $t_1 < t_2$ *and* $s_1 \geq s_2$. Otherwise both paths are incomparable. This leads to weaker dominance during search than for the earliest arrival problem, and consequently to more non-dominated solutions which can be offered to customers. It is therefore remarkable that we still achieve quite a reasonable performance.

Our approach can easily be extended to further criteria. In order to exemplify this, the "reliability of transfers" is introduced as an additional criterion. The reliability of transfers is a property of a connection that captures the probability of catching all trains within it. Since possible train delays cannot be ignored, such a criterion is of practical importance.

The remainder of this paper is organized as follows. In Section 2, we introduce the time-dependent graph model and describe the adaptations needed in order to make it suited for fully realistic timetables. A modification of Dijkstra's algorithm that makes it capable of minimizing multiple criteria is introduced in Section 3. Several speed-up techniques that do not violate the optimality of the search results are proposed. The results of the experimental analysis of our time-dependent search system are presented in Section 4. We analyze the impact of the proposed speed-up techniques on performance. The final performance is then compared to a fully optimized search using a time-expanded graph. The last aspect of our discussion covers the relationship between performance and the number of search criteria. Finally, Section 5 summarizes and gives an outlook on future work.

2   Realistic Time-Dependent Graph Model

In this section we will describe a time-dependent graph model as introduced in [5,10,11]. We will start off with a very basic time-dependent model and extend it in the following.

We assume the timetable to consist of a set $\mathcal{T}$ of trains, a set $\mathcal{S}$ of stations, and a set $\mathcal{E}$ of elementary connections. An *elementary connection* $e \in \mathcal{E}$ describes a connection between two adjacent train stations without intermediate stops. Such a connection contains a departure station from$(e) \in \mathcal{S}$, an arrival station to$(e) \in \mathcal{S}$, a departure time d(e), and an arrival time $a(e)$. In addition to that, each elementary connection has several properties like train class, traffic days and train number. Each train $tr \in \mathcal{T}$ is an ordered list of elements of $\mathcal{E}$. A *train connection* is composed of an ordered list of elementary connections which must be consistent with the sequence of departure and arrival stations.

2.1   Basic Time-Dependent Model

For each station $S \in \mathcal{S}$ in the timetable there is a node $v(S) \in V$ in the basic time-dependent graph $G = (V, E)$. We call these nodes *station nodes*. There is an edge $e_{AB} = (v(A), v(B)) \in E$ if the set $\mathcal{E}_{AB} := \{e \in \mathcal{E} | \text{from}(e) = A \wedge \text{to}(e) = B\}$ is non-empty. The characteristics of all elementary connections in $\mathcal{E}_{AB}$ are attributed to this single edge e_{AB}. Each edge has multiple length functions, one for each optimization criterion. These length functions are time-dependent: depending on the time t at which the edge is to be used, different connections in $\mathcal{E}_{AB}$ may be favorable. In general, this is implemented with an iterator which computes edge lengths "on-the-fly" and returns all necessary variants with different characteristics.

If we only consider travel time and make the assumption that a connection $e_1 \in \mathcal{E}_{AB}$ *may not* overtake another connection $e_2 \in \mathcal{E}_{AB}$ in the sense that d$(e_1) \geq$ d(e_2) and a$(e_1) <$ a(e_2), then the connection with the earliest departure after time t is the one chosen from $\mathcal{E}_{AB}$. Its travel time length is precisely a$(\text{rel}(\mathcal{E}_{AB}, t)) - t$, where $\text{rel}(\mathcal{E}_{AB}, t) := \arg\min_{e \in \mathcal{E}_{AB}, \text{d}(e) \geq t} \text{d}(e)$ is the relevant connection in $\mathcal{E}_{AB}$ at time t.

2.2   Transfers

In the basic model, transfers between different trains are not modeled differently than two consecutive elementary connections with the same train. In order to allow for our search to count the number of transfers and in order to assign a duration to transfers, the model has to be extended as follows. We assume here for simplicity that a constant transfer time is provided for each station.

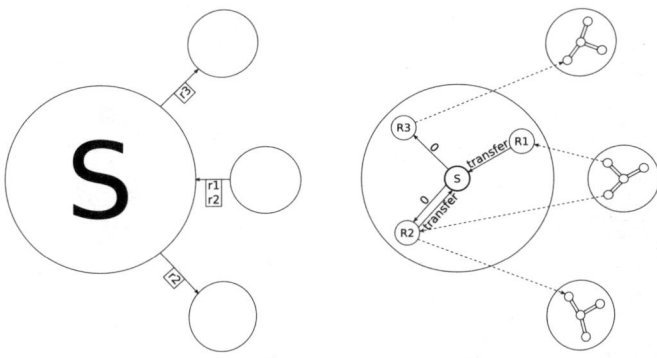

Fig. 1. Extension of a simple time-dependent graph (left) to support transfers. The timetable has three routes r1, r2, r3 so that the extended station (right) has three route nodes.

In order to still be able to take advantage of the fact that multiple elementary connections are modeled by a single edge, it is necessary to group train connections into *routes*. The set of routes forms a partition of $\mathcal{T}$ such that two connections are in the same route if and only if they share equal stations and properties. The departure and arrival times of two connections in the same route may differ as well as their traffic days. Using this partition, each station is represented by several *route nodes* in addition to its station node. The station node is used only to connect the route nodes and has no edges to nodes from other stations. The expanded model is depicted in Figure 1.

One route node is required for each route that arrives or departs at the station. For all connections in the same route, the corresponding route node plays the role of the station node in the basic model. The assumption that connections may not overtake each other can now be restricted to connections within a route. If we have overtaking elementary connections within a route, the route can simply be split up in order to separate the two elementary connections (and so we can get rid off this assumption). If the route has a connection that arrives at the station, an edge connecting the route node to the station node is introduced; if the route has a connection that departs from the station, an edge connecting the station node to the route node is introduced. One of these two edges needs to carry the transfer costs at the station and is called *transfer-edge*, the other has a transfer cost of 0. In the following we choose the edges from route nodes to station nodes as transfer-edges. This is called *exiting transfers* as opposed to *entering transfers*. We will see, that our choice is preferable due to performance advantages of the multi-criteria search.

2.3 Foot-Paths and Special Transfer Rules

We propose the following extensions to make the model fully realistic. In a real environment it is possible to walk from one station to another if the two

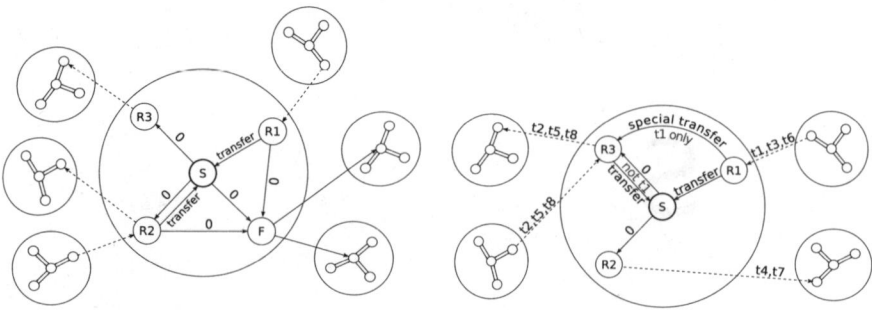

Fig. 2. (a) Illustration of a station with two foot-edges in the time-dependent model (b) Modifications to the graph for a station with a special transfer from train t_1 to train t_2

stations lie in geographic proximity. Realistic models therefore contain foot-paths to model this. Foot-paths are tuples (A, B, c) that represent a possibility to walk between stations A and B within c minutes. We assume, that c already contains all transfer costs at both A and B, so that no additional cost for switching trains arise. Foot-paths are special in that their length is constant in time. Figure 2 (a) shows the modifications that are needed in order to model a foot-path (A, B, c). It is not sufficient to simply add an edge from the station node of A to the station node of B with length c. This is because no additional transfer costs have to be paid when using a foot-path. Reducing c by the transfer cost at A, does not correctly model the costs when the journey starts at A. To circumvent these problems, an additional *foot-node* is added to the stations subgraph.

Another feature of realistic timetables are special transfer rules, that change the transfer time between two specific trains. The general transfer time of a station may be increased or decreased that way, depending on the real-world situation at the station. Two trains that use the same platform may for instance have a reduced transfer time. For each transfer rule several changes to the graph have to be made. Consider a special transfer time to get from train t_1 to train t_2 at station A. Let X denote the route node of A for t_1 and Y the route node of A for t_2. The station node for A is denoted by S. We assume that all special transfers are reasonable, i.e. it is not possible to reach a train departing before t_2 at Y if we arrived with t_1. However, there are cases in which it is explicitly made impossible to reach t_2 by setting the time of the special transfer higher than the usual transfer time. Figure 2 (b) shows the changes that have to be applied to the model when a special transfer rule is introduced. A new edge leads from X to Y carrying the special transfer cost. This edge may only be used after using t_1. The existing edges from S to Y and from Y to S have to be restricted so that they may *not* be used if t_1 is the last used train. This way Y cannot be reached from X without using the special transfer and the special transfer may not be used as shortcut to get to another route. For a proof of correctness of this model, we refer to the full version of this paper.

3   Multi-criteria Dijkstra and Speed-Up Techniques

In the following we briefly review an extension to Dijkstra's algorithm [16] that makes it capable of coping with several minimization criteria. Pseudocode is given in Algorithm 1. See Möhring [17] or the PhD-thesis of Theune [18] for a general description and correctness proofs.

3.1   Dijkstra's Algorithm for Multiple Criteria

The major difference when considering multiple search criteria is that nodes in the graph may be visited multiple times. The order in which nodes are visited in the classical algorithm guarantees that once a node is visited no better way of reaching it will be found later on. Multiple criteria allow for the possibility of different paths to a node that are not comparable – neither is strictly better.

Thus we need a way to remember all promising paths with which a node was reached. We do this by using multi-dimensional *labels*. Labels are associated with nodes and contain an entry for each criterion and a reference to its predecessor on the path. For every node in the graph we maintain a list of labels that are not dominated by any other label at the node. In the beginning, all label lists are empty. Then, start labels are created for all nodes with a timestamp within the query interval and are stored in a priority queue (lines 4-6). In the main loop of the algorithm, a lexicographically minimal label is extracted from the priority queue in each iteration (line 8). For the corresponding node of that label all outgoing edges are scanned and labels for their head nodes are created, provided that the edge is feasible (lines 9-11). Any new label is compared to all labels in the list corresponding to its node. It is only inserted into that list and into the priority queue if it is not dominated by any other label in the list. On the other hand, labels dominated by the new label are marked as invalid and removed from the node list (line 16).

3.2   Speed-Up Techniques

A good measure for the performance of this multi-criteria search algorithm is the number of labels created during a search. There are several techniques of reducing this number and thus increasing the algorithm's performance. As inserting labels into the priority queue is an expensive part of the search, the number of insertions can serve as a secondary measure of the algorithm's performance. We will now briefly discuss some important optimizations of the search, starting with some techniques that have been used in the time-expanded approach for some time [13]. In Section 3.2 we introduce two new and rather technical optimizations that have some impact on the search within the time-dependent graph.

Obtaining Lower Bounds. Some of the techniques described below make use of lower bounds for the distance of a node to the target node. These bounds can be available for some or all criteria. A general way of obtaining bounds is to

```
   Input: a timetable graph and a query
   Output: a set of Pareto-optimal labels at the terminal
 1 foreach node v do
 2 │    list<Label> labelListAt(v) := ∅;

 3 PriorityQueue pq := ∅;
 4 foreach node v in start interval do
 5 │    Label startLabel := createStartLabel(v);
 6 │    pq.insert(startLabel);

 7 while ! pq.isEmpty() do
 8 │    Label label := pq.extractLabel();
 9 │    foreach outgoing edge e=(v,w) of v=label.getNode() do
10 │    │    if isInfeasible(e) then continue; // ignore this edge
11 │    │    Label newLabel := createLabel(label, e);
12 │    │    if newLabel is dominated then  continue;
13 │    │    // newLabel is not dominated
14 │    │    pq.insert(newLabel);
15 │    │    labelListAt(w).insert(newLabel);
16 │    │    labelListAt(w).removeLabelsDominatedBy(newLabel);
```

Algorithm 1. Pseudocode for the generalized Dijkstra algorithm

simplify the graph enough to make it possible to search quickly. In a simplified auxiliary graph, a single-criterion backward search is performed in order to obtain lower bounds for all nodes and one criterion. In order to be able to perform a backward search, any time-dependency must be eliminated.

We have implemented two different versions of simplified graphs with different properties. The more efficient one uses the graph of the basic time-dependent model in which only travel time can be optimized and transfers are costless. Time-dependency is removed by replacing variable edge costs with their minimal cost over time. This graph is suited for obtaining lower bounds for travel time only. Another simplification procedure keeps the complete graph and only substitutes time-dependent edges with constant ones as in the first approach. The resulting graph is more complicated but yields tighter bounds and can also be used for transfers.

Dominance by Early Results. The basic version of the generalized Dijkstra algorithm tests only labels which reside at the same node for mutual domination. Therefore, sub-optimality of sub-paths can often only be detected at a later stage — at the latest at the terminal. This causes a significant amount of wasted work which we try to avoid.

A way to improve the behavior is to explicitly check newly created labels against results we already have. If they are already worse they may be discarded right away. The sooner our first results are obtained, the less avoidable labels are created this way. Lower bounds may be used when trying to dominate a label by earlier results. The criteria are therefore modified by adding the lower bound for the current node. That way labels that cannot lead to new Pareto optima are discarded as early as possible. If the lower bounds are tight enough, this can lead to major improvements once the first result has been found.

Goal-Directed Search. The lower bounds at a node can be used to add them as future costs to the cost values of a partial connection. If the extract operation from the priority queue is based on these modified values, this results in a goal-directed search as in the A* algorithm. It is important to note that this modification by itself leads to no improvement of running-time. It simply causes the search to find the first result earlier on. Together with the dominance by early results however, it leads to a major running-time improvement.

Avoid Hopping and Label Forwarding. Two phenomena that often arise when searching the time-dependent graph can be eliminated in order to improve performance. The first one is that labels propagate back to the node which they originated from. In this case the labels are immediately dominated. The search can easily be adapted to forbid *"hopping"*, i.e. the back-propagation of labels. The other phenomenon is due to the fact that all edges between station and route nodes in our graph have a cost of zero for all criteria. Because of this, newly created labels often have the same values for the single criteria as the label they originated from. Therefore, they are lexicographically minimal in the priority queue from the moment on they are inserted. We can thus avoid inserting them and simply hold them back until the current label has been processed completely. Before extracting further labels from the queue, the labels that are held back can be processed.

4   Computational Study

In the following, we analyze the performance of our multi-criteria search algorithm. We apply the above speed-up techniques and compare our prototype to a time-expanded approach. For the main part of our experiments we selected two relatively unrelated criteria, namely travel time and the number of transfers. Later we also show the influence on performance when adding an additional criterion to the search.

4.1   Train Network and Test Cases

The train network used in this study is derived from the train schedule of all trains within Germany of 2007 (56,994 trains, 8916 stations). The time-dependent graph has about 240,000 nodes and 670,000 edges while the corresponding time-expanded graph uses about 3,479,000 nodes and 5,633,000 edges. Three different sets of test cases were used. Each test case contains a source and a target station for the search, a date and a start time interval on that date. The first set of test cases is a synthetic one. It contains 1,000 randomly created tests that allow for arbitrary start time intervals (referred to as *random cases*). The second set also contains 1,000 randomly created tests which however have more realistic start time intervals of exactly one hour (*realistic cases*). The third set contains about 14,000 tests that were obtained from a snapshot of real connection queries provided by Deutsche Bahn AG (*real cases*).

4.2   Computational Environment

All computations were executed on an AMD Athlon(tm) 64 X2 dual core processor 4600+ with 2.4 GHz and 4 GB main memory running under Suse Linux 10.2. Our C++ code has been compiled with g++ 4.1.2 and compile option -O3.

4.3   Experiments

We first analyze the impact of single speed-up techniques. As a main indicator for performance we use several operation counts on representative operations, most importantly on the number of created labels, as well as on the number of labels which pass the domination tests and are inserted into the priority queue. We also provide CPU times, however, since our system is just a prototype to demonstrate feasibility of the approach, no serious effort was spent on fine-tuning the code in order to improve running time directly.

Impact of Exact Speed-Up Techniques. We start with a *base-line variant* which is the generalized Dijkstra algorithm on the fully realistic graph model without using any optimization techniques and choosing exiting transfers (cf. Section 2.2). Our first investigation compares this base-line variant with an optimized version which includes domination by early results as well as goal direction. The lower bounds are obtained from the basic time-independent graph (cf. Section 3.2). In addition to that, avoidance of hopping and label forwarding are used. Table 1 shows the combined impact of these techniques on performance. We observe an improvement of a factor of about six with respect to the number of created labels and a factor of 13 with respect to the number of insertions into the priority queue. A more careful analysis reveals the individual impact of the low level optimizations of avoiding the hopping of labels and their forwarding along costless edges (cf. Section 3.2). This can be seen in Table 2.

It can also be seen that the choice between entering and exiting transfers (cf. Section 2.2) makes a notable difference in performance. Together a factor of nearly two is achieved in the number of created labels and a factor of over three is achieved in the number of inserted labels. Note that the running times of the different sets of queries cannot be compared. The real cases use start time intervals of three hours while the realistic cases use one hour. This leads to an average number of about six non-dominated solutions for the real cases, but only an average of about two for the realistic cases. Therefore, different running times are to be expected. Although the average number of created labels is similar for both sets of instances, the actual distribution of the number of created labels has a significantly larger variance for the real cases.

Table 1. Comparison of the base-line variant with an optimized version

	1000 realistic cases		
	created labels	inserted labels	average time in seconds
base-line variant	1236744	636393	4.730
optimized version	207976	47967	1.050

Table 2. Performance improvement when avoiding hopping and forwarding labels

	1000 random cases			1000 realistic cases			14076 real cases		
	created labels	inserted labels	avg. time	created labels	inserted labels	avg. time	created labels	inserted labels	avg. time
entering transfers	1232592	545416	7.049s	385982	160200	1.606s	386764	176540	2.360s
exiting transfers	1072187	552012	5.990s	315565	160516	1.311s	343248	177193	2.098s
avoid hopping	682925	552014	5.453s	207984	160514	1.183s	212503	177192	1.932s
avoid hopping + label forwarding	682897	146766	4.690s	207976	47967	1.050s	212516	45114	1.570s

As explained in Section 3.2, there are several ways of obtaining lower bounds. The above results used the basic time-dependent graph. However, by using the more complex approach, lower bounds can be obtained for other criteria as well, like the number of transfers. Unfortunately, the lower bounds on the number of transfers do not improve the search sufficiently to overcome the effort of determining the bounds in the first place, as can be seen in Table 3.

An improvement can still be achieved with tighter bounds on the travel time. Compared to not using any heuristic, we obtain an improvement of factor about two. The most efficient variant of these bounds — the complex graph with travel time bounds only — will from here on be used as our *standard variant* for further comparisons.

Further Speed-Up by Realistic Assumptions. One of the strengths of our approach is the guaranteed optimality of the search results. We are not willing to sacrifice this advantage by using speed-up techniques that violate optimality. The only exception are optimizations that use realistic assumptions in order to limit the search to certain reasonable ranges for the criteria. The results of applying some of these techniques are shown in the following. There are two ways of restricting the allowed travel time. Firstly it can be restricted by a fixed upper limit like 24 hours. This helps a lot for long connections but does not help at all for short ones. A more adaptive restriction is to limit the allowed travel time to γ times the time of the fastest connection, where γ is a variable parameter of our algorithm. This improves the search a lot for short queries. Our results are summarized in Table 4. To limit the number of transfers did not show a notable effect on performance in our tests. A maximum of five allowed transfers did not yield a better performance, even though it makes some

Table 3. Performance when using several combinations of the simple and the complex graph in order to obtain lower bounds (realistic cases)

heuristic for lower bounds	created labels	inserted labels	average time in seconds
none	420803	92305	1.839
simple (time)	207976	47967	1.050
complex (time)	205260	45886	1.003
simple (time), complex (transfers)	207813	47939	1.106
complex (time & transfers)	205101	45866	1.159

Pareto-optimal connections impossible. Hence we dropped the limit on the number of transfers completely. A reasonable limitation can be put on the maximum waiting time at a station since long waiting periods are very unattractive for most passengers. This especially improves the search for connections running over night. The improvement can be seen in Table 5. Finally, the single limits can be applied together in different ways. We applied *conservative limits* of 24 hours for maximum travel time, five hours for maximum waiting time and $\gamma = 5$ and *tight limits* of ten hours for maximum travel time, three hours for maximum waiting time and $\gamma = 2$. The improvements can be found in Table 6. In summary, together with the exact speed-up techniques, a speed-up factor of about 20 over the base-line version has been achieved with respect to the number of created labels and a factor of 138 with respect to the number of insertions.

Comparison with a Time-Expanded Approach. In general, we expect a better performance of the time-dependent approach than of the time-expanded one. It is unclear however, whether this can be achieved in a multi-criteria setting. In order to answer this question, we compare the performance of our time-dependent approach with the time-expanded search incorporated in MOTIS. As the time-dependent system was developed as a proof of concept only, it makes not much sense to compare running times. We restrict our analysis to the comparison of the number of labels inserted into the priority queue. As can be seen in Table 7 the time-dependent approach creates much fewer labels. When using realistic assumptions, the time-dependent system adds 5.4 times less labels into the priority queue. However, it should be noted, that the time-dependent approach requires additional effort to compute actual edge lengths on-the-fly. Thus we expect (and empirically observe) similar running times for both approaches. As expected, the memory consumption of the time-expanded graph is a lot higher than that of the time-dependent one. In our tests, MOTIS needed 1.25 GB while the time-dependent graph used only 281 MB.

Adding an Additional Criterion: Reliability of Transfers. The above experiments were performed using travel time and the number of transfers as only search criteria. An interesting question is how the performance worsens when further criteria are introduced. This was explored by adding the "reliability of transfers" as a further criterion.

Table 4. Limiting the maximum travel time (realistic cases)

algorithmic variant	created labels	inserted labels	average time in seconds
standard	205260	45886	1.003
max. travel time = 24h	180910	30893	0.845
max. travel time = 15h	141602	16175	0.631
max. travel time = 10h	83162	6999	0.406
$\gamma = 5$	182535	32030	0.865
$\gamma = 3$	144678	17015	0.653
$\gamma = 2$	84125	6890	0.415

Table 5. Limiting the maximum waiting time (realistic cases)

algorithmic variant	created labels	inserted labels	average time in seconds
standard	205260	45886	1.003
max. waiting time = 5h	167914	19751	0.777
max. waiting time = 3h	151680	15441	0.637

Table 6. Performance improvement when combining limits (realistic cases)

algorithmic variant	created labels	inserted inserted	average time in seconds
standard	205260	45886	1.003
conservative limits	156515	17827	0.685
tight limits	63261	4605	0.335

The reliability of a single transfer is a function of the *buffer time* which is the available time exceeding the minimum transfer time at the station. This means that a passenger will catch the connecting train unless the incoming train is delayed by more than the buffer time. There are many plausible ways to map a buffer time t into a reliability measure. In this paper, we propose to define

$$\text{reliability} : t \mapsto s - \exp^{\ln(1-a) - \frac{1}{b} \cdot t},$$

with parameters $a = 0.6$, $b = 8$, $s = 0.99$ so that the maximal reliability of a single transfer is 99% and a buffer time of 0 minutes leads to 60% reliability. The reliability of connections with several transfers is defined as the product of the reliabilities of each single transfer. This yields a continuous reliability measure which we further transformed into a discrete one by subdividing the interval of [0,1] into 50, 20 and 10 equivalence classes of equal width. Table 8 summarizes the performance of the search when using different numbers of criteria. The addition of the number of transfers as second criterion leads to a slow-down of

Table 7. The number of labels inserted into the priority queue on average for both the time-dependent and the time-expanded search

real cases	inserted labels
time-expanded (optimized, conservative limits)	92538
time-expanded (optimized, tight limits)	64782
time-dependent (optimal, no limits)	44133
time-dependent (realistic assumptions, tight limits)	11913

Table 8. Relationship between the number of criteria and performance on 1000 realistic test cases. Different numbers of discretization steps are used for reliability.

criteria	created labels	inserted labels	average time in seconds	average number of Pareto optima
time	99284	19401	0.454	1.28
time, transfers	205260	45886	1.003	2.34
time, transfers, reliability (50 classes)	990664	160254	5.726	6.76
time, transfers, reliability (20 classes)	853742	149366	4.727	5.67
time, transfers, reliability (10 classes)	772822	142615	4.138	4.66

factor two, the addition of reliability of transfers as third criterion leads to a slow-down of another factor four if we use 10 equivalence classes.

5 Conclusions and Future Work

In this work we have presented our prototype for a time-dependent, multi-criteria search system that works in a fully realistic scenario. We have shown how to introduce the most important features of real timetables and how to improve performance significantly. We have provided the results of our experimental analysis that show that a speed-up factor of 20 with respect to the number of label creations and 138 with respect to the number of label insertions can be achieved under realistic assumptions. A comparison to the time-expanded approach was done, indicating that the new approach clearly is competitive. Finally we discussed the impact on performance when adding further criteria to the search.

In order to make the time-dependent approach able to replace current online search systems, its performance needs to be improved further. If possible, optimality should be maintained. It remains a challenge to design better speed-up techniques for multi-criteria search. Another goal is to extend our prototype to a dynamic scenario with train delays.

Acknowledgments

This work was partially supported by the DFG Focus Program Algorithm Engineering, grant Mu 1482/4-1. We wish to thank Deutsche Bahn AG for providing us timetable data for scientific use.

References

1. HaCon web-site (2007), `http://www.hacon.de/hafas/konzept.shtml`
2. Müller-Hannemann, M., Weihe, K.: On the cardinality of the Pareto set in bicriteria shortest path problems. Annals of Operations Research 147, 269–286 (2006)
3. Pallottino, S., Scutellà, M.G.: Shortest path algorithms in transportation models: Classical and innovative aspects. In: Equilibrium and Advanced Transportation Modelling, Kluwer Academic Publishers, Dordrecht (1998)
4. Schulz, F., Wagner, D., Weihe, K.: Dijkstra's algorithm on-line: An empirical case study from public railroad transport. ACM Journal of Experimental Algorithmics, Article 12, 5 (2000)
5. Pyrga, E., Schulz, F., Wagner, D., Zaroliagis, C.: Efficient models for timetable information in public transportation systems. ACM Journal of Experimental Algorithmics (JEA) 12, 2.4 (2007)
6. Cooke, K.L., Halsey, E.: The shortest route through a network with time-dependent internodal transit times. Journal of Mathematical Analysis and Applications 14, 493–498 (1966)
7. Orda, A., Rom, R.: Shortest-path and minimum-delay algorithms in networks with time-dependent edge-length. Journal of the ACM 37, 607–625 (1990)
8. Orda, A., Rom, R.: Minimum weight paths in time-dependent networks. Networks 21, 295–319 (1991)
9. Nachtigal, K.: Time depending shortest-path problems with applications to railway networks. European Journal of Operations Research 83, 154–166 (1995)
10. Brodal, G.S., Jacob, R.: Time-dependent networks as models to achieve fast exact time-table queries. In: Proceedings of the 3rd Workshop on Algorithmic Methods and Models for Optimization of Railways (ATMOS 2003). Electronic Notes in Theoretical Computer Science, vol. 92, pp. 3–15. Elsevier, Amsterdam (2004)
11. Pyrga, E., Schulz, F., Wagner, D., Zaroliagis, C.: Towards realistic modeling of time-table information through the time-dependent approach. In: Proceedings of the 3rd Workshop on Algorithmic Methods and Models for Optimization of Railways (ATMOS 2003). Electronic Notes in Theoretical Computer Science, vol. 92, pp. 85–103. Elsevier, Amsterdam (2004)
12. Müller-Hannemann, M., Schulz, F., Wagner, D., Zaroliagis, C.: Timetable information: Models and algorithms. In: Geraets, F., Kroon, L.G., Schoebel, A., Wagner, D., Zaroliagis, C.D. (eds.) Railway Optimization 2004. LNCS, vol. 4359, pp. 67–89. Springer, Heidelberg (2007)
13. Müller-Hannemann, M., Schnee, M.: Finding all attractive train connections by multi-criteria Pareto search. In: Geraets, F., Kroon, L.G., Schoebel, A., Wagner, D., Zaroliagis, C.D. (eds.) Railway Optimization 2004. LNCS, vol. 4359, pp. 246–263. Springer, Heidelberg (2007)
14. Bauer, R., Delling, D., Wagner, D.: Experimental study on speed-up techniques for timetable information systems. In: ATMOS 2007 (2007)
15. Bauer, R., Delling, D.: SHARC: Fast and Robust Unidirectional Routing. In: Proceedings of the 9th Workshop on Algorithm Engineering and Experiments (ALENEX 2008) (2008)
16. Dijkstra, E.W.: A note on two problems in connexion with graphs. Numerische Mathematik 1, 269–271 (1959)
17. Möhring, R.H.: Verteilte Verbindungssuche im öffentlichen Personenverkehr: Graphentheoretische Modelle und Algorithmen. In: Angewandte Mathematik - insbesondere Informatik, Vieweg, pp. 192–220 (1999)
18. Theune, D.: Robuste und effiziente Methoden zur Lösung von Wegproblemen. Teubner Verlag, Stuttgart (1995)

Author Index

Printing: Mercedes-Druck, Berlin
Binding: Stein+Lehmann, Berlin

Lecture Notes in Computer Science

Sublibrary 1: Theoretical Computer Science and General Issues

For information about Vols. 1– 4703
please contact your bookseller or Springer

Vol. 4910: V. Geffert, J. Karhumäki, A. Bertoni, B. Preneel, P. Návrat, M. Bieliková (Eds.), SOFSEM 2008: Theory and Practice of Computer Science. XV, 792 pages. 2008.

Vol. 4905: F. Logozzo, D.A. Peled, L.D. Zuck (Eds.), Verification, Model Checking, and Abstract Interpretation. X, 325 pages. 2008.

Vol. 4904: S. Rao, M. Chatterjee, P. Jayanti, C.S.R. Murthy, S.K. Saha (Eds.), Distributed Computing and Networking. XVIII, 588 pages. 2007.

Vol. 4878: E. Tovar, P. Tsigas, H. Fouchal (Eds.), Principles of Distributed Systems. XIII, 457 pages. 2007.

Vol. 4875: S.-H. Hong, T. Nishizeki, W. Quan (Eds.), Graph Drawing. XIII, 402 pages. 2008.

Vol. 4873: S. Aluru, M. Parashar, R. Badrinath, V.K. Prasanna (Eds.), High Performance Computing – HiPC 2007. XXIV, 663 pages. 2007.

Vol. 4863: A. Bonato, F.R.K. Chung (Eds.), Algorithms and Models for the Web-Graph. X, 217 pages. 2007.

Vol. 4860: G. Eleftherakis, P. Kefalas, G. Păun, G. Rozenberg, A. Salomaa (Eds.), Membrane Computing. IX, 453 pages. 2007.

Vol. 4855: V. Arvind, S. Prasad (Eds.), FSTTCS 2007: Foundations of Software Technology and Theoretical Computer Science. XIV, 558 pages. 2007.

Vol. 4854: L. Bougé, M. Forsell, J.L. Träff, A. Streit, W. Ziegler, M. Alexander, S. Childs (Eds.), Euro-Par 2007 Workshops: Parallel Processing. XVII, 236 pages. 2008.

Vol. 4851: S. Boztaş, H.-F.(F.) Lu (Eds.), Applied Algebra, Algebraic Algorithms and Error-Correcting Codes. XII, 368 pages. 2007.

Vol. 4848: M.H. Garzon, H. Yan (Eds.), DNA Computing. XI, 292 pages. 2008.

Vol. 4847: M. Xu, Y. Zhan, J. Cao, Y. Liu (Eds.), Advanced Parallel Processing Technologies. XIX, 767 pages. 2007.

Vol. 4846: I. Cervesato (Ed.), Advances in Computer Science – ASIAN 2007. XI, 313 pages. 2007.

Vol. 4838: T. Masuzawa, S. Tixeuil (Eds.), Stabilization, Safety, and Security of Distributed Systems. XIII, 409 pages. 2007.

Vol. 4835: T. Tokuyama (Ed.), Algorithms and Computation. XVII, 929 pages. 2007.

Vol. 4818: I. Lirkov, S. Margenov, J. Waśniewski (Eds.), Large-Scale Scientific Computing. XIV, 755 pages. 2008.

Vol. 4800: A. Avron, N. Dershowitz, A. Rabinovich (Eds.), Pillars of Computer Science. XXI, 683 pages. 2008.

Vol. 4783: J. Holub, J. Žďárek (Eds.), Implementation and Application of Automata. XIII, 324 pages. 2007.

Vol. 4782: R. Perrott, B.M. Chapman, J. Subhlok, R.F. de Mello, L.T. Yang (Eds.), High Performance Computing and Communications. XIX, 823 pages. 2007.

Vol. 4771: T. Bartz-Beielstein, M.J. Blesa Aguilera, C. Blum, B. Naujoks, A. Roli, G. Rudolph, M. Sampels (Eds.), Hybrid Metaheuristics. X, 202 pages. 2007.

Vol. 4770: V.G. Ganzha, E.W. Mayr, E.V. Vorozhtsov (Eds.), Computer Algebra in Scientific Computing. XIII, 460 pages. 2007.

Vol. 4769: A. Brandstädt, D. Kratsch, H. Müller (Eds.), Graph-Theoretic Concepts in Computer Science. XIII, 341 pages. 2007.

Vol. 4763: J.-F. Raskin, P.S. Thiagarajan (Eds.), Formal Modeling and Analysis of Timed Systems. X, 369 pages. 2007.

Vol. 4759: J. Labarta, K. Joe, T. Sato (Eds.), High-Performance Computing. XV, 524 pages. 2008.

Vol. 4750: M.L. Gavrilova, C.J.K. Tan (Eds.), Transactions on Computational Science I. XI, 181 pages. 2008.

Vol. 4746: A. Bondavalli, F. Brasileiro, S. Rajsbaum (Eds.), Dependable Computing. XV, 239 pages. 2007.

Vol. 4743: P. Thulasiraman, X. He, T.L. Xu, M.K. Denko, R.K. Thulasiram, L.T. Yang (Eds.), Frontiers of High Performance Computing and Networking ISPA 2007 Workshops. XXIX, 536 pages. 2007.

Vol. 4742: I. Stojmenovic, R.K. Thulasiram, L.T. Yang, W. Jia, M. Guo, R.F. de Mello (Eds.), Parallel and Distributed Processing and Applications. XX, 995 pages. 2007.

Vol. 4739: R. Moreno Díaz, F. Pichler, A. Quesada Arencibia (Eds.), Computer Aided Systems Theory – EUROCAST 2007. XIX, 1233 pages. 2007.

Vol. 4736: S. Winter, M. Duckham, L. Kulik, B. Kuipers (Eds.), Spatial Information Theory. XV, 455 pages. 2007.

Vol. 4732: K. Schneider, J. Brandt (Eds.), Theorem Proving in Higher Order Logics. IX, 401 pages. 2007.

Vol. 4731: A. Pelc (Ed.), Distributed Computing. XVI, 510 pages. 2007.

Vol. 4728: S. Bozapalidis, G. Rahonis (Eds.), Algebraic Informatics. VIII, 291 pages. 2007.

Vol. 4726: N. Ziviani, R. Baeza-Yates (Eds.), String Processing and Information Retrieval. XII, 311 pages. 2007.

Vol. 4719: R. Backhouse, J. Gibbons, R. Hinze, J. Jeuring (Eds.), Datatype-Generic Programming. XI, 369 pages. 2007.

Vol. 4711: C.B. Jones, Z. Liu, J. Woodcock (Eds.), Theoretical Aspects of Computing – ICTAC 2007. XI, 483 pages. 2007.

Vol. 4710: C.W. George, Z. Liu, J. Woodcock (Eds.), Domain Modeling and the Duration Calculus. XI, 237 pages. 2007.

Vol. 4708: L. Kučera, A. Kučera (Eds.), Mathematical Foundations of Computer Science 2007. XVIII, 764 pages. 2007.

Vol. 4707: O. Gervasi, M.L. Gavrilova (Eds.), Computational Science and Its Applications – ICCSA 2007, Part III. XXIV, 1205 pages. 2007.

Vol. 4706: O. Gervasi, M.L. Gavrilova (Eds.), Computational Science and Its Applications – ICCSA 2007, Part II. XXIII, 1129 pages. 2007.

Vol. 4705: O. Gervasi, M.L. Gavrilova (Eds.), Computational Science and Its Applications – ICCSA 2007, Part I. XLIV, 1169 pages. 2007.